Topics in Algebraic Graph Theory

The rapidly expanding area of algebraic graph theory uses two different branches of algebra to explore various aspects of graph theory: linear algebra (for spectral theory) and group theory (for studying graph symmetry). These areas have links with other areas of mathematics, such as logic and harmonic analysis, and are increasingly being used in such areas as computer networks where symmetry is an important feature. Other books cover portions of this material, but this book is unusual in covering both of these aspects and there are no other books with such a wide scope.

This book contains ten expository chapters written by acknowledged international experts in the field. Their well-written contributions have been carefully edited to enhance readability and to standardize the chapter structure, terminology and notation throughout the book. To help the reader, there is an extensive introductory chapter that covers the basic background material in graph theory, linear algebra and group theory. Each chapter concludes with an extensive list of references.

LOWELL W. BEINEKE is Schrey Professor of Mathematics at Indiana University-Purdue University Fort Wayne. His graph theory interests include topological graph theory, line graphs, tournaments, decompositions and vulnerability. With Robin J. Wilson he has edited *Selected Topics in Graph Theory* (3 volumes), *Applications of Graph Theory* and *Graph Connections*. He is currently the Editor of the *College Mathematics Journal*.

ROBIN J. WILSON is Head of the Pure Mathematics Department at the Open University and Gresham Professor of Geometry, London. He has written and edited many books on graph theory and combinatorics and on the history of mathematics, including *Introduction to Graph Theory* and *Four Colours Suffice*. His interests include graph coloring, spectral graph theory and the history of graph theory and combinatorics.

PETER J. CAMERON, internationally recognized for his substantial contributions to the area, served as academic consultant for this volume. He is a professor of mathematics at Queen Mary, University of London, and his mathematical interests are in permutation groups and their operands (logical, algebraic and combinatorial).

ENCYCLOPEDIA OF MATHEMATICS AND ITS APPLICATIONS

Volume 102

The titles below, and earlier volumes in the series, are available from booksellers or from Cambridge University Press at www.cambridge.org

50 F. Borceux *Handbook of Categorical Algebra I*
51 F. Borceux *Handbook of Categorical Algebra II*
52 F. Borceux *Handbook of Categorical Algebra III*
53 V. F. Kolchin *Random Graphs*
54 A. Katok and B. Hasselblatt *Introduction to the Modern Theory of Dynamical Systems*
55 V. N. Sachkov *Combinatorial Methods in Discrete Mathematics*
56 V. N. Sachkov *Probabilistic Methods in Discrete Mathematics*
57 P. M. Cohn *Skew Fields*
58 R. Gardner *Geometric Tomography*
59 G. A. Baker Jr. and P. Graves-Morris *Pade Approximants,* 2ed
60 J. Krajicek *Bounded Arithmetic, Propositional Logic and Complexity Theory*
61 H. Groemer *Geometric Applications of Fourier Series and Spherical Harmonics*
62 H. O. Fattorini *Infinite Dimensional Optimization and Control Theory*
63 A. C. Thompson *Minkowski Geometry*
64 R. B. Bapat and T. E. S. Raghavan *Nonnegative Matrices with Applications*
65 K. Engel *Sperner Theory*
66 D. Cvetkovic, P. Rowlinson, and S. Simic *Eigenspaces of Graphs*
67 F. Bergeron, G. Labelle, and P. Leroux *Combinatorial Species and Tree-Like Structures*
68 R. Goodman and N. Wallach *Representations and Invariants of the Classical Groups*
69 T. Beth, D. Jungnickel, and H. Lenz *Design Theory I,* 2ed
70 A. Pietsch and J. Wenzel *Orthonormal Systems for Banach Space Geometry*
71 G. E. Andrews, R. Askey, and R. Roy *Special Functions*
72 R. Ticciati *Quantum Field Theory for Mathematicians*
73 M. Stern *Semimodular Lattices*
74 I. Lasiecka and R. Triggiani *Control Theory for Partial Differential Equations I*
75 I. Lasiecka and R. Triggiani *Control Theory for Partial Differential Equations II*
76 A. A. Ivanov *Geometry of Sporadic Groups I*
77 A. Schinzel *Polynomials with Special Regard to Reducibility*
78 H. Lenz, T. Beth, and D. Jungnickel *Design Theory II,* 2ed
79 T. Palmer *Banach Algebras and the General Theory of *-Algebras II*
80 O. Stormark *Lie's Structural Approach to PDE Systems*
81 C. F. Dunkl and Y. Xu *Orthogonal Polynomials of Several Variables*
82 J. P. Mayberry *The Foundations of Mathematics in the Theory of Sets*
83 C. Foias et al. *Navier–Stokes Equations and Turbulence*
84 B. Polster and G. Steinke *Geometries on Surfaces*
85 R. B. Paris and D. Karninski *Asymptotics and Mellin–Barnes Integrals*
86 R. McEliece *The Theory of Information and Coding,* 2ed
87 B. Magurn *Algebraic Introduction to K-Theory*
88 T. Mora *Systems of Polynomial Equations I*
89 K. Bichteler *Stochastic Integration with Jumps*
90 M. Lothaire *Algebraic Combinatorics on Words*
91 A. A. Ivanov and S. V. Shpectorov *Geometry of Sporadic Groups II*
92 P. McMullen and E. Schulte *Abstract Regular Polytopes*
93 G. Gierz et al. *Continuous Lattices and Domains*
94 S. Finch *Mathematical Constants*
95 Y. Jabri *The Mountain Pass Theorem*
96 G. Gasper, and M. Rahman *Basic Hypergeometric Series,* 2ed
97 M. C. Pedicchio and W. Tholen *Categorical Foundations*
99 T. Mora *Solving Polynomial Equations Systems II*

Arthur Cayley (1821-1895)

From the *Illustrated London News,* 15 September 1883

Topics in Algebraic Graph Theory

Edited by

LOWELL W. BEINEKE
Indiana University-Purdue University
Fort Wayne

ROBIN J. WILSON
The Open University

Academic Consultant

PETER J. CAMERON
Queen Mary,
University of London

CAMBRIDGE
UNIVERSITY PRESS

CAMBRIDGE UNIVERSITY PRESS
Cambridge, New York, Melbourne, Madrid, Cape Town, Singapore,
São Paulo, Delhi, Dubai, Tokyo, Mexico City

Cambridge University Press
The Edinburgh Building, Cambridge CB2 8RU, UK

Published in the United States of America by Cambridge University Press, New York

www.cambridge.org
Information on this title: www.cambridge.org/9780521801973

First published 2005
Reprinted 2007

A catalogue record for this publication is available from the British Library

Library of Congress Cataloguing in Publication Data
Topics in algebraic graph theory / edited by Lowell W. Beineke and Robin J. Wilson,
academic consultant Peter J. Cameron.
p. cm. – (Encyclopedia of mathematics and its applications)
Includes bibliographical references and index.
ISBN 0-521-80197-4
1. Graph theory. I. Beineke, Lowell W. II. Wilson, Robin J. III. Series.
QA166.T64 2004
515'.5 – dc22 2004045915

ISBN 978-0-521-80197-3 Hardback

Contents

Preface

The field of graph theory has undergone tremendous growth during the past century. As recently as fifty years ago, the graph theory community had few members and most were in Europe and North America; today there are hundreds of graph theorists and they span the globe. By the mid-1970s, the field had reached the point where we perceived the need for a collection of surveys of the areas of graph theory: the result was our three-volume series *Selected Topics in Graph Theory*, comprising articles written by distinguished experts in a common style. During the past quarter-century, the transformation of the subject has continued, with individual areas (such as algebraic graph theory) expanding to the point of having important sub-branches themselves. This inspired us to conceive of a new series of books, each a collection of articles within a particular area written by experts within that area. This is the first of these books.

One innovative feature of this volume is the engagement of an academic consultant (Peter Cameron) to advise us on topics to be included and authors to be invited. We believe that this has been successful, the result being chapters covering the full range of areas within algebraic graph theory written by authors from around the world. Another important feature is that we have imposed uniform terminology and notation throughout, as far as possible, in the belief that this will aid readers in going from one chapter to another. For a similar reason we have not attempted to remove a small amount of overlap between the various chapters.

We hope that these features will make the book easier to use in an advanced course or seminar. We heartily thank the authors for cooperating on this, even though it sometimes required their abandoning some of their favourite conventions – for example, most graph theorists use G to denote a graph, whereas for algebraic graph theorists G usually denotes a group: the graph theorists won on this one. We also asked our contributors to undergo the ordeal of having their early versions subjected to detailed critical reading. We believe that the final product is thereby significantly better than it might otherwise have been, simply a collection of individually authored chapters. We want to express our sincere appreciation to all of our contributors for their cooperation.

We extend special thanks to Peter Cameron for his willingness to share his expertise as academic consultant – his advice has been invaluable. We are also grateful to Cambridge University Press for publishing this work; in particular, we thank Roger Astley for his advice, support and cooperation. Finally, we extend our appreciation to several universities for the different ways in which they have assisted with this endeavour: Purdue University awarded the first editor (LWB) sabbatical leave during which he was a guest of the Mathematical Institute at Oxford University, while the second editor (RJW) has had the cooperation of the Open University and Keble College, Oxford.

LWB, RJW
February 2004

Foreword

The topic of this volume is the connection between graph theory and algebra – more specifically, how algebra helps in the study of graphs, and how graph theory repays the debt.

There are two main connections between graph theory and algebra. These arise from two algebraic objects associated with a graph: its *adjacency matrix* and its *automorphism group*. Of course, they are related: the automorphism group can be regarded as the collection of all permutation matrices that commute with the adjacency matrix. Nevertheless, the two connections involve different algebraic techniques – linear algebra and group theory, respectively – and have a somewhat different flavour.

The introductory chapter presents the basic background material on graphs, linear algebra and groups that is used in the rest of the book.

Turning first to linear algebra, the first four chapters survey different aspects of the interaction between graphs and matrices. The adjacency matrix of a graph is a real symmetric matrix, and hence is diagonalizable: the eigenvalues of this matrix are closely connected with a variety of graph parameters. For directed graphs, the adjacency matrix is not necessarily diagonalizable, and things are much less straightforward. Moreover, in both cases the adjacency matrix is non-negative, and so the theory of such matrices developed by Perron and Frobenius can be applied.

The first chapter describes the basic results, including connections with counting closed walks and with clique and colouring problems, while the second gives a number of rather surprising applications. The third chapter looks in much more detail at the geometry of the eigenvectors of the adjacency matrix and surveys the theory of graph angles and star partitions which has been developed largely by the authors of this chapter.

For some questions about graphs, especially those connected with random walks and of approximating manifolds by graphs, a different matrix is more appropriate: this is the *Laplacian matrix*, the matrix $D - A$, where A is the adjacency matrix and D is the diagonal matrix whose diagonal entries are the vertex-degrees. This matrix is positive semi-definite with a zero eigenvalue; the smallest non-zero eigenvalue gives crucial information about the behaviour of a random walk on the graph. This is discussed in the fourth chapter.

In Chapter 5 the focus switches to automorphisms of graphs. Almost all graphs admit no non-trivial automorphisms; but often it is the case that graphs with symmetry have other important properties. Such graphs as the Petersen graph and the skeletons of regular polyhedra bear witness to this. Chapter 5 presents the basics about automorphisms.

Typically we find that the more symmetry we require of a graph, the more group theory contributes to the analysis of the graph. A first level of symmetry we might require is *vertex-transitivity*, the property that the automorphism group acts transitively on vertices. Among such graphs, the most important and widely studied are the *Cayley graphs*. A Cayley graph of a group has the group elements as vertices: the graph is obtained by specifying the neighbours of the identity element and obtaining the remaining edges by translation. Cayley graphs and their properties are described in Chapter 6. A stronger symmetry requirement is that the automorphism group acts transitively on the set of adjacent pairs of vertices. Such graphs are called *symmetric*, and they form the subject of Chapter 7.

The two algebraic strands, linear algebra and group theory, meet in the study of graphs satisfying 'higher regularity conditions': distance-regular graphs (and their special case, strongly regular graphs) and distance-transitive graphs. A graph is *strongly regular* if it is regular and if the number of common neighbours of two vertices v and w depends only on whether v and w are adjacent. It turns out that a regular connected graph that is not complete is strongly regular if and only if its adjacency matrix has just three distinct eigenvalues (the smallest possible number). Furthermore, if the automorphism group of a graph acts transitively on the sets of ordered pairs of vertices and of non-adjacent vertices, then the graph is strongly regular. Such graphs are the subject of Chapter 8.

This last condition leads naturally to the definition of a *distance-transitive* graph, one where the automorphism group acts transitively on the set of ordered pairs of vertices at each fixed distance up to the diameter of the graph. This is a very strong symmetry condition, and the goal of researchers in the area is the determination of all such graphs. The current state of this research programme is described in Chapter 9. The arguments here are very group-theoretic, using (for example) the classification of the finite simple groups.

If a graph is symmetric, then it is possible to exploit the symmetry to simplify searching for structures within the graph. Moreover, computational tools in group theory have reached a high level of sophistication. The final chapter discusses these matters with a detailed look at how such computations can be done.

PETER J. CAMERON

Introduction

LOWELL BEINEKE, ROBIN WILSON and PETER CAMERON

This introductory chapter is divided into three parts. The first presents the basic ideas of graph theory. The second concerns linear algebra (for Chapters 1–4), while the third concerns group theory (for Chapters 5–10).

1. Graph theory

This section presents the basic definitions, terminology and notations of graph theory, along with some fundamental results. Further information can be found in the many standard books on the subject – for example, West [4] or (for a simpler treatment) Wilson [5].

Graphs

A *graph* G is a pair of sets (V, E), where V is a finite non-empty set of elements called *vertices*, and E is a set of unordered pairs of distinct vertices called *edges*. The sets V and E are the *vertex-set* and the *edge-set* of G, and are often denoted by $V(G)$ and $E(G)$, respectively. An example of a graph is shown in Fig. 1.

The number of vertices in a graph is the *order* of the graph; usually it is denoted by n and the number of edges by m. Standard notation for the vertex-set is $V = \{v_1, v_2, \ldots, v_n\}$ and for the edge-set is $E = \{e_1, e_2, \ldots, e_m\}$. Arbitrary vertices are frequently represented by $u, v, w, \ldots$ and edges by $e, f, \ldots$.

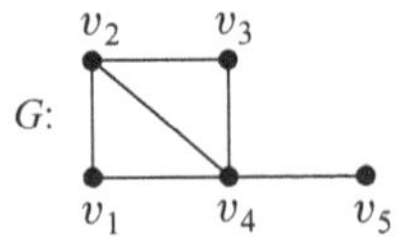

$V = \{v_1, v_2, v_3, v_4, v_5\}$

$E = \{v_1v_2, v_1v_4, v_2v_3, v_2v_4, v_3v_4, v_4v_5\}$

Fig. 1.

Variations of graphs

By definition, our graphs are *simple*, meaning that two vertices are connected by at most one edge. If several edges, called *multiple edges*, are allowed between two vertices, we have a *multigraph*. Sometimes, *loops* – edges joining vertices to themselves – are also permitted. In a *weighted graph*, the edges are assigned numerical values called *weights*. Finally, if the vertex-set is allowed to be infinite, then G is an *infinite graph*.

Perhaps the most important variation is that of *directed graphs*; these are discussed at the end of this section.

Adjacency and degrees

For convenience, the edge $\{v, w\}$ is commonly written as vw. We say that this edge *joins* v and w and that it is *incident with* v and w. In this case, v and w are *adjacent vertices*, or *neighbours*. The set of neighbours of a vertex v is its *neighbourhood* $N(v)$. Two edges are *adjacent edges* if they have a vertex in common.

The number of neighbours of a vertex v is called its *degree*, denoted by $\deg v$. Observe that the sum of the degrees in a graph is twice the number of edges. If all the degrees of G are equal, then G is *regular*, or is *k-regular* if that common degree is k. The maximum degree in a graph is often denoted by Δ.

Walks

A *walk* in a graph is a sequence of vertices and edges $v_0, e_1, v_1, \ldots, e_k, v_k$, in which each edge $e_i = v_{i-1}v_i$. This walk goes *from* v_0 *to* v_k or *connects* v_0 and v_k, and is called a v_0-v_k walk. It is frequently shortened to $v_0v_1 \ldots v_k$, since the edges may be inferred from this. Its *length* is k, the number of occurrences of edges. If $v_k = v_0$, the walk is *closed*.

Some important types of walk are the following:

- a *path* is a walk in which no vertex is repeated;
- a *trail* is a walk in which no edge is repeated;
- a *cycle* is a non-trivial closed trail in which no vertex is repeated.

Distance

In a connected graph, the *distance* between two vertices v and w is the minimum length of a path from v to w, and is denoted by $d(v, w)$. It is easy to see that distance satisfies the properties of a metric: for all vertices u, v and w,

- $d(v, w) \geq 0$, with equality if and only if $v = w$;
- $d(v, w) = d(w, v)$;
- $d(u, w) \leq d(u, v) + d(v, w)$

The *diameter* of a graph G is the maximum distance between two vertices of G. If G has cycles, the *girth* of G is the length of a shortest cycle, and the *circumference* is the length of a longest cycle.

Subgraphs

If G and H are graphs with $V(H) \subseteq V(G)$ and $E(H) \subseteq E(G)$, then H is a *subgraph* of G. If, moreover, $V(H) = V(G)$, then H is a *spanning* subgraph. The subgraph *induced* by a non-empty set S of vertices in G is that subgraph H with vertex-set S whose edge-set consists of those edges of G that join two vertices in S; it is denoted by $\langle S \rangle$ or $G[S]$. A subgraph H of G is *induced* if $H = \langle V(H) \rangle$. In Fig. 2, H_1 is a spanning subgraph of G, and H_2 is an induced subgraph.

Given a graph G, the *deletion of a vertex* v results in the subgraph obtained by excluding v and all edges incident with it. It is denoted by $G - v$ and is the subgraph induced by $V - \{v\}$. More generally, if $S \subset V$, we write $G - S$ for the graph obtained from G by deleting all of the vertices of S; that is, $G - S = \langle V - S \rangle$.

The *deletion of an edge* e results in the subgraph $G - e$ obtained by excluding e from E; for $F \subseteq E$, $G - F$ denotes the spanning subgraph with edge-set $E - F$.

Connectedness and connectivity

A graph G is *connected* if there is a path connecting each pair of vertices. A *(connected) component* of G is a maximal connected subgraph of G.

A vertex v of a graph G is a *cut-vertex* if $G - v$ has more components than G. A connected graph with no cut-vertices is 2-*connected* or *non-separable*. The following statements are equivalent for a graph G with at least three vertices:

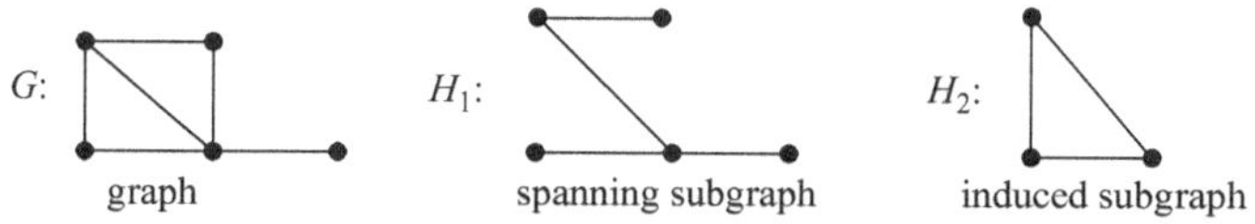

Fig. 2.

- G is non-separable;
- every pair of vertices lie on a cycle;
- every vertex and edge lie on a cycle;
- every pair of edges lie on a cycle;
- for any three vertices u, v, and w, there is a v-w path containing u;
- for any three vertices u, v, and w, there is a v-w path not containing u;
- for any two vertices v and w and any edge e, there is a v-w path containing e.

More generally, a graph G is *k-connected* if there is no set S with fewer than k vertices for which $G - S$ is a connected non-trivial graph. Menger characterized such graphs.

Menger's theorem *A graph G is k-connected if and only if, for each pair of vertices v and w, there is a set of k v-w paths that pairwise have only v and w in common.*

The *connectivity* $\kappa(G)$ of a graph G is the maximum value of k for which G is k-connected.

There are similar concepts and results for edges. A *cut-edge* (or *bridge*) is any edge whose deletion produces one more component than before. A non-trivial graph G is *k-edge-connected* if the result of removing fewer than k edges is always connected, and the *edge-connectivity* $\lambda(G)$ is the maximum value of k for which G is k-edge-connected. We note that Menger's theorem also has an edge version.

Bipartite graphs

If the vertices of a graph G can be partitioned into two non-empty sets so that no edge joins two vertices in the same set, then G is *bipartite*. The two sets are called *partite sets*, and if they have orders r and s, G may be called an $r \times s$ bipartite graph. The most important property of bipartite graphs is that they are the graphs that contain no cycles of odd length.

Trees

A *tree* is a connected graph that has no cycles. They have been characterized in many ways, a few of which we give here. For a graph G of order n:

- G is connected and has no cycles;
- G is connected and has $n - 1$ edges;
- G has no cycles and has $n - 1$ edges.

Any graph without cycles is a *forest*; note that each component of a forest is a tree.

Special graphs

We now introduce some individual types of graphs:

- the *complete graph* K_n has n vertices, each of which is adjacent to all of the others;
- the *null graph* N_n has n vertices and no edges;
- the *path graph* P_n consists of the vertices and edges of a path of length $n - 1$;
- the *cycle graph* C_n consists of the vertices and edges of a cycle of length n;
- the *complete bipartite graph* $K_{r,s}$ is the $r \times s$ bipartite graph in which each vertex is adjacent to all those in the other partite set;
- in the *complete k-partite graph*, $K_{r_1,r_2,\ldots,r_n}$ the vertices are in k sets (having orders $r_1, r_2, \ldots, r_k$) and each vertex is adjacent to all the others, except those in the same set. If the k sets all have order r, the graph is denoted by $K_{k(r)}$. The graph $K_{k(2)}$ is sometimes called the *k-dimensional octahedral graph* or *cocktail party graph*, also denoted by CP(k); $K_{3(2)}$ is the graph of an octahedron.
- the *d-dimensional cube* (or *d-cube*) Q_d is the graph whose vertices can be labelled with the 2^d binary d-tuples, in such a way that two vertices are adjacent when their labels differ in exactly one position. It is regular of degree d, and is isomorphic to the lattice of subgraphs of a set of d elements.

Examples of these graphs are given in Fig. 3.

Operations on graphs

There are several ways to get new graphs from old. We list some of the most important here.

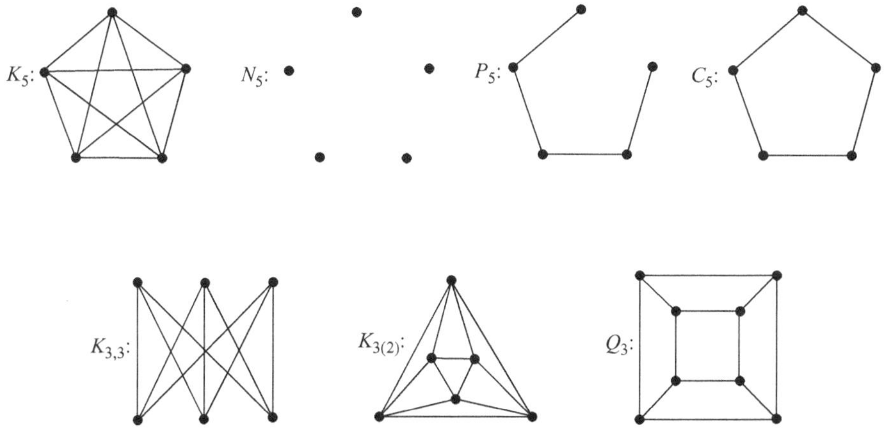

Fig. 3.

- The *complement* $\overline{G}$ of a graph G has the same vertices as G, but two vertices are adjacent in $\overline{G}$ if and only if they are not adjacent in G.

For the other operations, we assume that G and H are graphs with disjoint vertex-sets, $V(G) = \{v_1, v_2, \ldots, v_n\}$ and $V(H) = \{w_1, w_2, \ldots, w_t\}$:

- the *union* $G \cup H$ has vertex-set $V(G) \cup V(H)$ and edge-set $E(G) \cup E(H)$. The union of k graphs isomorphic to G is denoted by kG.
- the *join* $G + H$ is obtained from $G \cup H$ by adding all of the edges from vertices in G to those in H.
- the *(Cartesian) product* $G \square H$ or $G \times H$ has vertex-set $V(G) \times V(H)$, and (v_i, w_j) is adjacent to (v_h, w_k) if either (a) v_i is adjacent to v_h in G and $w_j = w_k$, or (b) $v_i = v_h$ and w_j is adjacent to w_k in H. In less formal terms, $G \square H$ can be obtained by taking n copies of H and joining corresponding vertices in different copies whenever there is an edge in G. Note that, for d-cubes, $Q_{d+1} = K_2 \square Q_d$ (with $Q_1 = K_2$).

Examples of these binary operations are given in Fig. 4.

There are two basic operations involving an edge of a graph. The *insertion of a vertex into an edge e* means that the edge $e = vw$ is replaced by a new vertex u and the two edges vu and uw. Two graphs are *homeomorphic* if each can be obtained from a third graph by a sequence of vertex insertions. The *contraction of the edge vw* means that v and w are replaced by a new vertex u that is adjacent to the other neighbours of v and w. If a graph H can be obtained from G by a sequence of edge contractions and the deletion of isolated vertices, then G is said to be *contractible* to H. Finally, H is a *minor* of G if it can be obtained from G by a sequence of edge-deletions and edge-contractions and the removal

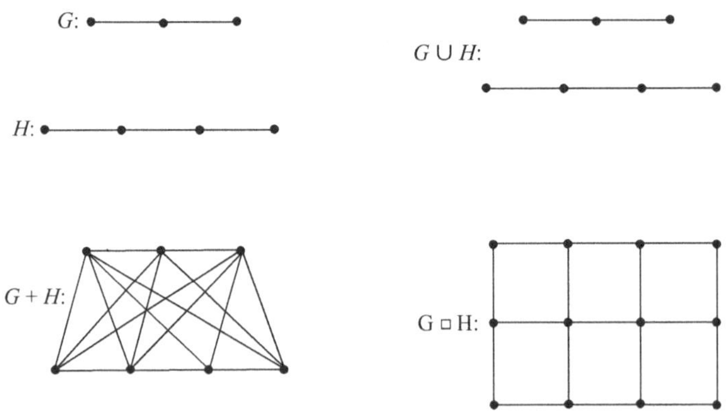

Fig. 4.

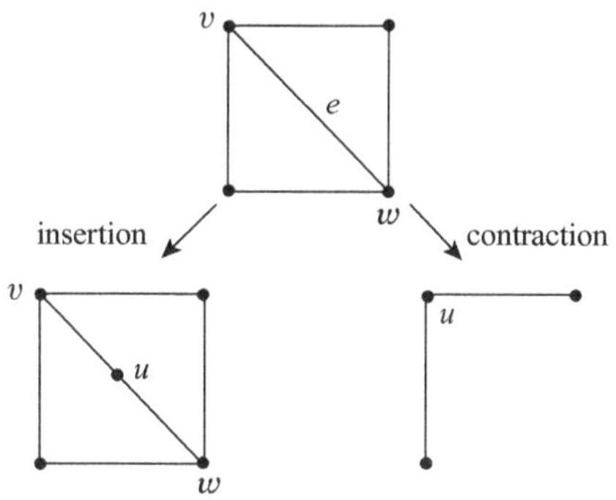

Fig. 5.

of isolated vertices. The operations of insertion and contraction are illustrated in Fig. 5.

Traversability

A connected graph G is *Eulerian* if it has a closed trail containing all of the edges of G; such a trail is called an *Eulerian trail*. The following are equivalent for a connected graph G:

- G is Eulerian;
- the degree of each vertex of G is even;
- the edge-set of G can be partitioned into cycles.

A graph G is *Hamiltonian* if it has a spanning cycle, and *traceable* if it has a spanning path. No 'good' characterizations of these graphs are known.

Planarity

A *planar graph* is one that can be embedded in the plane in such a way that no two edges meet except at a vertex incident with both. If a graph G is embedded in this way, then the points of the plane not on G are partitioned into open sets called *faces* or *regions*. Euler discovered the basic relationship between the numbers of vertices, edges and faces.

Euler's polyhedron formula *Let G be a connected graph embedded in the plane with n vertices, m edges and f faces. Then $n - m + f = 2$.*

It follows from this result that a planar graph with n vertices ($n \geq 3$) has at most $3(n-2)$ edges, and at most $2(n-2)$ edges if it is bipartite. From this it follows that the two graphs K_5 and $K_{3,3}$ are non-planar. Kuratowski proved that these two graphs are the only barriers to planarity.

Kuratowski's theorem *The following statements are equivalent for a graph G:*
- *G is planar;*
- *G has no subgraph that is homeomorphic to K_5 or $K_{3,3}$;*
- *G has no subgraph that is contractible to K_5 or $K_{3,3}$.*

Graph colourings

A graph G is *k-colourable* if, from a set of k colours, it is possible to assign a colour to each vertex in such a way that adjacent vertices always have different colours. The *chromatic number* $\chi(G)$ is the least value of k for which G is *k-chromatic*. It is easy to see that a graph is 2-colourable if and only if it is bipartite, but there is no 'good' way to determine which graphs are k-colourable for $k \geq 3$. Brooks's theorem provides one of the best-known bounds on the chromatic number of a graph.

Brooks's theorem *If G is a graph with maximum degree Δ that is neither an odd cycle nor a complete graph, then $\chi(G) \leq \Delta$.*

There are similar concepts for colouring edges. A graph G is *k-edge-colourable* if, from a set of k colours, it is possible to assign a colour to each edge in such a way that adjacent edges always have different colours. The *edge-chromatic number* $\chi'(G)$ is the least k for which G is k-edge-colourable. Vizing proved that the range of values of $\chi'(G)$ is very limited.

Vizing's theorem *If G is a graph with maximum degree Δ, then*

$$\Delta \leq \chi'(G) \leq \Delta + 1.$$

Line graphs

The *line graph* $L(G)$ of a graph G has the edges of G as its vertices, with two of these vertices adjacent if and only if the corresponding edges are adjacent in G. An example is given in Fig. 6.

A graph is a line graph if and only if its edges can be partitioned into complete subgraphs in such a way that no vertex is in more than two of these subgraphs.

Fig. 6.

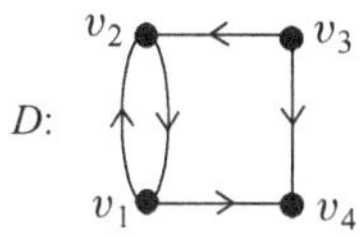

$V = \{v_1, v_2, v_3, v_4\}$
$E = \{\overrightarrow{v_1v_2}, \overrightarrow{v_1v_4}, \overrightarrow{v_2v_1}, \overrightarrow{v_3v_2}, \overrightarrow{v_3v_4}\}$.

Fig. 7.

Line graphs are also characterized by the property of having none of nine particular graphs as a forbidden subgraph.

Directed graphs

Digraphs are directed analogues of graphs, and thus have many similarities, as well as some important differences.

A *digraph* (or *directed graph*) D is a pair of sets (V, E) where V is a finite non-empty set of elements called *vertices*, and E is a set of ordered pairs of distinct elements of V called *arcs* or *directed edges*. Note that the elements of E are now ordered, which gives each of them a direction. An example of a digraph is given in Fig. 7.

Because of the similarities between graphs and digraphs, we mention only the main differences here and do not redefine those concepts that carry over easily.

An arc (v, w) of a digraph may be written as $\overrightarrow{vw}$, and is said to go *from* v *to* w, or to *go out of* v and *go into* w.

Walks, paths, trails and cycles are understood to be directed, unless otherwise indicated.

The *out-degree* $d^+(v)$ of a vertex v in a digraph is the number of arcs that go out of it, and the *in-degree* $d^-(v)$ is the number of arcs that go into it.

A digraph D is *strongly connected*, or *strong*, if there is a path from each vertex to each of the others. A *strong component* is a maximal strongly connected subgraph. Connectivity and edge-connectivity are defined in terms of strong connectedness.

A *tournament* is a digraph in which every pair of vertices are joined by exactly one arc. One interesting aspect of tournaments is their Hamiltonian properties:

- every tournament has a spanning path;
- a tournament has a Hamiltonian cycle if and only if it is strong.

2. Linear algebra

In this section we present the main results on vector spaces and matrices that are used in Chapters 1–4. For further details, see [3].

The space $\mathbf{R}^n$

The *real n-dimensional space* $\mathbf{R}^n$ consists of all n-tuples of real numbers $\mathbf{x} = (x_1, x_2, \ldots, x_n)$; in particular, the plane $\mathbf{R}^2$ consists of all pairs (x_1, x_2), and three-dimensional space $\mathbf{R}^3$ consists of all triples (x_1, x_2, x_3). The elements $\mathbf{x}$ are *vectors*, and the numbers x_i are the *coordinates* or *components* of $\mathbf{x}$.

When $\mathbf{x} = (x_1, x_2, \ldots, x_n)$ and $\mathbf{y} = (y_1, y_2, \ldots, y_n)$ are vectors in $\mathbf{R}^n$, we can form their *sum* $\mathbf{x} + \mathbf{y} = (x_1 + y_1, x_2 + y_2, \ldots, x_n + y_n)$, and if α is a scalar (real number), we can form the *scalar multiple* $\alpha\mathbf{x} = (\alpha x_1, \alpha x_2, \ldots, \alpha x_n)$.

The *zero vector* is the vector $\mathbf{0} = (0, 0, \ldots, 0)$, and the *additive inverse* of $\mathbf{x} = (x_1, x_2, \ldots, x_n)$ is the vector $-\mathbf{x} = (-x_1, -x_2, \ldots, -x_n)$.

We can similarly define the *complex n-dimensional space* $\mathbf{C}^n$, in which the vectors are all n-tuples of *complex* numbers $\mathbf{z} = (z_1, z_2, \ldots, z_n)$; in this case, we take the multiplying scalars α to be complex numbers.

Metric properties

When $\mathbf{x} = (x_1, x_2, \ldots, x_n)$ and $\mathbf{y} = (y_1, y_2, \ldots, y_n)$ are vectors in $\mathbf{R}^n$, their *dot product* is the scalar $\mathbf{x} \cdot \mathbf{y} = x_1y_1 + x_2y_2 + \cdots + x_ny_n$. The dot product is sometimes called the *inner product* and denoted by $\langle \mathbf{x}, \mathbf{y} \rangle$.

The *length* or *norm* $\|\mathbf{x}\|$ of a vector $\mathbf{x} = (x_1, x_2, \ldots, x_n)$ is

$$(\mathbf{x} \cdot \mathbf{x})^{1/2} = \left(x_1^2 + x_2^2 + \cdots + x_n^n\right)^{1/2}.$$

A *unit vector* is a vector $\mathbf{u}$ for which $\|\mathbf{u}\| = 1$, and for any non-zero vector $\mathbf{x}$, the vector $\mathbf{x}/\|\mathbf{x}\|$ is a unit vector.

When $\mathbf{x} = (x_1, x_2, \ldots, x_n)$ and $\mathbf{y} = (y_1, y_2, \ldots, y_n)$, the *distance* between $\mathbf{x}$ and $\mathbf{y}$ is $d(\mathbf{x}, \mathbf{y}) = \|\mathbf{x} - \mathbf{y}\|$. The distance function d satisfies the usual properties of a metric: *for any* $\mathbf{x}, \mathbf{y}, \mathbf{z} \in \mathbf{R}^n$,

- $d(\mathbf{x}, \mathbf{y}) \geq 0$, *and* $d(\mathbf{x}, \mathbf{y}) = 0$ *if and only if* $\mathbf{x} = \mathbf{y}$;
- $d(\mathbf{x}, \mathbf{y}) = d(\mathbf{y}, \mathbf{x})$;
- $d(\mathbf{x}, \mathbf{z}) \leq d(\mathbf{x}, \mathbf{y}) + d(\mathbf{y}, \mathbf{z})$ (*triangle inequality*).

The following result is usually called the *Cauchy-Schwarz inequality*:

Cauchy-Schwarz inequality *For any* $\mathbf{x}, \mathbf{y} \in \mathbf{R}^n$, $|\mathbf{x} \cdot \mathbf{y}| \leq \|\mathbf{x}\| \cdot \|\mathbf{y}\|$.

We define the angle θ between the non-zero vectors $\mathbf{x}$ and $\mathbf{y}$ by

$$\cos\theta = \mathbf{x}\cdot\mathbf{y}/\|\mathbf{x}\|\cdot\|\mathbf{y}\|.$$

Two vectors $\mathbf{x}$ and $\mathbf{y}$ are *orthogonal* if the angle between them is 0 – that is, if $\mathbf{x}\cdot\mathbf{y} = 0$. In this case, we have the following celebrated result.

Pythagoras's theorem *If* $\mathbf{x}$ *and* $\mathbf{y}$ *are orthogonal, then* $\|\mathbf{x}+\mathbf{y}\|^2 = \|\mathbf{x}\|^2 + \|\mathbf{y}\|^2$.

An *orthogonal set of vectors* is a set of vectors each pair of which is orthogonal. An *orthonormal set* is an orthogonal set in which each vector has length 1.

In a complex space $\mathbf{C}^n$ most of the above concepts are defined as above. One exception is that the dot product of two complex vectors $\mathbf{z} = (z_1, z_2, \ldots, z_n)$ and $\mathbf{w} = (w_1, w_2, \ldots, w_n)$ is now defined by $\mathbf{z}\cdot\mathbf{w} = z_1\overline{w}_1 + z_2\overline{w}_2 + \cdots + z_n\overline{w}_n$, where $\overline{w}$ is the complex conjugate of w.

Vector spaces

A *real vector space* V is a set of elements, called *vectors*, with rules of addition and scalar multiplication that satisfy the following conditions:

Addition
A1: For all $\mathbf{x}, \mathbf{y} \in V, \mathbf{x}+\mathbf{y} \in V$;
A2: For all $\mathbf{x}, \mathbf{y}, \mathbf{z} \in V, (\mathbf{x}+\mathbf{y})+\mathbf{z} = \mathbf{x}+(\mathbf{y}+\mathbf{z})$;
A3: There is an element $\mathbf{0} \in V$ *satisfying* $\mathbf{x}+\mathbf{0} = \mathbf{x}$, *for all* $\mathbf{x} \in V$;
A4: For each $\mathbf{x} \in V$, *there is an element* $-\mathbf{x}$ *satisfying* $\mathbf{x}+(-\mathbf{x}) = \mathbf{0}$;
A5: For all $\mathbf{x}, \mathbf{y} \in \mathbf{V}, \mathbf{x}+\mathbf{y} = \mathbf{y}+\mathbf{x}$.

Scalar multiplication
M1: For all $\mathbf{x} \in V$ *and* $\alpha \in \mathbf{R}, \alpha\mathbf{x} \in V$;
M2: For all $\mathbf{x} \in V, 1\mathbf{x} = \mathbf{x}$;
M3: For all $\alpha, \beta \in \mathbf{R}, \alpha(\beta\mathbf{x}) = (\alpha\beta)\mathbf{x}$;

Distributive laws
D1: For all $\alpha, \beta \in \mathbf{R}$ *and* $\mathbf{x} \in V, (\alpha+\beta)\mathbf{x} = \alpha\mathbf{x}+\beta\mathbf{x}$;
D2: For all $\alpha \in \mathbf{R}$ *and* $\mathbf{x}, \mathbf{y} \in V, \alpha(\mathbf{x}+\mathbf{y}) = \alpha\mathbf{x}+\alpha\mathbf{y}$.

Examples of real vector spaces are $\mathbf{R}^n$, $\mathbf{C}^n$, the set of all real polynomials, the set of all real infinite sequences, and the set of all functions $f : \mathbf{R} \to \mathbf{R}$, each with the appropriate definitions of addition and scalar multiplication.

Complex vector spaces are defined similarly, except that the scalars are elements of $\mathbf{C}$, rather than $\mathbf{R}$. More generally, the scalars can come from any field, such as the set $\mathbf{Q}$ of rational numbers, the integers $\mathbf{Z}_p$ modulo p, where p is a prime number, or the finite field $\mathbf{F}_q$, where q is a power of a prime.

Subspaces

A non-empty subset W of a vector space V is a *subspace* of V if W is itself a vector space with respect to the operations of addition and scalar multiplication in V. For example, the subspaces of $\mathbf{R}^3$ are $\{\mathbf{0}\}$, the lines and planes through $\mathbf{0}$, and $\mathbf{R}^3$ itself.

When X and Y are subspaces of a vector space V, their *intersection* $X \cap Y$ is also a subspace of V, as is their *sum* $X + Y = \{x + y : x \in X, y \in Y\}$.

When $V = X + Y$ and $X \cap Y = \{\mathbf{0}\}$, we call V the *direct sum* of X and Y, and write $V = X \oplus Y$.

Bases

Let $S = \{\mathbf{x}_1, \mathbf{x}_2, \ldots, \mathbf{x}_r\}$ be a set of vectors in a vector space V. Then any vector of the form

$$\alpha_1\mathbf{x}_1 + \alpha_2\mathbf{x}_2 + \cdots + \alpha_r\mathbf{x}_r,$$

where $\alpha_1, \alpha_2, \ldots, \alpha_r$ are scalars, is a *linear combination* of $\mathbf{x}_1, \mathbf{x}_2, \ldots, \mathbf{x}_r$. The set of all linear combinations of $\mathbf{x}_1, \mathbf{x}_2, \ldots, \mathbf{x}_r$ is a subspace of V called the *span* of S, denoted by $\langle S\rangle$ or $\langle \mathbf{x}_1, \mathbf{x}_2, \ldots, \mathbf{x}_r\rangle$. When $\langle S\rangle = V$, the set S *spans* V, or is a *spanning set* for V.

The set $S = \{\mathbf{x}_1, \mathbf{x}_2, \ldots, \mathbf{x}_r\}$ is *linearly dependent* if one of the vectors $\mathbf{x}_i$ is a linear combination of the others – in this case, there are scalars $\alpha_1, \alpha_2, \ldots, \alpha_r$, *not all zero*, for which

$$\alpha_1\mathbf{x}_1 + \alpha_2\mathbf{x}_2 + \cdots + \alpha_r\mathbf{x}_r = \mathbf{0}.$$

The set S is *linearly independent* if it is not linearly dependent – that is,

$$\alpha_1\mathbf{x}_1 + \alpha_2\mathbf{x}_2 + \cdots + \alpha_r\mathbf{x}_r = \mathbf{0}$$

holds only when $\alpha_1 = \alpha_2 = \cdots = \alpha_r = \mathbf{0}$.

A *basis* B is a linearly independent spanning set for V. In this case, each vector $\mathbf{x}$ of V can be written as a linear combination of the vectors in B *in exactly one way*; for example, the *standard basis* for $\mathbf{R}^3$ is $\{(1, 0, 0), (0, 1, 0), (0, 0, 1)\}$ and a basis for the set of all real polynomials is $\{1, x, x^2, \ldots\}$.

Dimension

A vector space V with a finite basis is *finite-dimensional*. In this situation, any two bases for V have the same number of elements. This number is the *dimension* of V, denoted by $\dim V$; for example, $\mathbf{R}^3$ has dimension 3. The dimension of a subspace of V is defined similarly.

When X and Y are subspaces of V, we have the *dimension theorem*:

$$\dim(X+Y) = \dim X + \dim Y - \dim(X \cap Y).$$

When $X \cap Y = \{\mathbf{0}\}$, this becomes

$$\dim(X \oplus Y) = \dim X + \dim Y.$$

Euclidean spaces

Let V be a real vector space, and suppose that with each pair of vectors $\mathbf{x}$ and $\mathbf{y}$ in V is associated a scalar $\langle \mathbf{x}, \mathbf{y} \rangle$. This is an *inner product* on V if it satisfies the following properties: *for any* $\mathbf{x}, \mathbf{y}, \mathbf{z} \in V$,

- $\langle \mathbf{x}, \mathbf{x} \rangle \geq 0$, *and* $\langle \mathbf{x}, \mathbf{x} \rangle = 0$ *if and only if* $\mathbf{x} = 0$;
- $\langle \mathbf{x}, \mathbf{y} \rangle = \langle \mathbf{y}, \mathbf{x} \rangle$;
- $\langle \alpha\mathbf{x} + \beta\mathbf{y}, \mathbf{z} \rangle = \alpha\langle \mathbf{x}, \mathbf{z} \rangle + \beta\langle \mathbf{y}, \mathbf{z} \rangle$.

The vector space V, together with this inner product, is called a *real inner product space*, or *Euclidean space*. Examples of Euclidean spaces are $\mathbf{R}^3$ with the dot product as inner product, and the space V of real-valued continuous functions on the interval $[-1, 1]$ with the inner product defined for $\mathbf{f}, \mathbf{g}$ in V by $\langle \mathbf{f}, \mathbf{g} \rangle = \int_{-1}^{1} \mathbf{f}(t)\mathbf{g}(t)\,dt$. Analogously to the dot product, we can define the metrical notions of length, distance and angle in any Euclidean space, and we can derive analogues of the Cauchy-Schwarz inequality and Pythagoras's theorem.

An *orthogonal basis* for a Euclidean space is a basis in which any two distinct basis vectors are orthogonal. If, further, each basis vector has length 1, then the basis is an *orthonormal basis*. If V is a Euclidean space, the *orthogonal complement* $W^\perp$ of a subspace W is the set of all vectors in V that are orthogonal to all vectors in W – that is,

$$W^\perp = \{\mathbf{v} \in V : \langle \mathbf{v}, \mathbf{w} \rangle = 0 \text{ for all } \mathbf{w} \in W\}.$$

Linear transformations

When V and W are real vector spaces, a function $T : V \to W$ is a *linear transformation* if, for all $\mathbf{v}_1, \mathbf{v}_2 \in V$ and $\alpha, \beta \in \mathbf{R}$,

$$T(\alpha\mathbf{v}_1 + \beta\mathbf{v}_2) = \alpha T(\mathbf{v}_1) + \beta T(\mathbf{v}_2).$$

If $V = W$, then T is sometimes called a *linear operator* on V.

The linear transformation T is *onto*, or *surjective*, when $T(V) = W$, and is *one-one*, or *injective*, if $T(\mathbf{v}_1) = T(\mathbf{v}_2)$ only when $\mathbf{v}_1 = \mathbf{v}_2$.

The *image* of T is the subspace of W defined by

$$\text{im}(T) = \{\mathbf{w} \in W : \mathbf{w} = T(\mathbf{v}), \text{ for some } \mathbf{v} \in V\};$$

note that T is onto if and only if $\text{im}(T) = W$.

The *kernel*, or *null space*, of T is the subspace of V defined by

$$\ker(T) = \{\mathbf{v} \in V : T(\mathbf{v}) = \mathbf{0}_W\};$$

note that T is one-one if and only if $\ker(T) = \{\mathbf{0}_V\}$.

Defining the *rank* and *nullity* of T by

$$\text{rank}(T) = \dim \text{im}(T) \quad \text{and} \quad \text{nullity}(T) = \dim \ker(T),$$

we obtain the *rank-nullity formula*:

$$\text{rank}(T) + \text{nullity}(T) = \dim V.$$

Algebra of linear transformations

When $S : U \to V$ and $T : V \to W$ are linear transformations, we can form their *composition* $T \circ S : U \to W$, defined by

$$(T \circ S)(\mathbf{u}) = T(S(\mathbf{u})), \qquad \text{for all } \mathbf{u} \in U.$$

The composition of linear transformations is associative.

The linear transformation $T : V \to W$ is *invertible*, or *non-singular*, if there is a linear transformation T^{-1}, called the *inverse* of T, for which $T^{-1} \circ T$ is the identity transformation on V and $T \circ T^{-1}$ is the identity transformation on W. Note that a linear transformation is invertible if and only if it is one-one and onto.

The matrix of a linear transformation

Let $T : V \to W$ be a linear transformation, let $\{\mathbf{e}_1, \mathbf{e}_2, \ldots, \mathbf{e}_n\}$ be a basis for V and let $\{\mathbf{f}_1, \mathbf{f}_2, \ldots, \mathbf{f}_m\}$ be a basis for W. For each $i = 1, 2, \ldots, n$, we can write

$$T(\mathbf{e}_i) = a_{1i}\mathbf{f}_1 + a_{2i}\mathbf{f}_2 + \cdots + a_{mi}\mathbf{f}_m,$$

for some scalars $a_{1i}, a_{2i}, \ldots, a_{mi}$. The rectangular array of scalars

$$\mathbf{A} = \begin{pmatrix} a_{11} & a_{12} & \cdots & a_{1n} \\ a_{21} & a_{22} & \cdots & a_{2n} \\ \cdot & \cdot & \cdot & \cdot \\ a_{m1} & a_{m2} & \cdots & a_{mn} \end{pmatrix}$$

is the *matrix of T associated with the given bases*. The scalar a_{ij} is the *ij-entry* of $\mathbf{A}$, and we abbreviate the above array by writing $\mathbf{A} = (a_{ij})$.

Since the matrix $\mathbf{A}$ has m rows and n columns, we call $\mathbf{A}$ an $m \times n$ *matrix*; a matrix for which $m = n$ is a *square matrix* of *order n*. The *diagonal* of a square matrix $\mathbf{A} = (a_{ij})$ consists of the entries a_{ii} down the leading (top-left to bottom-right) diagonal.

Operations on matrices

Given a matrix $\mathbf{A} = (a_{ij})$ and a scalar κ, we can define the *scalar multiple* $\kappa\mathbf{A} = (\kappa a_{ij})$; note that $1\mathbf{A} = \mathbf{A}$.

Given two $m \times n$ matrices $\mathbf{A} = (a_{ij})$ and $\mathbf{B} = (b_{ij})$, we define their *sum* $\mathbf{A} + \mathbf{B}$ to be the matrix $\mathbf{C} = (c_{ij})$, where $c_{ij} = a_{ij} + b_{ij}$; matrix addition is commutative and associative.

Given an $m \times k$ matrix $\mathbf{A} = (a_{ij})$ and a $k \times n$ matrix $\mathbf{B} = (b_{ij})$, we define their *product* $\mathbf{AB}$ to be the $m \times n$ matrix $\mathbf{C} = (c_{ij})$, where $c_{ij} = \Sigma_k a_{ik} b_{kj}$; matrix multiplication is associative, but not commutative in general. The matrix product $\mathbf{AA}$ is written $\mathbf{A}^2$, with similar notation for higher powers of $\mathbf{A}$.

Given two matrices $\mathbf{A} = (a_{ij})$ and $\mathbf{B} = (b_{ij})$ of the same size, we define their *Kronecker product* $\mathbf{A} \otimes \mathbf{B}$ to be the matrix $\mathbf{C} = (c_{ij})$, where $c_{ij} = a_{ij} b_{ij}$.

Given an $m \times n$ matrix $\mathbf{A} = (a_{ij})$, we can interchange the rows and columns to form the $n \times m$ matrix $\mathbf{A}^T = (a_{ji})$ called its *transpose*. It follows that

$$(\mathbf{A}^T)^T = \mathbf{A} \quad \text{and} \quad (\kappa\mathbf{A})^T = \kappa(\mathbf{A})^T, \quad \text{for any scalar } \kappa,$$

and that, for matrices $\mathbf{A}$ and $\mathbf{B}$ of appropriate sizes,

$$(\mathbf{A} + \mathbf{B})^T = \mathbf{A}^T + \mathbf{B}^T \quad \text{and} \quad (\mathbf{AB})^T = \mathbf{B}^T\mathbf{A}^T.$$

The *trace* $\mathrm{tr}(\mathbf{A})$ of a square matrix $\mathbf{A}$ is the sum of the diagonal entries of $\mathbf{A}$; for matrices $\mathbf{A}$ and $\mathbf{B}$ of appropriate sizes,

$$\mathrm{tr}(\mathbf{A} + \mathbf{B}) = \mathrm{tr}(\mathbf{A}) + \mathrm{tr}(\mathbf{B}) \quad \text{and} \quad \mathrm{tr}(\mathbf{AB}) = \mathrm{tr}(\mathbf{BA}).$$

Types of matrices

A *zero matrix* $\mathbf{0}$ is a matrix in which each entry is 0; for matrices $\mathbf{A}$ and $\mathbf{0}$ of the same size, $\mathbf{A} + \mathbf{0} = \mathbf{0} + \mathbf{A} = \mathbf{A}$. An *all-1 matrix* $\mathbf{J}$ is a matrix in which each entry is 1.

A square matrix $\mathbf{A} = (a_{ij})$ is *symmetric* if $\mathbf{A}^T = \mathbf{A}$ (so $a_{ij} = a_{ji}$ for all i and j), and *anti-symmetric* or *skew-symmetric* if $\mathbf{A}^T = -\mathbf{A}$ (so $a_{ij} = -a_{ji}$ for all i and j).

A *diagonal matrix* $\mathbf{A}$ is a square matrix in which every non-diagonal entry is 0. The *identity matrix* $\mathbf{I}$ is the diagonal matrix with 1s on its diagonal; for square matrices $\mathbf{A}$ and $\mathbf{I}$ of the same order, $\mathbf{AI} = \mathbf{IA} = \mathbf{A}$. A *permutation matrix* is a matrix obtained from $\mathbf{I}$ by permuting the rows or columns.

An *upper triangular matrix* is a square matrix in which every entry below and to the left of the diagonal is 0; a *lower triangular matrix* is defined similarly.

A *block matrix* is a matrix arranged in submatrices $\mathbf{B}_{ij}$ called *blocks*, as follows:

$$\begin{pmatrix} \mathbf{B}_{11} & \mathbf{B}_{12} & \cdots & \mathbf{B}_{1s} \\ \mathbf{B}_{21} & \mathbf{B}_{22} & \cdots & \mathbf{B}_{2s} \\ \cdot & \cdot & \cdot & \cdot \\ \mathbf{B}_{r1} & \mathbf{B}_{r2} & \cdots & \mathbf{B}_{rs} \end{pmatrix}$$

A *circulant matrix* $\mathbf{A} = (a_{ij})$ is an $n \times n$ matrix in which each successive row is obtained by moving the preceding row by one position to the right; thus, for each i and j, $a_{ij} = a_{i+1,j+1}$, where the subscripts are taken modulo n.

A square matrix $\mathbf{A}$ is *invertible* if there is a matrix $\mathbf{B}$ for which $\mathbf{AB} = \mathbf{BA} = \mathbf{I}$; the matrix $\mathbf{B}$ is the *inverse* of $\mathbf{A}$, denoted by $\mathbf{A}^{-1}$. Note that $(\mathbf{A}^{-1})^{-1} = \mathbf{A}$, and that, for square matrices $\mathbf{A}$ and $\mathbf{B}$ of the same order, $(\mathbf{AB})^{-1} = \mathbf{B}^{-1}\mathbf{A}^{-1}$.

Determinants

Let $\mathbf{A} = (a_{ij})$ be an $n \times n$ square matrix. The *determinant* of $\mathbf{A}$, written $\det \mathbf{A}$ or $|\mathbf{A}|$, is the sum

$$\sum (\operatorname{sgn} s)\, a_{1s(1)}\, a_{2s(2)} \ldots a_{ns(n)},$$

taken over all permutations s of $\{1, 2, \ldots, n\}$, where $\operatorname{sgn} s$ is 1 if s is an even permutation, and -1 if s is odd; for example, when $n = 3$, the permutations 123, 231 and 312 are even, and 132, 213 and 321 are odd, and so

$$\begin{pmatrix} a_{11} & a_{12} & a_{13} \\ a_{21} & a_{22} & a_{23} \\ a_{31} & a_{32} & a_{33} \end{pmatrix} = \begin{array}{l} a_{11}a_{22}a_{33} + a_{12}a_{23}a_{31} + a_{13}a_{21}a_{32} \\ \quad - a_{11}a_{23}a_{32} - a_{12}a_{21}a_{33} - a_{13}a_{22}a_{31}. \end{array}$$

Note that, *for any square matrices* $\mathbf{A}$ *and* $\mathbf{B}$,

- $\det \mathbf{A} = \det \mathbf{A}^T$;
- $\det \mathbf{AB} = \det \mathbf{A} \det \mathbf{B}$;
- *if* $\det \mathbf{A} \neq 0$, *then* $\det \mathbf{A}^{-1} = (\det \mathbf{A})^{-1}$.

Given a square matrix $\mathbf{A}$, let the *minor* M_{ij} be the determinant of the matrix obtained from $\mathbf{A}$ by deleting the ith row and the jth column, and let the *cofactor* $A_{ij} = (-1)^{i+j} M_{ij}$. Then

$$\det \mathbf{A} = \sum_i a_{ij} A_{ij} = \sum_j a_{ij} A_{ij}.$$

The matrix adj $\mathbf{A} = (A_{ij})$ is the *adjoint matrix*, and if $\det \mathbf{A} \neq 0$, then

$$\mathbf{A}^{-1} = (\det \mathbf{A})^{-1}(\operatorname{adj} \mathbf{A}).$$

Change of basis

Let $T: V \to W$ be a linear transformation, let $\{\mathbf{e}_1, \mathbf{e}_2, \ldots, \mathbf{e}_m\}$ and $\{\mathbf{e}'_1, \mathbf{e}'_2, \ldots, \mathbf{e}'_m\}$ be bases for V, and let $\{\mathbf{f}_1, \mathbf{f}_2, \ldots, \mathbf{f}_n\}$ and $\{\mathbf{f}'_1, \mathbf{f}'_2, \ldots, \mathbf{f}'_n\}$ be bases for W. If the matrix of T associated with the bases $\{\mathbf{e}_1, \mathbf{e}_2, \ldots, \mathbf{e}_m\}$ and $\{\mathbf{f}_1, \mathbf{f}_2, \ldots, \mathbf{f}_n\}$ is $\mathbf{A}$, then the matrix of T associated with the bases $\{\mathbf{e}'_1, \mathbf{e}'_2, \ldots, \mathbf{e}'_m\}$ and $\{\mathbf{f}'_1, \mathbf{f}'_2, \ldots, \mathbf{f}'_n\}$ is $\mathbf{Q}^{-1}\mathbf{AP}$, where $\mathbf{P} = (p_{ij})$ and $\mathbf{Q} = (q_{ij})$ are the transition matrices defined by $\mathbf{e}'_j = \sum_i p_{ij}\mathbf{e}_i$ and $\mathbf{f}'_j = \sum_i q_{ij}\mathbf{f}_i$.

Now let $T: V \to V$ be a linear transformation, and let $\{\mathbf{e}_1, \mathbf{e}_2, \ldots, \mathbf{e}_m\}$ and $\{\mathbf{e}'_1, \mathbf{e}'_2, \ldots, \mathbf{e}'_m\}$ be bases for both V and W. If the matrix of T associated with the basis $\{\mathbf{e}_1, \mathbf{e}_2, \ldots, \mathbf{e}_m\}$ in V and W is $\mathbf{A}$, then, by the above result, the matrix of T associated with the basis $\{\mathbf{e}'_1, \mathbf{e}'_2, \ldots, \mathbf{e}'_m\}$ in V and W is $\mathbf{P}^{-1}\mathbf{AP}$, where $\mathbf{P} = (p_{ij})$ is the transition matrix defined by $\mathbf{e}'_j = \sum_i p_{ij}\mathbf{e}_i$. Two matrices $\mathbf{A}$ and $\mathbf{B}$ are *similar* if $\mathbf{B} = \mathbf{P}^{-1}\mathbf{AP}$, for some invertible matrix $\mathbf{P}$.

Eigenvalues and eigenvectors

If $\mathbf{A}$ is a square matrix, then the polynomial $\det(\lambda\mathbf{I} - \mathbf{A})$ is the *characteristic polynomial* of $\mathbf{A}$ and the equation $\det(\lambda\mathbf{I} - \mathbf{A}) = 0$ is its *characteristic equation*. The roots of this equation are the *eigenvalues* of $\mathbf{A}$; a repeated root is a *multiple eigenvalue*, and a non-repeated root is a *simple eigenvalue*. The sum of the eigenvalues of $\mathbf{A}$ is equal to the trace of $\mathbf{A}$. When $\mathbf{A}$ is a symmetric matrix, the eigenvalues of $\mathbf{A}$ are all real. The set of eigenvalues of $\mathbf{A}$ is the *spectrum* of $\mathbf{A}$.

Every square matrix satisfies a number of polynomial equations; for example, if $\mathbf{A}$ is an $n \times n$ matrix, then the matrices $\mathbf{I}, \mathbf{A}, \mathbf{A}^2, \ldots, \mathbf{A}^{n^2}$ are linearly dependent and so there is a polynomial equation connecting them. The unique monic polynomial of lowest degree satisfied by $\mathbf{A}$ is the *minimal polynomial* of $\mathbf{A}$. The *Cayley-Hamilton theorem* states that *every matrix satisfies its characteristic equation*. It follows that the minimum polynomial divides the characteristic polynomial; moreover, the minimum polynomial and the characteristic equation share the same irreducible factors.

Let $T: V \to V$ be a linear transformation. Regardless of the basis used, all matrices associated with T have the same eigenvalues: these are the scalars λ for which $T(\mathbf{v}) = \lambda\mathbf{v}$, for some non-zero vector $\mathbf{v}$, and each such vector is an associated *eigenvector*. For a given eigenvalue λ the set of all such eigenvectors, together with $\mathbf{0}$, is a subspace of V, called the *eigenspace* associated with λ.

Diagonalizing matrices

When $\mathbf{v}_1, \mathbf{v}_2, \ldots, \mathbf{v}_r$ are eigenvectors associated with *distinct* eigenvalues $\lambda_1, \lambda_2, \ldots, \lambda_r$, the set $\{\mathbf{v}_1, \mathbf{v}_2, \ldots, \mathbf{v}_r\}$ is linearly independent. Eigenvectors associated with the same eigenvalue λ may be linearly dependent or independent. If we can find a basis of eigenvectors, and if $\mathbf{P}$ is the matrix with these eigenvectors as columns, then $\mathbf{P}^{-1}\mathbf{AP}$ is a diagonal matrix. Conversely, if $\mathbf{A}$ is any square matrix and if we can find a matrix $\mathbf{P}$ for which $\mathbf{P}^{-1}\mathbf{AP}$ is a diagonal matrix, then there is a basis of eigenvectors, and these eigenvectors form the columns of $\mathbf{P}$; in this case, we say that $\mathbf{A}$ is *diagonalizable*. Not all square matrices are diagonalizable: for example, the matrix

$$\begin{pmatrix} 1 & 1 \\ 0 & 1 \end{pmatrix}$$

is not diagonalizable, since every eigenvector is a multiple of $(1 \quad 0)$ and so there is no basis of eigenvectors.

Every symmetric matrix $\mathbf{A}$ has an orthonormal basis of eigenvectors, and so is diagonalizable. Moreover, the corresponding transition matrix $\mathbf{P}$ is then an *orthogonal matrix* ($\mathbf{P}^{-1} = \mathbf{P}^T$), the matrix $\mathbf{P}^T\mathbf{AP}$ is diagonal, and the matrix $\mathbf{A}$ is called *orthogonally diagonalizable*.

Even when we cannot diagonalize a given matrix, we can always choose a basis so that the resulting matrix can be decomposed into *block diagonal form*, with square matrices $\mathbf{B}_1, \mathbf{B}_2, \ldots, \mathbf{B}_k$ arranged down the diagonal, and zeros elsewhere. There are various existence and uniqueness theorems concerning such decompositions.

Quadratic forms

A *quadratic form* is an expression of the form $q(\mathbf{x}) = \mathbf{x}^T\mathbf{Ax}$, where $\mathbf{x}$ is a column vector and $\mathbf{A}$ is a symmetric matrix; for example,

$$q(\mathbf{x}) = 2x_1^2 + 3x_2^2 - 4x_3^2 + x_1x_2 - 6x_1x_3$$

is a quadratic form corresponding to the symmetric matrix

$$\mathbf{A} = \begin{pmatrix} 2 & 1/2 & -3 \\ 1/2 & 3 & 0 \\ -3 & 0 & -4 \end{pmatrix}.$$

A quadratic form q is *positive definite* if $q(\mathbf{x}) > 0$, and *positive semidefinite* if $q(\mathbf{x}) \geq 0$, for every non-zero vector $\mathbf{x}$.

3. Group theory

This section introduces some basic material about permutation groups and automorphism groups of graphs and explains the notation used for groups. Further information can be found in the books by Cameron [1] and Dixon and Mortimer [2] and the many references cited therein.

Groups

A *group* is a set $\mathcal{G}$ with a binary operation $\circ$ satisfying the conditions:

- for all $g, h, k \in \mathcal{G}$, $(g \circ h) \circ k = g \circ (h \circ k)$ (*associative law*);
- there exists an element $1 \in \mathcal{G}$ (the *identity*) such that $1 \circ g = g \circ 1 = g$ for all $g \in \mathcal{G}$;
- for each $g \in \mathcal{G}$, there is an element $g^{-1} \in G$ (the *inverse* of g) such that $g \circ g^{-1} = g^{-1} \circ g = 1$.

We usually write $g \circ h$ more briefly as gh.

If in addition we have the condition:

- for all $g, h \in \mathcal{G}$, $g \circ h = h \circ g$ (*commutative law*)

then the group $\mathcal{G}$ is *Abelian* (or *commutative*).

Groups are important here because the set of automorphisms of a graph (with the operation of composition of mappings) is a group. In many cases, the group encodes important information about the graph; and in general, the use of symmetry can be used to do combinatorial searches in the graph more efficiently.

Homomorphisms

A *homomorphism* from a group $\mathcal{G}_1$ to a group $\mathcal{G}_2$ is a function $\phi: \mathcal{G}_1 \to \mathcal{G}_2$ for which $\phi(gh) = \phi(g)\phi(h)$, for all $g, h \in \mathcal{G}_1$. Note two things about this definition. First, in the defining equation $\phi(gh) = \phi(g)\phi(h)$, the operation gh on the left is the group operation in $\mathcal{G}_1$, while the operation $\phi(g)\phi(h)$ on the right is the group operation in $\mathcal{G}_2$. Second, it is common among algebraists to write homomorphisms on the right, so instead of $\phi(g)$, we write $g\phi$. The advantage of this is that the composition of ϕ with a homomorphism $\psi: \mathcal{G}_2 \to \mathcal{G}_3$ is now $\phi\psi$, where $g(\phi\psi) = (g\phi)\psi$.

An *isomorphism* from $\mathcal{G}_1$ to $\mathcal{G}_2$ is a homomorphism that is one-to-one and onto. If there is an isomorphism from $\mathcal{G}_1$ to $\mathcal{G}_2$, then the groups $\mathcal{G}_1$ and $\mathcal{G}_2$ are *isomorphic*. Algebraists regard isomorphic groups as being essentially the same, and do not carefully distinguish between them.

Subgroups

A *subgroup* of a group is a subset that forms a group in its own right with respect to the same operation. If $\mathcal{H}$ is a subgroup of $\mathcal{G}$ (sometimes written $\mathcal{H} < \mathcal{G}$), then there are two natural partitions of $\mathcal{G}$:

- the partition into sets $\mathcal{H}x = \{hx : h \in \mathcal{H}\}$, called *right cosets* of $\mathcal{H}$;
- the partition into sets $x\mathcal{H} = \{xh : h \in \mathcal{H}\}$, called *left cosets* of $\mathcal{H}$.

The number of right (or left) cosets is the *index* of $\mathcal{H}$ in $\mathcal{G}$.

Let $\phi : \mathcal{G}_1 \to \mathcal{G}_2$ be a homomorphism, and let 1 denote the identity element of $\mathcal{G}_2$. Then the set

$$\mathcal{N} = \{g \in \mathcal{G}_1 : g\phi = 1\}$$

is the *kernel* of ϕ. The kernel of a homomorphism is a subgroup of $\mathcal{G}_1$, and has the additional property that

$$g^{-1}ng \in \mathcal{N}, \qquad \text{for all } g \in \mathcal{G}_1, n \in \mathcal{N}.$$

Such a subgroup is called a *normal subgroup*. It turns out that every normal subgroup is the kernel of a homomorphism, and that a subgroup is normal if and only if the partitions of the group into right and left cosets of the subgroup coincide (so that we can simply speak of *cosets*).

Let $\mathcal{N}$ be a normal subgroup of $\mathcal{G}$. Then we define $\mathcal{G}/\mathcal{N}$ to be the set of cosets of $\mathcal{N}$ in $\mathcal{G}$, and define an operation on $\mathcal{G}/\mathcal{N}$ by the rule

$$(\mathcal{N}x)(\mathcal{N}y) = \mathcal{N}xy.$$

It can be shown that this operation is indeed well defined (that is, independent of the chosen representatives x and y of the cosets) and defines a group, which is called the *quotient group* or *factor group*, also denoted by $\mathcal{G}/\mathcal{N}$.

Every group $\mathcal{G}$ has two trivial normal subgroups: the whole group $\mathcal{G}$ and the identity $\{1\}$. A group with no other normal subgroups is *simple*; any other group is *composite*.

Composition series

An important structural result about finite groups is the *Jordan–Hölder theorem*.

Jordan-Hölder theorem *Let $\mathcal{G}$ be a finite group. Then the following hold:*

(a) there is a chain

$$\{1\} = \mathcal{G}_r < \mathcal{G}_{r-1} < \cdots < \mathcal{G}_1 < \mathcal{G}_0 = \mathcal{G}$$

of subgroups of $\mathcal{G}$ such that, for $1 \le i \le r$, $\mathcal{G}_i$ is a normal subgroup of $\mathcal{G}_{i-1}$ and the factor group $\mathcal{G}_{i-1}/\mathcal{G}_i$ is simple.

(b) for any two such chains, the multisets of isomorphism types of simple factor groups are the same.

The series in (*a*) is a *composition series* for $\mathcal{G}$, and the simple factor groups are the *composition factors*.

This theorem indicates that the simple groups are the 'building blocks' from which arbitrary groups can be constructed. Thus finite group theory falls into two parts, concerned with answering the questions

- what are the finite simple groups?
- how can they be put together to build arbitrary groups?

Finite simple groups

The first question above has recently been completely solved. It is not possible here to give any hint of the proof, and the result can be stated only in broad terms.

There are four classes of finite simple groups:

- If p is a prime number, then the *cyclic group* $\mathbf{Z}_p$ of order p (which can be realized as the group of rotations of a regular p-gon) is simple; indeed, it has no subgroups at all except for itself and the identity.
- For $n \ge 2$, the set of permutations of $\{1, 2, \ldots, n\}$ that have even parity (that is, are products of even numbers of transpositions) forms a normal subgroup $\mathcal{A}_n$ of $\mathcal{S}_n$ with index 2, called the *alternating group*. For $n \ge 5$, the alternating group is simple.
- There is a large collection of *groups of Lie type*, closely related to groups of matrices over finite fields. The simplest to describe are the groups $\mathrm{PSL}_n(q)$, obtained from the group $\mathrm{SL}_n(q)$ of all $n \times n$ matrices over the finite field $\mathbf{F}_q$ of q elements which have determinant 1. The set of scalar matrices of determinant 1 forms a normal subgroup $\mathcal{Z}$ of $\mathrm{SL}_n(q)$, and the group $\mathrm{PSL}_n(q)$ is the quotient group $\mathrm{SL}_n(q)/\mathcal{Z}$. The group $\mathrm{PSL}_n(q)$ is simple for $n \ge 2$ in all except two cases, $\mathrm{PSL}_2(2)$ and $\mathrm{PSL}_2(3)$ (which happen to be isomorphic to $\mathcal{S}_3$ and $\mathcal{A}_4$, respectively). The other groups of this type are harder to describe.
- There are just twenty-six further finite simple groups; these are the *sporadic groups*.

The *Classification of finite simple groups* asserts the following.

Classification theorem *Any finite simple group is cyclic of prime order, or an alternating group, or a group of Lie type, or one of the twenty-six sporadic groups.*

Permutation groups

A *permutation* of the set Ω is a bijective mapping $g\colon \Omega \to \Omega$. We write the image of the point $v \in \Omega$ under the permutation g as vg, rather than $g(v)$. The *composition* g_1g_2 of two permutations g_1 and g_2 is the permutation obtained by applying g_1 and then g_2 that is,

$$v(g_1g_2) = (vg_1)g_2 \quad \text{for each } v \in \Omega.$$

A *permutation group* on Ω is a set $\mathcal{G}$ of permutations of Ω satisfying the following conditions:

- $\mathcal{G}$ is closed under composition: if $g_1, g_2 \in \mathcal{G}$ then $g_1g_2 \in \mathcal{G}$;
- $\mathcal{G}$ contains the *identity permutation* 1, defined by $v1 = v$ for $v \in \Omega$.
- $\mathcal{G}$ is closed under inversion, where the *inverse of* g is the permutation g^{-1} defined by the rule that $vg^{-1} = w$ if $wg = v$.

The *degree* of the permutation group $\mathcal{G}$ is the cardinality of the set Ω.

The simplest example of a permutation group is the set of all permutations of a set Ω. This is the *symmetric group*, denoted by $\mathrm{Sym}(\Omega)$. More generally, an *action* of $\mathcal{G}$ on Ω is a homomorphism from $\mathcal{G}$ to $\mathrm{Sym}(\Omega)$. The image of the homomorphism is then a permutation group. The action is *faithful* if its kernel is $\{1\}$ – that is, if distinct group elements map to distinct permutations. If the action is faithful, then $\mathcal{G}$ is isomorphic to a permutation group on Ω.

Products

Let $\mathcal{G}_1$ and $\mathcal{G}_2$ be permutation groups on Ω_1 and Ω_2 respectively. We define two products:

The *direct product* $\mathcal{G}_1 \times \mathcal{G}_2$ acts on the disjoint union $\Omega_1 \cup \Omega_2$. Each element of the group is an ordered pair (g_1, g_2), with $g_1 \in \mathcal{G}_1$, $g_2 \in \mathcal{G}_2$, and acts by the rule

$$v(g_1, g_2) = \begin{cases} vg_1 & \text{if } v \in \Omega_1, \\ vg_2 & \text{if } v \in \Omega_2. \end{cases}$$

It also has an action as a permutation group on the Cartesian product $\Omega_1 \times \Omega_2$, where

$$(v_1, v_2)(g_1, g_2) = (v_1g_1, v_2g_2), \qquad \text{for } v_1 \in \Omega_1, v_2 \in \Omega_2.$$

The *wreath product* $\mathcal{G}_1 \wr \mathcal{G}_2$ is a permutation group on $\Omega_1 \times \Omega_2$. Its elements are compositions of two types of permutations:

- for each function $f : \Omega_2 \to \mathcal{G}_1$, the permutation given by

$$(v_1, v_2)f = (v_1 f(v_2), v_2);$$

- for each element $g \in \mathcal{G}_2$, the permutation given by

$$(v_1, v_2)g = (v_1, v_2 g).$$

It also has an action as a permutation group on the set of all functions from Ω_2 to Ω_1. We do not define this here, but note that it arises in connection with the automorphism groups of the Hamming graphs in Chapter 7, Section 5.

The same (abstract) group may act as a permutation group on many different sets. Two actions of $\mathcal{G}$ on sets Ω_1 and Ω_2 are *isomorphic* if there is a bijection f from Ω_1 to Ω_2 that 'commutes with the action of $\mathcal{G}$' – that is,

$$(vf)g = (vg)f, \text{ for all } v \in \Omega_1 \text{ and } g \in \mathcal{G}.$$

More generally, the permutation groups $\mathcal{G}_1$ on Ω_1 and $\mathcal{G}_2$ on Ω_2 are *equivalent* if there are a bijection f from Ω_1 to Ω_2 and an isomorphism ϕ from $\mathcal{G}_1$ to $\mathcal{G}_2$ such that

$$(vf)(g\phi) = (vg)f, \text{ for all } v \in \Omega_1 \text{ and } g \in \mathcal{G}_1.$$

Note that, although isomorphic actions are equivalent, two actions of the same group may be equivalent without being isomorphic.

Automorphism groups of graphs

Let $G = (V, E)$ be a simple graph, possibly directed and possibly containing loops. An *automorphism* of G is a permutation g of V with the property that $\{vg, wg\}$ is an edge if and only if $\{v, w\}$ is an edge – or, if G is a digraph, that (vg, wg) is an arc if and only if (v, w) is an arc. Now the set of all automorphisms of G is a permutation group $\mathrm{Aut}(G)$, called the *automorphism group* of G.

The definition of an automorphism of a multigraph is a little more complicated. The most straightforward approach is to interpret a multigraph as a weighted graph. If $a_{v,w}$ denotes the multiplicity of vw as an edge of G, then an automorphism is a permutation of V satisfying $a_{vg,wg} = a_{v,w}$. Again, the set of automorphisms is a group.

We note the following results:

Theorem

- *A simple undirected graph and its complement have the same automorphism group.*

- *The automorphism group of the complete graph K_n or the null graph N_n is the symmetric group $\mathcal{S}_n$.*
- *The 5-cycle C_5 has ten automorphisms, realized geometrically as the rotations and reflections of a regular pentagon.*

This last group is the *dihedral group* D_{10}. More generally, $\mathrm{Aut}(C_n)$ is the dihedral group D_{2n}, for $n \geq 3$.

We can describe the automorphism group of a disconnected graph in terms of the group products introduced in the last section.

Theorem *Let G be a graph whose connected components are n_1 copies of G_1, n_2 copies of $G_2, \ldots, n_r$ copies of G_r, where $G_1, G_2, \ldots, G_r$ are pairwise non-isomorphic connected graphs. Then*

$$\mathrm{Aut}(G) = (\mathrm{Aut}(G_1) \wr \mathcal{S}_{n_1}) \times (\mathrm{Aut}(G_2) \wr \mathcal{S}_{n_2}) \times \cdots \times (\mathrm{Aut}(G_r) \wr \mathcal{S}_{n_r}).$$

This theorem illustrates the notions of both direct and wreath product. First, if $r = 1$, then any automorphism is a composition of n_1 independently chosen automorphisms of G_1 (acting on the components) and an arbitrary permutation of the components – that is, the group is $\mathrm{Aut}(G_1) \wr \mathcal{S}_{n_1}$. For general values of r, automorphisms of the subgraphs consisting of copies of each G_i can be chosen independently and combined, giving the direct product.

Orbits and stabilizers

Let $\mathcal{G}$ be a permutation group on Ω. The relation $\sim$ on Ω, defined by

$$v \sim w \quad \text{if} \quad w = vg \quad \text{for some } g \in \mathcal{G},$$

is an equivalence relation, and its equivalence classes are the *orbits* of $\mathcal{G}$. $\mathcal{G}$ is *transitive* if it has just one orbit; thus, $\mathcal{G}$ is transitive if, for any $v, w \in \Omega$, there exists $g \in \Omega$ such that $vg = w$.

The *stabilizer* $\mathcal{G}_v$ of a point $v \in \Omega$ is the set

$$\mathcal{H} = \{g \in \mathcal{G} : vg = v\};$$

it is a subgroup of $\mathcal{G}$. Moreover, if w is a point in the same orbit as v, then the set

$$\{g \in \mathcal{G} : vg = w\}$$

is a right coset of $\mathcal{H}$ in $\mathcal{G}$. This correspondence is a bijection between the orbit of v and the set $\mathcal{G}/\mathcal{H}$ of right cosets of $\mathcal{H}$ in $\mathcal{G}$. It is, moreover, an isomorphism between the actions of $\mathcal{G}$ on the orbit of v and on the set $\mathcal{G}/\mathcal{H}$, where the group acts by right multiplication on the latter set. It is thus possible to identify Ω with the union of the coset spaces of a family of subgroups of $\mathcal{G}$, one subgroup for each orbit of $\mathcal{G}$.

Group actions

Given a subgroup $\mathcal{H}$ of a group $\mathcal{G}$, there is an action of $\mathcal{G}$ as a permutation group on the set of right cosets of $\mathcal{H}$: this action is transitive. However, the induced permutation group may not be isomorphic to $\mathcal{G}$, since there may be different elements of $\mathcal{G}$ inducing the same permutation (or, equivalently, a non-identity element of $\mathcal{G}$ inducing the identity permutation). Indeed, the stabilizer of the coset $\mathcal{H}g$ is the *conjugate* $g^{-1}\mathcal{H}g$ of $\mathcal{H}$, so the kernel of the action (the subgroup fixing all cosets) is the *core* of $\mathcal{H}$, the intersection of all the conjugates. We say that $\mathcal{H}$ is *core-free* if

$$\bigcap_{g \in \mathcal{G}} g^{-1}\mathcal{H}g = \{1\};$$

thus, the action is faithful if and only if $\mathcal{H}$ is core-free.

In particular, if $\mathcal{H}$ is the identity subgroup, then it is core-free. In this case, the cosets are just the singleton subsets of $\mathcal{G}$, and we lose nothing by identifying them with the elements of $\mathcal{G}$. We obtain the action of $\mathcal{G}$ on itself by right multiplication, where the element g induces the permutation $x \mapsto xg$ of $\mathcal{G}$. This action was used by Cayley to show that every group is isomorphic to a permutation group.

Transitivity

Let $G = (V, E)$ be a graph or digraph, and let $\mathcal{G}$ be a group of automorphisms of G (a subgroup of the automorphism group Aut(G)). G is *$\mathcal{G}$-vertex-transitive*, or $\mathcal{G}$ acts *vertex-transitively* on G, if $\mathcal{G}$ is transitive on the set V; edge-transitivity is defined similarly. The graph G is *$\mathcal{G}$-symmetric* if $\mathcal{G}$ acts transitively on the set of ordered pairs (v, w) of adjacent vertices. (If G is a digraph, this is equivalent to edge-transitivity, but for an undirected graph it is a little stronger, since it implies the existence of an element of $\mathcal{G}$ that interchanges the two ends of an edge.) Finally, G is *vertex-transitive* (or *edge-transitive*, or *symmetric*) if the appropriate transitivity holds with $\mathcal{G} = \text{Aut}(G)$. The term *arc-transitive* is sometimes used as a synonym for 'symmetric'. Also, if G is $\mathcal{G}$-symmetric, we call $(G, \mathcal{G})$ a *symmetric pair*. For example, the pentagon C_5 is a symmetric graph, and (C_5, D_{10}) is a symmetric pair. The cyclic subgroup of order 5 acts vertex-transitively, but not symmetrically, on C_5.

Orbitals and rank

In the 1960s, Sims introduced graph-theoretic methods into the study of permutation groups, as follows.

Let $\mathcal{G}$ be a permutation group on Ω. Then there is a natural action of $\mathcal{G}$ on $\Omega \times \Omega$, the *coordinate-wise action*:

$$(v, w)g = (vg, wg),$$

for $g \in \mathcal{G}$, $v, w \in \Omega$. An orbit of $\mathcal{G}$ in this action is an *orbital*, and the number of orbitals is the *rank*. Orbitals are of two types:

- a *diagonal orbital* is one of the form $O = \{(v, v): v \in A\}$, where A is an orbit of $\mathcal{G}$ in Ω;
- a *non-diagonal orbital* is one consisting of ordered pairs of distinct elements.

If the degree is greater than 1, there is at least one orbital of each type, and so the rank is at least 2. It is equal to 2 if and only if $\mathcal{G}$ is *2-transitive* (or *doubly transitive*) – that is, for any two pairs (v_1, v_2) and (w_1, w_2) of distinct elements of Ω, there is an element $g \in \mathcal{G}$ such that $v_1 g = w_1$ and $v_2 g = w_2$.

There is a natural *pairing* of the orbitals: the orbital O^* paired with O is $\{(w, v): (v, w) \in O\}$. If $O = O^*$, the orbital O is *self-paired*; trivially, a diagonal orbital is self-paired.

Sims defined the *orbital graph* $G(O)$ associated with an orbital O as an undirected graph if O is self-paired and a directed graph otherwise, as follows: the vertex-set is Ω, and the edges are all pairs vw for which $(v, w) \in O$. If O is diagonal, then $G(O)$ just has a loop at every vertex in the corresponding orbit; and if O is not self-paired, then $G(O^*)$ is the converse digraph of $G(O)$.

Graphs admitting a given group

Sims' construction of orbital graphs can in a sense be reversed. Let $\mathcal{G}$ be a permutation group on Ω. Which simple graphs or digraphs G with vertex-set Ω admit $\mathcal{G}$ as a group of automorphisms? It is clear that, for any such graph, the set of ordered pairs of adjacent vertices is preserved by $\mathcal{G}$, and so is a union of orbitals. Conversely, any union of orbitals is the arc set of a digraph G for which $\mathcal{G}$ is a subgroup of Aut(G).

The construction is very flexible. If we want to exclude loops, then we consider only non-diagonal orbitals. If we insist that the graphs are undirected, then we take only unions that contain both or neither of each pair O, O^* of orbitals. It is also very practical: instead of having to list all the edges in order to specify a graph, we need give only a set of orbit representatives. This fact is exploited in Chapter 10.

This construction can be interpreted within the group. If A is an orbit of $\mathcal{G}$ on Ω, and if $\mathcal{H} = \mathcal{G}_v$ is the stabilizer of the point $v \in A$, then (as noted earlier) there is a bijection between A and the set of right cosets of $\mathcal{H}$ in $\mathcal{G}$, as follows: to the point $w \in A$ corresponds the set $\{g \in \mathcal{G} \colon vg = w\}$. So we can identify Ω with the set of all such cosets. Now each orbital O of $\mathcal{G}$ is contained in $A_1 \times A_2$, for some orbits A_1, A_2 of $\mathcal{G}$ on Ω. If $\mathcal{H}_i$ is the stabilizer of a point $v_i \in A_i$, for $i = 1, 2$, then the orbit O consists of all pairs $(\mathcal{H}_1 x, \mathcal{H}_2 y)$ of cosets such that $x \in \mathcal{H}_1 g \mathcal{H}_2 y$ – that is, $\mathcal{H}_1 x \subseteq \mathcal{H}_1 g \mathcal{H}_2 y$, for some element $g \in \mathcal{G}$. Now the *double coset* $\mathcal{H}_1 g \mathcal{H}_2$ corresponds to this orbit. The construction can be reversed: to the double coset $\mathcal{H}_1 g \mathcal{H}_2$ corresponds the above set of pairs of cosets (a subset of $A_1 \times A_2$), and this set is an orbit of $\mathcal{G}$.

So any graph on the vertex-set Ω which admits $\mathcal{G}$ as a group of automorphisms can be represented by a family of subgroups of $\mathcal{G}$ (the stabilizers of vertices in the $\mathcal{G}$-orbits in Ω), together with a family of double cosets of these subgroups, corresponding to the $\mathcal{G}$-orbits on $\Omega \times \Omega$ that are edges of the graph.

The description becomes simpler when $\mathcal{G}$ is vertex-transitive, and even more so when $\mathcal{G}$ is symmetric. First note that, if $\mathcal{G}$ is transitive on Ω, then the stabilizer $\mathcal{H}$ of a point of Ω is core-free in $\mathcal{G}$. So any vertex-transitive graph has the form

$$G(\mathcal{G}, \mathcal{H}, \mathcal{S}),$$

where $\mathcal{H}$ is a core-free subgroup of $\mathcal{G}$, and $\mathcal{S}$ is the union of a set of double cosets of the form $\mathcal{H} x \mathcal{H}$; in fact $\mathcal{S} = \{g \in \mathcal{G} \colon vg \sim v\}$, where $\mathcal{H} = \mathcal{G}_v$. The graph is undirected if, for all $x \in \mathcal{S}$, we have $x^{-1} \in \mathcal{S}$. The connected component containing v consists of the vertices whose corresponding cosets belong to the subgroup generated by $\mathcal{S}$. Such a graph is a *coset graph*.

Now the graph $G(\mathcal{G}, \mathcal{H}, \mathcal{S})$ is $\mathcal{G}$-symmetric if and only if $\mathcal{S}$ consists of just one double coset $\mathcal{H} g \mathcal{H}$. It is also undirected if and only if $g^{-1} \in \mathcal{H} g \mathcal{H}$. This leads to the following result of Sabidussi:

Theorem *Let G be an undirected $\mathcal{G}$-symmetric graph. Then G is isomorphic to $(\mathcal{G}, \mathcal{H}, \mathcal{H} g \mathcal{H})$, for some subgroup $\mathcal{H}$ and some element g of $\mathcal{G}$, where $\mathcal{H} = \mathcal{G}_v$ and $vg \sim v$ for some vertex v, and $g^{-1} \in \mathcal{H} g \mathcal{H}$. This isomorphism is an equivalence of $\mathcal{G}$-actions. Moreover, G is connected if and only if $\langle \mathcal{H}, g \rangle = \mathcal{G}$.*

Note that we may assume, without loss of generality, that the order of g is a power of 2 and that g interchanges v and vg.

The fact that the given symmetric graph G is isomorphic to a coset graph for the given group $\mathcal{G}$ suggests the need for a clear notion of equivalence of symmetric pairs. An *equivalence* between symmetric pairs $(G, \mathcal{G})$ and $(G', \mathcal{G}')$ consists of a graph isomorphism $f \colon G \to G'$ and a group isomorphism $\phi : \mathcal{G} \to \mathcal{G}'$ such that, for all vertices u of G and all $g \in \mathcal{G}$, $((u)f)g = (u(g\phi))f$. We then write

$(G, \mathcal{G}) \cong (G', \mathcal{G}')$. Thus, every symmetric pair $(G, \mathcal{G})$ is equivalent to the symmetric pair $(G^*, \mathcal{G})$, where G^* is the coset graph $G(\mathcal{G}, \mathcal{H}, \mathcal{H}g\mathcal{H})$.

Primitivity and double transitivity

Let $\mathcal{G}$ be a transitive permutation group on Ω. $\mathcal{G}$ is *primitive* on Ω if the only equivalence relations on Ω that are invariant under the action of $\mathcal{G}$ are the trivial ones – the relation of equality, and the 'universal' relation with a single equivalence class. Now the following assertions hold:

- a 2-transitive group is primitive;
- $\mathcal{G}$ is primitive on Ω if and only if a stabilizer $\mathcal{H} = \mathcal{G}_v$ is a maximal proper subgroup of $\mathcal{G}$, for $v \in \Omega$;
- If $\mathcal{G}$ is primitive on Ω, and $\mathcal{N}$ is a non-trivial normal subgroup of Ω, then $\mathcal{N}$ is transitive on Ω.

Motivated by the second assertion, we call a permutation group $\mathcal{G}$ on Ω *quasi-primitive* if all of its non-trivial normal subgroups are transitive. Thus, a primitive group is quasiprimitive.

If $\mathcal{G}$ is transitive on Ω, then there is a single diagonal orbital $\{(v, v)\colon v \in \Omega\}$; the corresponding orbital graph consists of a loop at each vertex. Sims proved the following result:

Theorem *The transitive permutation group $\mathcal{G}$ on Ω is primitive if and only if every non-diagonal orbital graph is connected.*

For, let $G(O)$ be the orbital graph corresponding to the orbital O containing (v, w). It can be shown that, if $\mathcal{H} = \mathcal{G}_v$ and $vg = w$, then the connected component of $G(O)$ containing v is the orbit of the subgroup $\langle \mathcal{H}, g\rangle$. So, if $\mathcal{G}$ is primitive, then $\mathcal{H}$ is a maximal subgroup of $\mathcal{G}$, and so $\langle H, g\rangle = \mathcal{G}$ for all $g \notin \mathcal{H}$; thus all non-diagonal orbital graphs are connected. Conversely, if $\mathcal{G}$ is not primitive, then there is a proper subgroup $\mathcal{K}$ of $\mathcal{G}$ with $\mathcal{H} < \mathcal{K}$. So, if $g \in \mathcal{K}$, $g \notin \mathcal{H}$, and $w = vg$, then the orbital graph containing the edge (v, w) is neither diagonal nor connected.

Many classifications of particular classes of (quasi)primitive groups relevant to graph theory have been obtained, using the Classification of finite simple groups. Among these are:

- the 2-transitive groups;
- the primitive permutation groups with rank 3;
- the primitive permutation groups of odd degree, other than those of 'affine type'.

One of the main tools in applying this powerful theorem is the O'Nan-Scott Theorem, which describes the structure of the *socle* (the product of the minimal normal subgroups) of a (quasi)primitive permutation group. See Chapter 7, Section 3, for an account of this theorem.

References

1. P. J. Cameron, *Permutation Groups*, London Math. Soc. Student Texts **45**, Cambridge Univ. Press, 1999.
2. J. D. Dixon and B. Mortimer, *Permutation Groups*, Springer, 1996.
3. S. Lipschutz, *Linear Algebra*, Schaum's Outline Series, McGraw-Hill, 1974.
4. D. B. West, *Introduction to Graph Theory* (2nd ed.), Prentice-Hall, 2001.
5. R. J. Wilson, *Introduction to Graph Theory* (4th ed.), Pearson, 1996.

1
Eigenvalues of graphs

MICHAEL DOOB

This chapter gives a survey of some of the relationships between the properties of a graph and the spectrum of its adjacency matrix. First we give some working examples and use them to illustrate the necessary matrix theory background. Next, the eigenvalues of a graph are related to the path structure. Then we examine the construction of eigenvalues via graph labellings. The implications of bounding the eigenvalues (from both above and below) are also explored. Finally, we examine the question of whether a graph is determined by its spectrum. In addition, we note some relationships between a graph and other matrices related to the adjacency matrix.

1. Introduction

The *eigenvalues* of a graph are the eigenvalues of its usual (0-1)-adjacency matrix. The relationships between the algebraic properties of these eigenvalues and the usual (topological and geometric) properties of graphs have been studied quite intensively. At first blush it seems implausible that such eigenvalues would have any relationship with such properties of a graph. In this chapter we shall show that this is not the case. We shall also lay the groundwork for some deeper results appearing in subsequent chapters by giving a general background for the theory and presenting some useful computational tools.

Eigenvalues of other matrices (Laplacian, Seidel matrices, etc.) associated with a graph are also of interest, and when we talk of such eigenvalues we will specifically note the type of matrix under consideration. Relationships between these matrices and the adjacency matrix will be discussed.

A good introduction to the some of the earlier results concerning eigenvalues of graphs is given in [25]. More advanced work is discussed in [10] and [15]. A list of all the graphs with up to seven vertices, with their eigenvalues and characteristic polynomials, appears in [22].

2. Some examples

It is always good to have some particular graphs with which to work, so first we give some examples and their eigenvalues. We start with a specific example, the graph Q_3 in Fig. 1.

The adjacency matrix $\mathbf{A}$ is a square matrix of order 8:

$$\mathbf{A} = \begin{pmatrix} 0 & 0 & 0 & 0 & 0 & 1 & 1 & 1 \\ 0 & 0 & 0 & 0 & 1 & 0 & 1 & 1 \\ 0 & 0 & 0 & 0 & 1 & 1 & 0 & 1 \\ 0 & 0 & 0 & 0 & 1 & 1 & 1 & 0 \\ 0 & 1 & 1 & 1 & 0 & 0 & 0 & 0 \\ 1 & 0 & 1 & 1 & 0 & 0 & 0 & 0 \\ 1 & 1 & 0 & 1 & 0 & 0 & 0 & 0 \\ 1 & 1 & 1 & 0 & 0 & 0 & 0 & 0 \end{pmatrix}.$$

The characteristic polynomial of this matrix is easily computed:

$$\det(x\mathbf{I} - \mathbf{A}) = x^8 - 12x^6 + 30x^4 - 28x^2 + 9 = (x-3)(x+3)(x-1)^3(x+1)^3,$$

and so the eigenvalues of Q_3 are 3, −3, 1 and −1, with respective multiplicities 1, 1, 3 and 3.

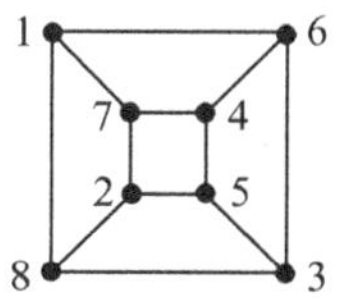

Fig. 1.

Table 1.

Graph	Eigenvalues	Respective multiplicities
K_n	$n-1, -1$	$1, n-1$
$K_{r,s}$	$\sqrt{rs}, 0, -\sqrt{rs}$	$1, r+s-2, 1$
C_n	$2\cos\left(\frac{2k\pi}{n}\right), k = 1, 2, \ldots, n$	$2, 1, 1, \ldots, 1, 2$ for n even $1, 1, \ldots, 1, 2$ for n odd
P_n	$2\cos\left(\frac{k\pi}{n+1}\right), k = 1, 2, \ldots, n$	$1, 1, \ldots, 1$
$L(K_r)$	$2r-4, r-4, -2$	$1, r-1, r(r-3)/2$
$L(K_{r,r})$	$2r-2, r-2, -2$	$1, 2(r-1), (r-1)^2$
$L(K_{r,s})$	$r+s-2, r-2, s-2, -2$	$1, s-1, r-1, (r-1)(s-1)$
Q_d	$d-2k, k = 0, 1, \ldots, d$	$\binom{d}{k}, k = 0, 1, \ldots, d$
CP(r) or $K_{r(2)}$	$2r-2, 0, -2$	$1, r, r-1$

Table 1 gives some more examples. By the end of this chapter, we shall have verified all of them. In this table, note that two graphs have the same spectrum only if they are isomorphic. This is not true in general, as the following examples show. The first pair of non-isomorphic graphs, shown in Fig. 2, is the smallest one with the same eigenvalues: they are 2, 0 and -2, with respective multiplicities 1, 3 and 1. We see from this example that, in general, connectivity cannot be determined from the eigenvalues alone.

The next graphs, shown in Fig. 3, are the smallest *connected* pair with the same eigenvalues. These two graphs have diameters 2 and 4. Hence, in general, the diameter of a graph cannot be determined from the eigenvalues.

The trees in Fig. 4, are the smallest pair with the same eigenvalues: $\pm\frac{1}{2} \pm \frac{\sqrt{13}}{2}$, each of multiplicity 1, and 0 with multiplicity 4. These trees have very different

Fig. 2. Characteristic polynomial: $p(x) = x^5 - 4x^3$

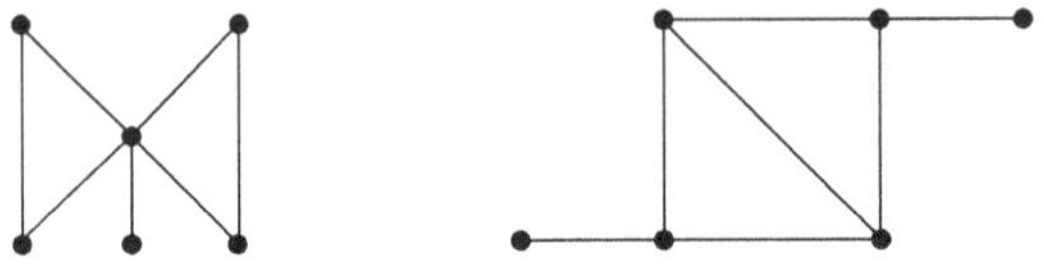

Fig. 3. Characteristic polynomial: $p(x) = x^6 - 7x^4 - 4x^3 + 7x^2 + 4x - 1$

Fig. 4. Characteristic polynomial: $p(x) = x^8 - 7x^6 + 9x^4$

degrees. Hence, in general, we cannot determine much about the degree sequence from the eigenvalues of a graph.

From these examples we see that there are several properties of graphs that are not determined by their eigenvalues; more will be said about this in Section 9. On the other hand, there are other properties (for example, the number of edges and the number of triangles in a graph) that are determined by the eigenvalues. Some of these will be given in Section 4. The sorting of graph properties into those determined by the eigenvalues or not continues to be of research interest.

3. A little matrix theory

Since the adjacency matrix of a graph is real and symmetric, we may use known results from the theory of matrices to our advantage. We will state some of them without proofs, which may be found in [14] or [20]. The first is sometimes called the *principal axis theorem*.

Theorem 3.1 (Principal axis theorem) *If* $\mathbf{A}$ *is a real symmetric matrix of order n, then* $\mathbf{A}$ *has n real eigenvalues and a corresponding orthonormal set of eigenvectors.*

If $\mathbf{U}$ is a matrix with an orthonormal set of eigenvectors as columns, then $\mathbf{U}^T = \mathbf{U}^{-1}$ and $\mathbf{U}^T\mathbf{A}\mathbf{U} = \mathbf{D}$, where $\mathbf{D}$ is the diagonal matrix whose entries are the corresponding eigenvalues. This gives a diagonalization theorem.

Theorem 3.2 *If* $\mathbf{A}$ *is a real symmetric matrix, then there exists a matrix* $\mathbf{U}$ *such that* $\mathbf{U}^T\mathbf{A}\mathbf{U} = \mathbf{D}$. *In addition, the minimum polynomial is* $\prod(x - \lambda_i)$, *where the product is taken over the distinct eigenvalues.*

The next theorem, the *Perron-Frobenius theorem*, gives information about the largest eigenvalue of a matrix with non-negative entries.

Theorem 3.3 (Perron-Frobenius theorem) *If* $\mathbf{A}$ *is a non-negative matrix with eigenvalues* $\lambda_1 \geq \lambda_2 \geq \cdots \geq \lambda_n$, *then* $|\lambda_1| \geq |\lambda_k|$, *for* $k = 1, 2, \ldots, n$, *and the eigenvalue* λ_1 *has an eigenvector with all entries non-negative. If* $\mathbf{A}$ *is* *indecomposable, then the eigenvalue λ_1 is simple ($\lambda_1 > \lambda_2$), and the eigenvector has all entries positive.*

The adjacency matrix of a graph is indecomposable precisely when the graph is connected. Hence the largest eigenvalue of a connected graph is simple.

It is a straightforward observation that the spectrum of a graph is the union of the spectra of its connected components. So, unless otherwise stated, we assume that all graphs under consideration are connected.

The following result is known as the *Interlacing theorem.*

Theorem 3.4 (Interlacing theorem) *Let $\mathbf{A}$ be a real symmetric matrix with eigenvalues $\lambda_1 \geq \lambda_2 \geq \cdots \geq \lambda_n$, and let $\mu_1 \geq \mu_2 \geq \cdots \geq \mu_{n-1}$ be the eigenvalues of a principal submatrix of $\mathbf{A}$. Then $\lambda_i \geq \mu_i \geq \lambda_{i+1}$, for $i = 1, 2, \ldots, n-1$.*

This theorem is of major importance in the study of graph eigenvalues. A principal submatrix corresponds to an induced subgraph with one fewer vertex. This result can be visualized as interlacing the eigenvalues on the real axis, as shown in Fig. 5.

The Interlacing theorem and an easy induction yield the following result for induced subgraphs:

Corollary 3.5 *If H is an induced subgraph of G, if $\mu_1 \geq \mu_2 \geq \cdots \geq \mu_m$ are the eigenvalues of H, and if $\lambda_1 \geq \lambda_2 \geq \cdots \geq \lambda_n$ are the eigenvalues of G, then $\lambda_{i+n-m} \leq \mu_i \leq \lambda_i$, for $i = 1, 2, \ldots, m$.*

4. Eigenvalues and walks

The oldest, and perhaps the most fundamental, relationship between the eigenvalues of a graph and its geometric properties concerns walks – particularly, closed walks. Suppose that $\mathbf{A}$ is the adjacency matrix of a graph. Then the powers of $\mathbf{A}$ enumerate the walks in that graph.

Theorem 4.1 *If a graph G has adjacency matrix $\mathbf{A}$, then for $k = 0, 1, \ldots,$ the ij-entry of $\mathbf{A}^k$ is the number of v_i-v_j walks of length k.*

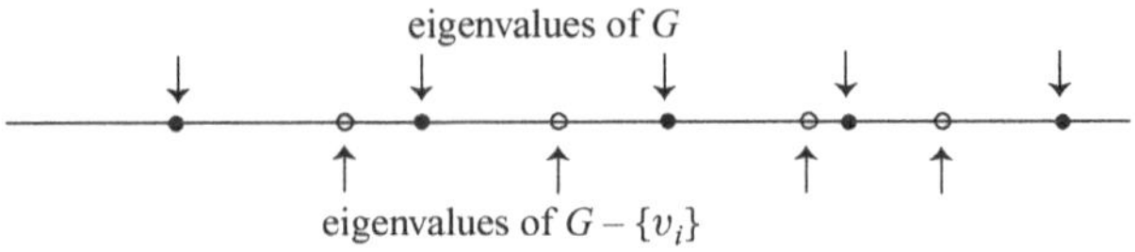

Fig. 5.

Proof The case $k = 0$ is trivial, but will be useful later in this section; the case $k = 1$ is the definition of the adjacency matrix. For larger values of k, the proof follows by an easy induction that uses nothing more than the definition of matrix multiplication. ■

This theorem is often applied to enumerate closed walks of length k. These correspond to diagonal entries in $\mathbf{A}^k$, and so the number of closed walks of length k is the sum of the diagonal entries, which in turn is equal to tr $(\mathbf{A}^k)$.

Corollary 4.2 *If* $\mathbf{A}$ *is the adjacency matrix of a graph* G *with degree sequence* $d_1, d_2, \ldots, d_n$, *then*
- $A^2_{ii} = \deg v_i$;
- *the number of edges of* G *is given by*

$$|E(G)| = \tfrac{1}{2}\sum_{i=1}^{n} \deg v_i = \tfrac{1}{2}\mathrm{tr}\,(\mathbf{A}^2);$$

- *the number of triangles in* G *is* $\frac{1}{6}\mathrm{tr}\,(\mathbf{A}^3)$.

From this we see that the number of edges and the number of triangles in a graph are determined by its eigenvalues. This argument does not extend to k-cycles, with $k > 3$: the first example of cospectral graphs in Section 2 shows that the number of 4-cycles cannot be determined by the eigenvalues alone. However, the degree sequence adds enough extra information to determine the 4-cycles.

Corollary 4.3 *If* $d_1, d_2, \ldots, d_n$ *is the degree sequence of* G, *then the number of 4-cycles in* G *is* $\frac{1}{8}(\mathrm{tr}\mathbf{A}^4 + \mathrm{tr}\mathbf{A}^2 - \Sigma_{i=1}^{n} d_i^2)$.

We can also use Theorem 4.1 to bound the diameter of a graph by the number of distinct eigenvalues.

Theorem 4.4 *If a graph* G *has diameter* D *and* t *distinct eigenvalues, then*

$$D \leq t - 1.$$

Proof Suppose not. Then there exist two vertices v_i and v_j such that the distance between them is exactly t. Thus $\mathbf{A}^t_{ij} > 0$, but $\mathbf{A}^k_{ij} = 0$ for $k = 1, 2, \ldots, t - 1$. So if $q(x)$ is any polynomial of degree t, the ij-entry of $q(\mathbf{A})$ must be non-zero. But, by Theorem 3.2, the minimum polynomial is of degree t; when evaluated at $\mathbf{A}$, this polynomial has every entry equal to 0, and hence we have a contradiction. ■

The graphs K_n, $K_{r,s}$, C_n, P_n, $L(K_r)$, $L(K_{r,r})$, Q_d and CP(r), in the table of examples in Section 2, all attain the bound of the last theorem, while $L(K_{r,s})$ does not whenever $r \neq s$. Distance-regular graphs (described later in this chapter) meet this bound too; however, there is no known general characterization of graphs that attain this bound.

Matrix equations involving the minimum and other polynomials are often used to reveal information about the graph, and *vice versa*. For example, the complete graph K_n must have $\mathbf{A}^2_{ij} = n - 2$ if $i \neq j$, and $\mathbf{A}^2_{ii} = n - 1$. Thus

$$\mathbf{A}^2 - (n-2)\mathbf{A} = \mathbf{I},$$

and

$$x^2 - (n-2)x - 1 = (x - (n-1))(x+1)$$

is the minimum polynomial. Since the trace of $\mathbf{A}$ is 0, the eigenvalues are $n - 1$ (simple) and -1 ($n - 1$ times). Conversely, if a graph with n vertices has the same eigenvalues as K_n, then the minimum polynomial forces the adjacency matrix to satisfy $\mathbf{A}^2 - (n-2)\mathbf{A} = \mathbf{I}$, which makes the graph regular with degree $n - 1$, and hence K_n. We say that K_n is *characterized* by its spectrum.

Another polynomial is used in the case of regular connected graphs. In this case, the adjacency matrix $\mathbf{A}$ and the all-1 matrix $\mathbf{J}$ commute. If the degree of the graph is r, then the all-1 vector $\mathbf{j}$ is an eigenvector of $\mathbf{A}$ (with eigenvalue r) and of $\mathbf{J}$ (with eigenvalue n). Any other eigenvector of $\mathbf{A}$ is orthogonal to $\mathbf{j}$, and so is also an eigenvector of $\mathbf{J}$ (with eigenvalue 0). If we consider the interpolating polynomial

$$h(x) = n \prod \frac{x - \lambda_i}{r - \lambda_i},$$

where the product is taken over all distinct eigenvalues that are not equal to r, then $h(r) = n$ and $h(\lambda) = 0$ for any other eigenvalue λ. Using our orthonormal basis of eigenvectors, we see that $h(\mathbf{A}) - \mathbf{J}$ is a symmetric matrix that annihilates any vector, and hence $h(\mathbf{A}) - \mathbf{J} = 0$, or $h(\mathbf{A}) = \mathbf{J}$. This polynomial is called the *Hoffman polynomial*.

Strongly regular graphs

A graph is *strongly regular* with parameters (n, r, λ, μ) if it has n vertices, is regular of degree r, has each pair of adjacent vertices mutually adjacent to λ other vertices, and has each pair of non-adjacent vertices mutually adjacent to μ other vertices. By convention, we exclude any graph in which each component is complete (that is, $\mu = 0$), so that all strongly regular graphs are connected. An example of a strongly regular graph is the Petersen graph: it clearly has parameters (10, 3, 0, 1). Strongly regular graphs are considered in detail in Chapter 8.

For any strongly regular graph there are only three types of entries in the adjacency matrix: those for which two vertices are adjacent, those for which two vertices are non-adjacent, and the diagonal. We know that the entry in $\mathbf{A}^2$ is λ, μ and r in each of these cases. From this it is clear that

$$\mathbf{A}^2 + (\mu - \lambda)\mathbf{A} + (\mu - r)\mathbf{I} = \mu\mathbf{J}.$$

Hence we have essentially the Hoffman polynomial, and we know that the eigenvalues of the graph are r and the two roots of

$$x^2 + (\mu - \lambda)x + (\mu - r).$$

This means that strongly regular graphs attain the diameter bound of Theorem 4.4.

Once we have the eigenvalues, it is trivial to compute their multiplicities. If the distinct eigenvalues of the graph are r, λ_2 and λ_3, with respective multiplicities 1, m_2 and m_3, then we have

$$m_2 + m_3 = n - 1 \quad \text{and} \quad m_2\lambda_2 + m_3\lambda_3 = -r,$$

since the adjacency matrix of any graph has trace 0. These multiplicities must be integers, and, since $\lambda_2 \neq \lambda_3$, they are determined by the equations displayed above. The integrality of m_2 and m_3 has been an especially useful tool for showing the non-existence of potential strongly regular graphs with certain parameters.

We have thus verified the spectra of $L(K_n)$, $L(K_{r,r})$, $K_{r,r}$ and $\mathrm{CP}(r)$, as presented in Section 2, since these graphs are strongly regular.

A regular graph of degree r and diameter d can have at most

$$1 + r + r(r - 1) + \cdots + r(r - 1)^d$$

vertices. A graph that attains this bound is called a *Moore graph* of diameter d. A Moore graph of diameter $d = 2$ is strongly regular with parameters $(r^2 + 1, r, 0, 1)$. One of the oldest existence questions in spectral graph theory arises from these graphs: for which r do these Moore graphs exist?

We follow the logic given above: the eigenvalues are r plus the two solutions of

$$x^2 + x - (r - 1) = 0$$

– that is,

$$\tfrac{1}{2}(-1 \pm \sqrt{4r - 3}).$$

Then

$$\lambda_2 + \lambda_3 = -1 \quad \text{and} \quad \lambda_2 - \lambda_3 = \sqrt{4r-3}.$$

In addition,

$$m_2 + m_3 = r^2 \quad \text{and} \quad m_2\lambda_2 + m_3\lambda_3 = -r.$$

Hence,

$$(m_2 - m_3)\sqrt{4r-3} = r(r-2).$$

If $\sqrt{4r-3}$ is not an integer, then $m_2 = m_3$ and $r = 2$, so that the graph is C_5. Otherwise, $\sqrt{4r-3}$ is an odd integer, $2t+1$, so that $2t+1$ divides

$$16r(r-2) = (2t+1)^2(4t^2+4t-1) - 15.$$

Hence $2t+1$ divides 15, and so $t = 0, 1, 2$ or 7 – which in turn implies that $r = 1, 3, 7$ or 57. The case $r = 3$ gives the Petersen graph, and the case $r = 7$ also yields a single graph, called the *Hoffman-Singleton graph*. What happens in the final case $r = 57$ (a question raised in 1960 in [17]) has remained unsolved, despite much effort over several decades. The Petersen graph contains two copies of C_5, the Hoffman-Singleton graph has 50 vertices and contains 10 copies of the Petersen graph. The graph with $r = 57$, if it exists, would have 3250 vertices.

Much more will be said about strongly regular graphs in Chapter 8.

Distance-regular graphs

Strongly regular graphs can be generalized to distance-regular graphs with diameter greater than 2. We define $n_i(v)$ to be the number of vertices at distance i from v, and for two vertices v and w at distance k from each other, $p_{ij}^k(v, w)$ is the number of vertices at distance i from v and distance j from w. If $n_i(v)$ is the same for all v, and if $p_{ij}^k(v, w)$ is the same for all vertices v and w at distance k from each other, then the graph is *distance-regular*. A strongly regular graph with parameters (n, r, λ, μ) is such a graph with $n_1(v) = r$, $n_2(v) = n - r - 1$, $p_{11}^1 = \lambda$ and $p_{11}^2 = \mu$. The d-dimensional cube Q_d is distance-regular with diameter d.

A distance-regular graph of diameter d has an associated intersection array, an idea introduced in [2]. It has parameters $c_i = p_{i-1,1}^i$, $b_i = p_{i+1,1}^i$ and $a_i = p_{i1}^i = r - b_i - c_i$, for $i = 0, 1, \ldots, d$. It turns out that the eigenvalues of $\mathbf{A}$ are the same as the eigenvalues of the tridiagonal matrix

$$\begin{pmatrix} a_0 & b_0 & 0 & 0 & \cdots & 0 & 0 \\ c_1 & a_1 & b_1 & 0 & \cdots & 0 & 0 \\ 0 & c_2 & a_2 & b_2 & \cdots & 0 & 0 \\ \vdots & \vdots & \vdots & \vdots & \ddots & \vdots & \vdots \\ 0 & 0 & 0 & 0 & \cdots & a_{d-1} & b_{d-1} \\ 0 & 0 & 0 & 0 & \cdots & c_d & a_d \end{pmatrix}.$$

Since the order of this matrix is only $d + 1$, there can be at most $d + 1$ eigenvalues. By Theorem 4.4, the graph must have at least $d + 1$ eigenvalues. Hence all distance-regular graphs attain the diameter bound of Theorem 4.4.

We conclude this section with two important distance-regular graphs.

The *Hamming graph* $H(d, l)$ has the d-tuples $(x_1, x_2, \ldots, x_d)$ as its vertices, where $1 \leq x_i \leq l$. Two vertices are adjacent if, as d-tuples, they agree in all but one coordinate. This graph has diameter d, since the distance between two vertices is the number of coordinates in which the corresponding d-tuples differ. The eigenvalues are $kl - d$, with multiplicity $\binom{d}{k}$, for $k = 0, 1, \ldots, d$. Notice that if $l = 2$, we have the cube graph Q_d.

The *Johnson graph* $J(d, l)$ has vertices corresponding to the $\binom{n}{d}$ subsets of $\{1, 2, \ldots, l\}$. Two vertices are adjacent if, as subsets, their intersection is of size $d - 1$. This graph has diameter d, since the distance between two vertices is k if and only if the corresponding subsets have intersection of size $d - k$. Notice that if $d = 2$, we have the graph $L(K_l)$.

The book [4] is an encyclopaedic reference on distance-regular graphs.

5. Eigenvalues and labellings of graphs

The columns of the adjacency matrix **A** correspond to the vertices of the graph. If **A** acts on a vector **x**, then the entries of that vector can have the same correspondence, so a vector may be used to label the vertices. When we use an eigenvector, the eigenvalues will reflect the geometric structure of the graph.

Theorem 5.1 *If* **A** *is the adjacency matrix of a graph G with vertices* $\{v_1, v_2, \ldots, v_n\}$, *and the vector* **x** *is viewed as a labelling of the vertices of G, then, for* $k = 1, 2, \ldots, n$, $(\mathbf{Ax})_k$ *is the sum of the labels on the vertices adjacent to* v_k.

This result is convenient for direct calculation. For example, consider the cycle C_n, and let $\varepsilon_k = e^{2i\pi k/n}$ (an nth root of unity). If the vertices in order around the cycle are $v_1, v_2, \ldots, v_n$, let $\mathbf{x} = (x_1, x_2, \ldots, x_n)$ be defined by letting $x_j = \varepsilon_k^j$.

Then $(\mathbf{A}\mathbf{x})_j = x_{j-1} + x_{j+1}$ (subscripts taken modulo n), and

$$\frac{x_{j-1} + x_{j+1}}{x_j} = \frac{e^{2i\pi(j-1)k/n} + e^{2i\pi(j+1)k/n}}{e^{2i\pi jk/n}} = e^{-2i\pi k/n} + e^{2i\pi k/n} = \varepsilon_k^{-1} + \varepsilon_k.$$

Hence, $\mathbf{A}\mathbf{x} = (\varepsilon_k^{-1} + \varepsilon_k)\mathbf{x}$ and, on letting k range from 1 to n, we obtain all the eigenvalues of C_n. Notice that the roots of unity ε_k and ε_{n-k} yield the same eigenvalue, and so, unless $k = n$ or $n/2$, this eigenvalue has multiplicity 2; otherwise, the multiplicity is 1.

Similar labellings can be used to find the eigenvalues of a path. Consider the path P_n and the cycle C_{2n+2}. Let $\mathbf{x}$ be any eigenvector of P_n and use that vector to label the first n vertices of C_{2n+2}. Label the next vertex 0, the next n vertices with $-\mathbf{x}$ in reverse order, and the last vertex 0. It is now easy to verify that we have an eigenvector of C_{2n+2}. If we shift the labels one step around the cycle (labelling the first vertex 0, the next n vertices by $\mathbf{x}$, the next one 0, and the final n by $-\mathbf{x}$ in reverse order), we get a second independent eigenvector. Hence the eigenvalues of P_n correspond to the double eigenvalues of C_{2n+2}. This technique first appeared in [25].

We can also use eigenvalue labellings to deduce an easy result about bipartite graphs. If G has a bipartition (S, T), and if $\mathbf{x}$ is an eigenvector with corresponding eigenvalue λ, then we can replace the label $\mathbf{x}_i$ with $-\mathbf{x}_i$ for each vertex in S to get an eigenvector with corresponding eigenvalue $-\lambda$. Hence, λ and $-\lambda$ have the same multiplicity, and the spectrum is symmetric about 0. Conversely, if G has a spectrum that is symmetric about 0, then the trace of $\mathbf{A}^{2k+1}$ is 0 and the graph has no odd cycles, and so G is bipartite. This yields one of the oldest results concerning the spectra of graphs.

Theorem 5.2 *A graph is bipartite if and only if its spectrum is symmetric about* 0.

Now suppose that v_i and v_j are non-adjacent vertices with the same neighbours, and suppose that $\mathbf{x}$ is an eigenvector with eigenvalue λ. If Σ is the sum of the labels on the neighbours of v_i and v_j, then clearly

$$\lambda x_i = (\mathbf{A}\mathbf{x})_i = \Sigma = (\mathbf{A}\mathbf{x})_j = \lambda x_j,$$

and hence $\lambda(x_i - x_j) = 0$. This yields the following result about the symmetry of the eigenvectors of a graph.

Corollary 5.3 *If v and w are non-adjacent vertices with the same neighbours, and if $\mathbf{x}$ is an eigenvector with eigenvalue λ, then either v and w have the same label or $\lambda = 0$.*

We can apply this result to $K_{r,s}$. If $\lambda \neq 0$, then any eigenvector $\mathbf{x}$ gives labels that can have only two possible values, say y and z. We can then say that $\lambda y = sz$ and $\lambda z = ry$, and in particular, $y \neq 0$ and $z \neq 0$. Hence $rsyz = szry = (\lambda y)(\lambda z) = \lambda^2 yz$. So $\lambda^2 = rs$, and the only non-zero eigenvalues are $\pm\sqrt{rs}$. Since $K_{r,s}$ is connected and has trace 0, these two non-zero values are simple eigenvalues.

A similar result can be obtained for adjacent vertices. Suppose that v_i and v_j are adjacent vertices with the same neighbours, and that $\mathbf{x}$ is a eigenvector with eigenvalue λ. Again, let Σ be the sum of the labels on the neighbours of v_i and v_j. Then

$$\Sigma + x_j = (\mathbf{Ax})_i = \lambda x_i \quad \text{and} \quad \Sigma + x_i = (\mathbf{Ax})_j = \lambda x_j,$$

so that $(\lambda + 1)(x_i - x_j) = 0$.

Corollary 5.4 *If v and w are adjacent vertices with the same neighbours, and if $\mathbf{x}$ is an eigenvector with eigenvalue λ, then either v and w have the same label or $\lambda = -1$.*

We can apply this result to the complete graph K_n with one edge deleted. Suppose that G has n vertices, all of which are pairwise joined except for v_1 and v_2, and suppose that $\mathbf{x}$ is an eigenvector with eigenvalue λ. For $\lambda = 0$, we see from Corollary 5.4 that $\mathbf{x}_3 = \mathbf{x}_4 = \cdots = \mathbf{x}_n$, but Corollary 5.3 tells us nothing. However, $(\mathbf{Ax})_1 = 0$ tells us that the common value is 0, and then $\mathbf{x}_1 = -\mathbf{x}_2$; thus $\lambda = 0$ has multiplicity 1. For $\lambda = -1$, we see from Corollary 5.3 that $\mathbf{x}_1 = \mathbf{x}_2$, but Corollary 5.4 tells us nothing. However,

$$-\mathbf{x}_1 = (\mathbf{Ax})_1 = \sum_{i=3}^{n} \mathbf{x}_i = 0 = (\mathbf{Ax})_2 = -\mathbf{x}_2,$$

and so $\mathbf{x}_1 = \mathbf{x}_2 = \sum_{i=1}^{n} \mathbf{x}_i$. On the other hand, $(\mathbf{Ax})_3 = -\mathbf{x}_3$ implies that $\sum_{i=1}^{n} \mathbf{x}_i = 0$. Hence the eigenspace is

$$\left\{ (\mathbf{x}_1, \mathbf{x}_2, \ldots, \mathbf{x}_n) : \mathbf{x}_1 = \mathbf{x}_2 = 0 = \sum_{i=3}^{n} \mathbf{x}_i \right\},$$

which has dimension $n - 3$. For any other eigenvalue λ, $\mathbf{x}_1 = \mathbf{x}_2 = a$ and $\mathbf{x}_3 = \mathbf{x}_4 = \cdots = \mathbf{x}_n = b$. Hence $\lambda a = (n-2)b$ and $\lambda b = (n-3)b + 2a$. Elimination gives

$$\lambda = \tfrac{1}{2}\left(n - 3 \pm \sqrt{(n+1)^2 - 8}\right).$$

Thus we have completely determined the spectrum.

Starting with a matrix $\mathbf{A}$, the *Rayleigh quotient* is defined using the usual inner product for any non-zero vector $\mathbf{x}$ as $\langle \mathbf{Ax}, \mathbf{x} \rangle / \langle \mathbf{x}, \mathbf{x} \rangle$. Since the Rayleigh quotients of $\mathbf{x}$ and $c\mathbf{x}$ are identical for $c \neq 0$, we can restrict our attention to vectors of unit length when convenient.

We first observe that if $\mathbf{A}$ is the adjacency matrix of a graph, then the Rayleigh quotient takes on precisely all values between the largest and smallest eigenvalues.

Theorem 5.5 *Let G be a graph with adjacency matrix* $\mathbf{A}$ *and with eigenvalues* $\lambda_1 \geq \lambda_2 \geq \cdots \geq \lambda_n$. *Then* $\langle \mathbf{A}\mathbf{x}, \mathbf{x}\rangle / \langle \mathbf{x}, \mathbf{x}\rangle$ *takes on precisely the values in the interval* $[\lambda_n, \lambda_1]$. *In addition,* $\langle \mathbf{A}\mathbf{x}, \mathbf{x}\rangle / \langle \mathbf{x}, \mathbf{x}\rangle = \lambda_1$ *or* λ_n *only if* $\mathbf{x}$ *is an eigenvector.*

Proof Let y be any non-zero vector, and let $\{\mathbf{x}_1, \mathbf{x}_2, \ldots, \mathbf{x}_n\}$ be an orthonormal basis of eigenvectors for $\mathbf{A}$. Then $\mathbf{y} = \sum r_i \mathbf{x}_i$ and

$$\frac{\langle \mathbf{A}\mathbf{y}, \mathbf{y}\rangle}{\langle \mathbf{y}, \mathbf{y}\rangle} = \frac{\left\langle \mathbf{A}\left(\sum r_i \mathbf{x}_i\right), \sum r_i \mathbf{x}_i \right\rangle}{\left\langle \sum r_j \mathbf{x}_j, \sum r_j \mathbf{x}_j \right\rangle} = \sum \frac{r_i^2 \lambda_i}{\sum r_j^2} = \sum \frac{r_i^2}{\sum r_j^2} \lambda_i .$$

The coefficients of λ_i are all non-negative and add to 1 – that is, the Rayleigh quotient is a convex combination of the eigenvalues. Hence, the Rayleigh quotient of every non-zero vector is in the interval $[\lambda_n, \lambda_1]$, each value in that interval is attained by the Rayleigh quotient for some vector, and the extreme values of the interval are attained only by eigenvectors. ■

Observe also that if λ_1 is the largest eigenvalue of a connected graph G, $\mathbf{x}$ is a corresponding eigenvector, and $\langle \mathbf{A}\mathbf{y}, \mathbf{y}\rangle / \langle \mathbf{y}, \mathbf{y}\rangle = \lambda_1$, then we can write $\mathbf{y}$ as a convex combination of eigenvectors of G. Since λ_1 is simple, this means that $\mathbf{y} = r\mathbf{x}$ for some real number $r \neq 0$, and so $\mathbf{y}$ has all of its coordinates positive or all of its coordinates negative.

Corollary 5.6 *If H is a proper induced subgraph of a connected graph G, if H has maximum eigenvalue* μ, *and if G has maximum eigenvalue* λ, *then* $\mu < \lambda$.

Proof We know that $\mu \leq \lambda$, from Theorem 3.4. If $\mathbf{z}$ is an eigenvector of H with eigenvalue μ, we can extend $\mathbf{z}$ to $\mathbf{y}$ by adding 0 as entries corresponding to vertices not in H; then $\langle \mathbf{A}\mathbf{y}, \mathbf{y}\rangle / \langle \mathbf{y}, \mathbf{y}\rangle = \mu$. If $\mu = \lambda$, then by the observation in the previous paragraph, all the coordinates of $\mathbf{y}$ are non-zero. This can happen only if $H = G$, and so $\mu < \lambda$. ■

Corollary 5.7 *If e is an edge of a connected graph G, then the largest eigenvalue of G is strictly greater than the largest eigenvalue of* $G - e$.

Proof Let $H = G - e$. If H is disconnected, then its maximum eigenvalue is attained on one of the components, which is a proper induced subgraph, so Corollary 5.6 applies. If H is connected, let $\mathbf{x}$ be an eigenvector of H corresponding to its largest eigenvalue μ. If the deleted edge joins v_i and v_j, and if $\mathbf{A}$ is the adjacency matrix for G, then the largest eigenvalue of G is bounded below by $\langle \mathbf{Ax}, \mathbf{x} \rangle / \langle \mathbf{x}, \mathbf{x} \rangle = \mu + 2x_i x_j$. Since $\mathbf{x}$ has only positive or only negative entries, the largest eigenvalue of G must be greater than that of H. ∎

The classic proof of the Perron-Frobenius theorem brackets the largest eigenvalue of a non-negative matrix by considering the values of $(\mathbf{Ax})_i / \mathbf{x}_i (i = 1, 2, \ldots, n)$, where $\mathbf{x}$ is a vector with non-negative coordinates. The largest eigenvalue λ_1 satisfies

$$\min_i \left\{ \frac{(\mathbf{Ax})_i}{\mathbf{x}_i} \right\} \leq \lambda_1 \leq \max_i \left\{ \frac{(\mathbf{Ax})_i}{\mathbf{x}_i} \right\}.$$

The proof actually shows that if we maximize the left value and minimize the right value over $\mathbf{x} \geq \mathbf{0}$, then we get the same value from both cases. If we let $\mathbf{j}$ be the all-1 vector and $\mathbf{A}$ be the adjacency matrix of a graph G with degrees $d_1, d_2, \ldots, d_n$, then we get

$$\min_i d_i \leq \lambda_1 \leq \max_i d_i.$$

On the other hand, if we use Theorem 5.5 with the same vector $\mathbf{j}$, we get $\frac{1}{n} \sum_{i=1}^{n} d_i \leq \lambda$. This results in the following theorem.

Theorem 5.8 *If G is a graph with degrees $d_1, d_2, \ldots, d_n$ and maximum eigenvalue λ_1, then*

$$\frac{1}{n} \sum_{i=1}^{n} d_i \leq \lambda_1 \leq \max_i d_i.$$

Equality is attained if and only if the graph is regular.

In other words, the maximum eigenvalue lies between the average degree and the maximum degree, and equals both values if and only if the graph is regular.

6. Lower bounds for the eigenvalues

Finding lower bounds for the eigenvalues of graphs has been a recurring theme in the study of graph spectra. In this section we use $\lambda(G)$ to denote the smallest eigenvalue.

Proposition 6.1 *Let G be a connected graph with least eigenvalue $\lambda(G)$. Then*

- *$\lambda(G) \leq 0$, with equality for a null graph;*
- *if G is not null, then $\lambda(G) \leq -1$, with equality if and only if G is complete;*

- *if G is neither complete nor null, then $\lambda(G) \leq -\sqrt{2}$, with equality if and only if $G = K_{1,2}$;*
- *if G is neither complete nor $K_{1,2}$, and if $\lambda(G) \geq -1.5$, then G is*

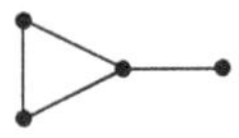

(In fact, $\lambda(G) = -\frac{2}{3}\sqrt{10}\sin(\frac{1}{3}\arccos(\frac{\sqrt{10}}{100}) + \frac{\pi}{6}) + \frac{1}{3} \approx -1.4811943$, for this graph.)

Proof Since the trace of any adjacency matrix is 0, $\lambda(G) \leq 0$. If a graph has an edge, then that edge is a two-vertex induced subgraph with least eigenvalue -1. Theorem 3.5 implies that $\lambda(G) \leq -1$. If G is not a complete graph, then $K_{1,2}$ is an induced subgraph with least eigenvalue $-\sqrt{2}$. Among the graphs G with four vertices, the one given in the last statement is the only one with least eigenvalue $\lambda(G) \geq -1.5$. ■

So we now know that the graphs G with $\lambda(G) \geq -1.5$ are pretty sporadic. The situation for $-2 \leq \lambda(G) \leq -1.5$ is just the opposite. First, we make an observation that motivated the original study of graphs for which $-2 \leq \lambda(G)$.

The *vertex-edge incidence matrix* of a graph is a (0, 1)-matrix with rows corresponding to the vertices, columns corresponding to the edges, and an entry equal to 1 if and only if its row and column correspond to an incident vertex and edge.

Theorem 6.2 *Any line graph L(G) satisfies $\lambda(L(G)) \geq -2$.*

Proof Let $\mathbf{K}$ be the vertex-edge incidence matrix of G. Evidently,

$$\mathbf{K}\mathbf{K}^T = 2\mathbf{I} + \mathbf{A}(L(G)).$$

Since $\mathbf{K}\mathbf{K}^T$ is positive semidefinite, it has non-negative eigenvalues. ■

Two non-isomorphic graphs with more than four vertices have non-isomorphic line graphs. So, in a sense, the number of graphs satisfying $-2 \leq \lambda(G) < 1.5$ is the same as the number of all line graphs.

Not all graphs satisfying $-2 \leq \lambda(G)$ are line graphs. For example, if G is a cocktail party graph CP(r), then $\lambda(G) = -2$, even though for any $r > 2$, G is not a line graph. So the determination of all graphs with $-2 \leq \lambda(G)$ is more complicated, but, as it turns out, leads to one of the most beautiful results in the study of graph spectra.

Suppose that G is a graph with n vertices and adjacency matrix $\mathbf{A}$, and assume that $-2 \leq \lambda(G)$. Then $\mathbf{A} + 2\mathbf{I}$ is positive semidefinite, and hence is of the form $\mathbf{K}\mathbf{K}^T$ for some appropriate matrix $\mathbf{K}$ with n rows; the number of columns is the

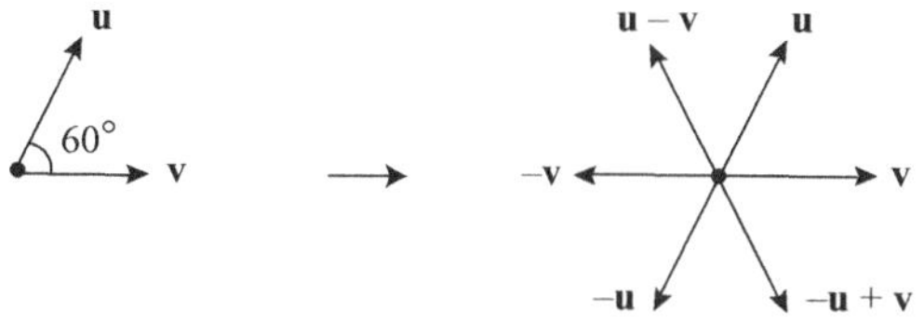

Fig. 6.

rank of $\mathbf{A} + 2\mathbf{I}$. Let $\mathbf{k}_1, \mathbf{k}_2, \ldots, \mathbf{k}_n$ be the n rows of $\mathbf{K}$. Then $\mathbf{k}_i \cdot \mathbf{k}_j = 1$ if v_i and v_j are adjacent, $\mathbf{k}_i \cdot \mathbf{k}_j = 0$ if they are not, and $\mathbf{k}_i \cdot \mathbf{k}_i = 2$. So we may think of these vectors as having length $\sqrt{2}$ and meeting at angles of 60° or 90°. Now if we have a set of vectors, and two vectors $\mathbf{u}$ and $\mathbf{v}$ meet at 60°, then we can add four other vectors $-\mathbf{u}$, $-\mathbf{v}$, $\mathbf{u} - \mathbf{v}$ and $-\mathbf{u} + \mathbf{v}$, to get three equiangular lines in the form of a 'star', as shown in Fig. 6. So if we start with vectors $\mathbf{k}_1, \mathbf{k}_2, \ldots, \mathbf{k}_n$, then we can add vectors when needed to complete the star. Since there can only be a finite number of vectors with each pair meeting at 60°, 90° or 120°, we can repeat this process a finite number of times until it eventually stops at some set Φ. This new set of vectors is called *star closed.*

Notice that Φ has the following properties:

- Φ is finite;
- if $\mathbf{x}$ and $c\mathbf{x} \in \Phi$, then $c = \pm 1$;
- if $\mathbf{x}, \mathbf{y} \in \Phi$, then $2\langle \mathbf{x}, \mathbf{y}\rangle / \langle \mathbf{x}, \mathbf{x}\rangle$ is an integer;
- if $\mathbf{x} \in \Phi$ and if H is the hyperplane orthogonal to $\mathbf{x}$, then Φ is closed under reflections by H: in other words, $\mathbf{x}, \mathbf{y} \in \Phi$ implies that $\mathbf{y} - 2\frac{\langle \mathbf{x}, \mathbf{y}\rangle}{\langle \mathbf{x}, \mathbf{x}\rangle}\mathbf{x}$ is also in Φ.

Sets of vectors satisfying these properties are called *root systems* and have been studied extensively, especially since they are used to classify real semisimple Lie algebras. Our set Φ has the additional property that all vectors have the same length. All such root systems are known: complete expositions can be found in [3, Ch. 6] and [7]. The complete list of such root systems is shown in Table 2, along with one set of generating vectors; see also Chapter 3, Section 5. Since $\mathbf{v} \in \Phi$ if and only if $-\mathbf{v} \in \Phi$, it is customary to count the number of lines in a root system. The standard basis is denoted by $\{\mathbf{e}_1, \mathbf{e}_2, \ldots, \mathbf{e}_n\}$, where $(\mathbf{e}_i)_j$ is 1 if $i = j$, and 0 otherwise. The beauty of this result is that the arguments are reversible: starting from the graph theory we can classify all the root systems. The details may be found in [6].

When the adjacency matrix of a graph satisfies $\mathbf{A} = \mathbf{K}\mathbf{K}^T - 2\mathbf{I}$, where the rows of $\mathbf{K}$ are taken from a root system, we say that the root system *represents* the graph.

Suppose that G is a bipartite graph with bipartition (X, Y). For an edge joining $v_i \in X$ and $v_j \in Y$, consider the vector $\mathbf{e}_i - \mathbf{e}_j$. Let $\mathbf{K}$ be the matrix with all such vectors as rows. Then $\mathbf{K}\mathbf{K}^T = 2\mathbf{I} + \mathbf{A}(L(G))$.

Table 2.

Name	Generating vectors	Number of lines
A_n	$\{\mathbf{e}_i - \mathbf{e}_j : 1 \le i < j \le n+1\}$	$\frac{1}{2}n(n+1)$
D_n	$\{\mathbf{e}_i \pm \mathbf{e}_j : 1 \le i < j \le n\}$	$n(n-1)$
E_6	$A_5 \cup \{\mathbf{e}_7 - \mathbf{e}_8\} \cup \left\{\frac{1}{2}\sum_{i=1}^{8} \varepsilon_i \mathbf{e}_i : \varepsilon_i = \pm 1, \sum_{i=1}^{6} \varepsilon_i = \varepsilon_7 + \varepsilon_8 = 0\right\}$	36
E_7	$A_7 \cup \left\{\frac{1}{2}\sum_{i=1}^{8} \varepsilon_i \mathbf{e}_i : \varepsilon_i = \pm 1, \sum_{i=1}^{8} \varepsilon_i = 0\right\}$	63
E_8	$D_8 \cup \left\{\frac{1}{2}\sum_{i=1}^{8} \varepsilon_i \mathbf{e}_i : \varepsilon_i = \pm 1, \prod_{i=1}^{8} \varepsilon_i = 1\right\}$	120

Conversely, if $\mathbf{KK}^T$ has the rows of $\mathbf{K}$ taken from the root system A_n, then any row is of the form $\mathbf{e}_i - \mathbf{e}_j$; for each such $\mathbf{e}_i - \mathbf{e}_j$, join v_i and v_j. This graph G has vertices corresponding to the columns of $\mathbf{K}$. Multiplying a row vector by -1 when necessary allows each column to have entries with the same sign. Let the columns with positive entries form one set of the bipartition, and the columns with negative entries form the other one. The graph is then bipartite and $\mathbf{KK}^T = 2\mathbf{I} + \mathbf{A}(L(G))$.

Theorem 6.3 *A graph is represented by A_n if and only if it is the line graph of a bipartite graph.*

It is easy to see that any line graph can be represented by D_n. If the edge joins v_i and v_j, let $\mathbf{e}_i + \mathbf{e}_j$ be a row of $\mathbf{K}$; then $\mathbf{KK}^T - 2\mathbf{I} = \mathbf{A}(L(G))$. Notice that the columns of $\mathbf{K}$ correspond to the vertices of G.

If we construct $\mathbf{K}$ by taking the vectors $\mathbf{e}_1 \pm \mathbf{e}_j$ $(j = 2, 3, \ldots, n+1)$ as $2n$ rows, then $\mathbf{KK}^T - 2\mathbf{I} = \mathbf{A}(\mathrm{CP}(n))$. Since $\mathrm{CP}(n)$ is not a line graph for $n > 2$, we see that graphs other than line graphs may be represented by D_n.

The *generalized line graph* $L(G, a_1, a_2, \ldots, a_n)$ is defined for a graph G with n vertices. It starts with disjoint copies of $L(G)$ and $\mathrm{CP}(a_1), \mathrm{CP}(a_2), \ldots, \mathrm{CP}(a_n)$. Then every vertex in $\mathrm{CP}(a_i)$ is joined to each vertex in $L(G)$ whose corresponding edge in G has v_i as an endpoint.

Now consider the following construction of $\mathbf{K}$. The rows for the line graph $L(G)$ are constructed as before: for each edge $v_i v_j$, let $\mathbf{e}_i + \mathbf{e}_j$ be a row of $\mathbf{K}$. For each $\mathrm{CP}(a_i)$, add the vectors $\mathbf{e}_i \pm \mathbf{e}_{n+a_1+\cdots+a_{i-1}+j}$, for $j = 1, 2, \ldots, a_i$. (The last subscript is just an artifact to increase the column count by 1 each time that a new pair of vectors is appended to $\mathbf{K}$.) Upon completion of the construction, we have

$$\mathbf{KK}^T = 2\mathbf{I} + \mathbf{A}(L(G, a_1, a_2, \ldots, a_n)).$$

Conversely, if we have a set of vectors R taken from D_n, where the inner product of any two of them is 0 or 1, we can construct a generalized line graph in the following way. For $1 \le i \ne j \le n$, it may be that two vectors of the form

$\pm\mathbf{e}_i \pm \mathbf{e}_j$ are in R: we take all such vectors and put them aside. For the remaining vectors, if $\pm\mathbf{e}_i \pm \mathbf{e}_j$ and $\pm\mathbf{e}_j \pm \mathbf{e}_k$ are two vectors in R, then $i \neq k$ and the $\pm\mathbf{e}_j$ terms have the same sign. Now use these remaining vectors as rows of $\mathbf{K}$. We can multiply a column by -1 without changing the inner product of any two rows. Since all entries in a column have the same sign, we may assume with no loss of generality that all of these vectors are of the form $\mathbf{e}_i + \mathbf{e}_j$. Now let the vertices of G correspond to the non-zero columns of $\mathbf{K}$, and join v_i and v_j if $\mathbf{e}_i + \mathbf{e}_j$ is a row of $\mathbf{K}$. Then $\mathbf{KK}^T = 2\mathbf{I} + \mathbf{A}(L(G))$ (so far).

Now we go back to the vectors we put aside. They come in pairs and are of the form $\mathbf{e}_i \pm \mathbf{e}_j$ or $\pm\mathbf{e}_i + \mathbf{e}_j$; we need consider only the case $\mathbf{e}_i \pm \mathbf{e}_j$ since the other case is symmetric. We append these vectors as rows of $\mathbf{K}$. Note that the $\pm\mathbf{e}_j$ are the only non-zero entries in the jth column since an inner product of -1 is forbidden. The column containing $\mathbf{e}_i$ may contain other entries, but if it does, then the column corresponds to a vertex in G. The rows corresponding to the newly added vectors that are non-zero in the ith column must induce a CP(a_i) subgraph in $\mathbf{KK}^T - 2\mathbf{I}$. Finally, we observe that each vertex in the CP(a_i) is adjacent to v_i. Thus

$$\mathbf{KK}^T = 2\mathbf{I} + \mathbf{A}(L(G, a_1, a_2, \ldots, a_n)).$$

Theorem 6.4 *A graph is represented by D_n if and only if it is a generalized line graph.*

Using the table of generating vectors for root systems given previously, we immediately deduce the inclusions $E_6 \subset E_7 \subset E_8$; hence, we have the following theorem.

Theorem 6.5 *If $\lambda(G) \geq -2$, then G is a generalized line graph or G is represented by E_8.*

Since E_8 is finite, all but a finite number of graphs with $\lambda(G) \geq -2$ are generalized line graphs. Of those that are not, it is known that the largest has 36 vertices. If we consider graphs with $\lambda(G) > -2$, then $\mathbf{KK}^T$ is non-singular, and hence $\mathbf{K}$ has six, seven or eight rows according as G is represented by E_6, E_7 or E_8. In fact, there are 20 such graphs on six vertices, 110 graphs on seven vertices, and 443 graphs on eight vertices. These were originally found by a mixture of theoretical analysis and computer search (see [13] and [11]). Further details appear in Chapter 2.

7. Upper bounds for the eigenvalues

In the previous section we saw how to use root systems to find graphs with eigenvalues bounded from below by -2. A similar analysis can be used to find graphs with eigenvalues bounded from above by 2.

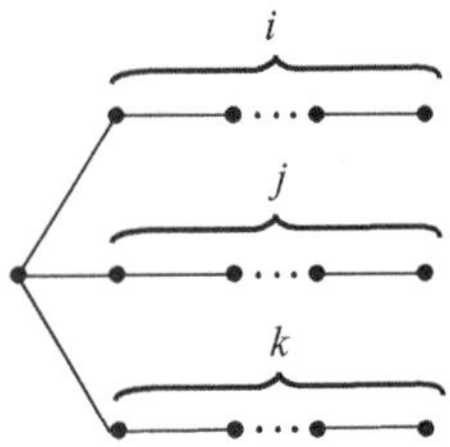

Fig. 7. The graph $T(i, j, k)$

To proceed, we define the graph $T(i, j, k)$, shown in Fig. 7: take three paths P_i, P_j and P_k and add a new vertex adjacent to one end-vertex of each path. The graph is then a tree with $i + j + k + 1$ vertices, three pendant vertices and one vertex of degree 3.

Now we form a matrix $\mathbf{K}$ from a set of vectors, where the inner product of any two vectors is 0 or -1. We can then define the adjacency matrix of G by the equation

$$\mathbf{KK}^T = 2\mathbf{I} - \mathbf{A}(G)$$

and we will have a graph whose eigenvalues are bounded from above by 2. Such sets of vectors have been fully studied: they are called *fundamental sets* of roots.

Theorem 7.1 *If G is a graph with largest eigenvalue $\lambda_1 = 2$, then G is one of the following graphs: C_n, $K_{1,4}$, $T(2, 2, 2)$, $T(3, 3, 1)$, $T(5, 2, 1)$ or*

Actually, it is an easy exercise in vertex labelling to show that each of these graphs has maximum eigenvalue 2, and to show that no other graphs can be so labelled (see [27]). Thus the results concerning fundamental sets of roots can be derived directly from graph theory.

From Corollary 5.7 we now know all graphs with largest eigenvalue less than 2.

Corollary 7.2 *If G is a graph with largest eigenvalue $\lambda_1 < 2$, then G is a path P_n, $T(1, 1, r)$, $T(1, 2, 4)$, $T(1, 2, 3)$ or $T(1, 2, 2)$.*

We can say more about graphs with largest eigenvalue greater than 2. In [9], graphs are described whose largest eigenvalue is bounded above by $(2 + \sqrt{5})^{1/2} \sim 2.058171$. It turns out that all such graphs are trees of the form $T(i, j, k)$ or $S(j, k, l)$, where the latter graph appears in Fig. 8. Notice that the graph given in the conclusion of Theorem 7.1 is $S(1, k, 1)$.

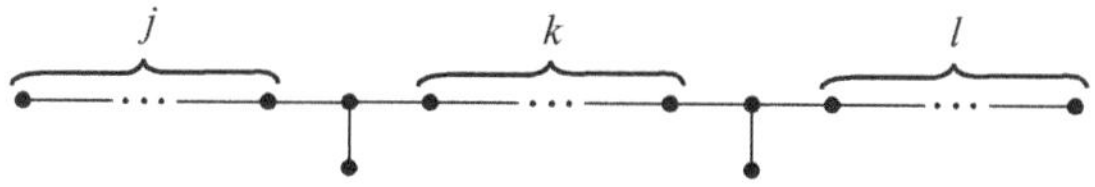

Fig. 8. The graph $S(j, k, l)$

Theorem 7.3 *If G is a graph with largest eigenvalue λ_1 and if $2 < \lambda_1 < (2+\sqrt{5})^{1/2}$, then G is one of the following:*

- $T(1, 2, k)$, *with* $k > 5$;
- $T(1, j, k)$, *with* $2 < j < k$;
- $T(2, 2, k)$, *with* $2 < k$;
- $T(2, 3, 3)$;
- $S(j, k, l)$, *for* $(j, l) \neq (1, 1)$ *and large enough* k.

The exact value of 'large enough' in this theorem has been determined (see [5]).

Finding bounds for the second largest eigenvalue λ_2 has also been of interest. Obviously $\lambda_2(K_n) = -1$, for $n > 1$. Smith [27] showed what happens for $\lambda_2 \leq 0$.

Theorem 7.4 $\lambda_2(G) \leq 0$ *if and only if G is a complete multipartite graph.*

Proof Consider the three graphs shown in Fig. 9. Each has two positive eigenvalues, so by Theorem 3.4 none can be a subgraph of a graph G with $\lambda_2(G) \leq 0$. Since the graph is connected, the only possibility is for G to be a complete multipartite graph. ∎

Graphs with $\lambda_2(G) \leq 1$ can be partially described in terms of their complements: these have eigenvalues bounded from below by -2, or have just one eigenvalue λ_n with $\lambda_n < -2$. Details can be found in [8].

Graphs with $\lambda_2(G) \leq \sqrt{2} - 1$ and $\lambda_2(G) \leq (\sqrt{5} - 1)/2$ have also been described in terms of minimal forbidden subgraphs – that is, graphs not satisfying the eigenvalue bound and minimal with respect to inclusion. It is known that the possible values of $\lambda_2(G)$ are dense in $(\sqrt{(2+\sqrt{5}}, \infty)$, but nowhere dense in $(-\infty, \sqrt{2} - 1)$. Further details appear in [12], [18] and [26].

There are several fascinating interrelations between $\lambda_2(G)$ and other areas of mathematical interest. In particular, this eigenvalue is related to expanders and superconcentrators [1] and to Ramanujan graphs [19].

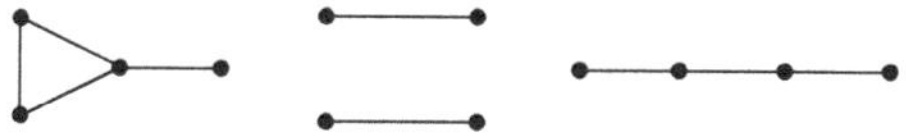

Fig. 9.

8. Other matrices related to graphs

There are other matrices obtained from graphs whose eigenvalues are interesting: as with the adjacency matrix, the rows and columns correspond to the vertices of the graph. One such is the *Seidel matrix*: it is 0 on the diagonal, -1 for adjacent vertices, and 1 for non-adjacent vertices. Another is the *Laplacian*: it has $\deg v_i$ in the ith diagonal position, -1 for adjacent vertices, and 0 for non-adjacent vertices. Finally, the complement of a graph has an adjacency matrix with 0 on the diagonal, 0 for vertices adjacent in the original graph, and 1 for vertices similarly non-adjacent.

For a regular graph G of degree r, each of these matrices can be written in the form $\alpha\mathbf{J} + \beta\mathbf{A} + \gamma\mathbf{I}$. For the Seidel matrix, $\alpha = 1$, $\beta = -2$ and $\gamma = -1$; for the Laplacian, $\alpha = 0$, $\beta = -1$ and $\gamma = r$; for the complement of G, $\alpha = 1$, $\beta = -1$ and $\gamma = -1$. It is easy to compute the eigenvalues of these matrices; to avoid trivialities, we assume that $\beta \neq 0$.

Theorem 8.1 *If G is regular of degree r and has eigenvalues $\lambda_1 = r > \lambda_2 \geq \cdots \geq \lambda_n$, then the matrix $\alpha\mathbf{J} + \beta\mathbf{A} + \gamma\mathbf{I}$ has eigenvalues $\alpha n + \beta r + \gamma$ and $\beta\lambda_i + \gamma$, for $i = 2, 3, \ldots n$.*

Proof A regular graph has r as an eigenvalue with $\mathbf{j}$ as an eigenvector; this vector gives the first eigenvalue. Any other eigenvector $\mathbf{x}$ of G is orthogonal to $\mathbf{j}$, so $\mathbf{Jx} = \mathbf{0}$ and the result follows. ■

For non-regular graphs this theorem is no longer true. However, the slippage in the multiplicity is at most 1.

The largest eigenvalue of a graph is called the *dominant* one; any other eigenvalue is called *subdominant*.

Theorem 8.2 *Let G be a graph with λ as a subdominant eigenvalue of multiplicity $m\lambda$. Then $\alpha\mathbf{J} + \beta\mathbf{A} + \gamma\mathbf{I}$ has $\beta\lambda + \gamma$ as an eigenvalue with multiplicity m, where $m_\lambda - 1 \leq m \leq m_\lambda + 1$.*

Proof Let H be the hyperplane orthogonal to vector $\mathbf{j}$, and let E_λ be the eigenspace of the eigenvalue λ. Then

$$\dim(H + E_\lambda) + \dim(H \cap E_\lambda) = \dim H + \dim E_\lambda.$$

If $E_\lambda \subseteq H$, then $\dim(H + E_\lambda) = n - 1$ and $\dim(H \cap E_\lambda) = m_\lambda$. This implies that the argument used for the regular case still applies, and so $m = m_\lambda$. Otherwise, we have $\dim(H + E_\lambda) = n$ and $\dim(H \cap E_\lambda) = m_\lambda - 1$. The argument used in the regular case now implies that $m \geq m_\lambda - 1$.

If $\mathbf{C} = \alpha\mathbf{J} + \beta\mathbf{A} + \gamma\mathbf{I}$, then

$$\mathbf{A} = -\beta^{-1}\alpha\mathbf{J} + \beta^{-1}\mathbf{C} - \beta^{-1}\gamma\mathbf{I}.$$

Using the argument of the previous paragraph on the new matrix, we get $m_\lambda \geq m - 1$. Hence $m_\lambda - 1 \leq m \leq m_\lambda + 1$. ■

This result applies immediately to complements and to the Seidel matrix of a non-regular graph. Although it does not apply to the Laplacian directly, we can still say something for bipartite graphs. If G is bipartite, we let $\mathbf{K}$ be a variant of the vertex-edge incidence matrix, as follows. Let (X, Y) be the partition of the vertices. For each edge $v_i v_j (v_i \in X,\ v_j \in Y)$, let $\mathbf{e}_i - \mathbf{e}_j$ be a column of $\mathbf{K}$: this is the same A_n representation that we used to prove Theorem 6.3. Then $\mathbf{KK}^T = \mathbf{L}$, the Laplacian of G. All the entries in a given row have the same sign. Hence, $\mathbf{K}^T\mathbf{K} = 2\mathbf{I} + \mathbf{A}(L(G))$. Now $\mathbf{L}$ has 0 as a simple eigenvalue (see Section 4), and $\mathbf{KK}^T$ and $\mathbf{K}^T\mathbf{K}$ have the same non-zero eigenvalues.

Theorem 8.3 *If G is a bipartite graph with line graph $L(G)$ and Laplacian $\mathbf{L}$, then $\mu (> 0)$ is an eigenvalue of $\mathbf{L}$ if and only if $\mu - 2\ (> -2)$ is an eigenvalue of $L(G)$.*

The validity of this theorem for paths was first observed by Haemers [16].

9. Cospectral graphs

From the moment it was realized that different graphs could have the same spectrum, the hunt for such graphs was on, and many families of cospectral graphs have been found. Broadly speaking, two types of constructions are used: one uses operations on graphs (complements, products, etc.) to produce new cospectral ones, while the other adroitly pastes different graphs together. We look at each of these.

Using graph operations

One easy way to construct cospectral graphs is due to Hoffman (see [21]). Take two non-isomorphic cospectral regular graphs G_1 and G_2, and consider the graph formed by taking k copies of G_1 and $s - k$ copies of G_2. Let H_k be the complement of this graph. By Theorem 8.1, the graphs H_k $(k = 0, 1, \ldots, s)$ are cospectral. Thus we can have arbitrarily large sets of non-isomorphic cospectral graphs.

Given graphs G and H with vertex-sets $\{v_1, v_2, \ldots, v_n\}$ and $\{w_1, w_2, \ldots, w_m\}$, the *Cartesian product* $G \,\square\, H$ of G and H has as its vertices the pairs $\{(v_i, w_j) : i = 1, 2, \ldots, n,\ j = 1, 2, \ldots, m\}$. Two vertices (v, w) and (v', w') are joined if either $v = v'$ and w and w' are joined in H, or $w = w'$ and v and v' are joined in G. We use $\otimes$ to denote the *Kronecker product* of matrices. This product is associative and so may be extended to the product of more than two factors; for example, the d-dimensional cube is the Cartesian product of d copies of K_2.

It is easy to see that

$$\mathbf{A}(G \,\square\, H) = (\mathbf{A}(G) \otimes \mathbf{I}) + (\mathbf{I} \otimes \mathbf{A}(H)).$$

Let $\mathbf{x}$ be an eigenvector of G with eigenvalue λ, and $\mathbf{y}$ be an eigenvector of H with eigenvalue μ. Then, using the properties of the Kronecker product,

$$\begin{aligned}
&((\mathbf{A}(G) \otimes \mathbf{I}) + (\mathbf{I} \otimes \mathbf{A}(H)))(\mathbf{x} \otimes \mathbf{y}) \\
&\quad = (\mathbf{A}(G) \otimes \mathbf{I})(\mathbf{x} \otimes \mathbf{y}) + (\mathbf{I} \otimes \mathbf{A}(H))(\mathbf{x} \otimes \mathbf{y}) \\
&\quad = (\mathbf{A}(G)\mathbf{x} \otimes \mathbf{I}\mathbf{y}) + (\mathbf{I}\mathbf{x} \otimes \mathbf{A}(H)\mathbf{y}) \\
&\quad = (\lambda\mathbf{x} \otimes \mathbf{y}) + (\mathbf{x} \otimes \mu\mathbf{y}) \\
&\quad = \lambda(\mathbf{x} \otimes \mathbf{y}) + \mu(\mathbf{x} \otimes \mathbf{y}) = (\lambda + \mu)(\mathbf{x} \otimes \mathbf{y}).
\end{aligned}$$

Thus, the eigenvalues of $G \,\square\, H$ are just the pairwise sums of the eigenvalues of G and H.

An attractive feature of this product is that the factorization is essentially unique (see [22]). We may take two cospectral graphs G_1 and G_2 and form H_k, the Cartesian product of k copies of G_1 and $s - k$ copies of G_2. Then the graphs H_k $(k = 0, 1, \ldots, s)$ are all cospectral.

The *Shrikhande graph* is defined in [4]. It is strongly regular with the same parameters as $L(K_{4,4})$. When G_1 is the Shrikhande graph and G_2 is $L(K_{4,4})$, then the graphs H_k $(k = 1, 2, \ldots, s)$ are distance-regular with the same parameters, and therefore cospectral; they are known as *Doob graphs*. These graphs imply that arbitrarily large families of cospectral distance-regular graphs with large diameter exist. Similar constructions for creating cospectral distance-regular graphs can be found in [4].

This concept can be extended to *NEPS (non-extended p-sum) graphs*. For this, we start with n graphs $G_1, G_2, \ldots, G_n$ and a set of vectors $\mathcal{B} = \{(\beta_1, \beta_2, \ldots, \beta_n) : \beta_i = 0 \text{ or } 1\}$. The vertex-set of the NEPS graph is the Cartesian product of the vertex-sets of $G_1, G_2, \ldots, G_n$, just as in the Cartesian product of graphs. Two vertices are adjacent if, as n-tuples, there is a vector $(\beta_1, \beta_2, \ldots, \beta_n) \in \mathcal{B}$ so that the coordinates of the n-tuples agree exactly where $\beta_i = 1$. When $\mathcal{B} = \{\mathbf{e}_1, \mathbf{e}_2, \ldots, \mathbf{e}_n\}$, the NEPS is the ordinary Cartesian product. If $\mathbf{A}_1, \mathbf{A}_2, \ldots, \mathbf{A}_n$ are the respective adjacency matrices of $G_1, G_2, \ldots, G_n$, and $\mathbf{A}$ is the adjacency matrix of the NEPS product, then

$$\mathbf{A} = \sum_{(\beta_1, \ldots, \beta_n) \in \mathcal{B}} \mathbf{A}_1^{\beta_1} \otimes \mathbf{A}_2^{\beta_2} \otimes \cdots \otimes \mathbf{A}_n^{\beta_n}.$$

If $\mathbf{x}_1, \mathbf{x}_2, \ldots, \mathbf{x}_n$ are eigenvectors of $\mathbf{A}_1, \mathbf{A}_2, \ldots, \mathbf{A}_n$ with eigenvalues $\lambda_1, \lambda_2, \ldots, \lambda_n$, then

$$\begin{aligned}\mathbf{A}(\mathbf{x}_1 \otimes \mathbf{x}_2 \otimes \cdots \otimes \mathbf{x}_n) &= \sum_{(\beta_1,\ldots,\beta_n)\in\mathcal{B}} \mathbf{A}_1^{\beta_1} \otimes \cdots \otimes \mathbf{A}_n^{\beta_n}(\mathbf{x}_1 \otimes \cdots \otimes \mathbf{x}_n) \\ &= \sum_{(\beta_1,\ldots,\beta_n)\in\mathcal{B}} \left(\lambda_1^{\beta_1}\mathbf{x}_1 \otimes \lambda_2^{\beta_2}\mathbf{x}_2 \otimes \cdots \otimes \lambda_n^{\beta_n}\mathbf{x}_n\right) \\ &= \sum_{(\beta_1,\ldots,\beta_n)\in\mathcal{B}} \lambda_1^{\beta_1}\lambda_2^{\beta_2}\ldots\lambda_n^{\beta_n}(\mathbf{x}_1 \otimes \mathbf{x}_2 \otimes \cdots \otimes \mathbf{x}_n).\end{aligned}$$

It follows that we can compute the eigenvalues of the NEPS graphs from the eigenvalues of the factors. More use of the NEPS graphs will be made in Chapter 3, Section 7.

Pasting graphs together

A second method of constructing cospectral graphs is by pasting smaller graphs together. One way is to take two graphs, designate a special vertex (or root) in each of them, and then identify these two vertices. We denote this new graph by $G \cdot H$, the roots being understood from the context. Further, we denote by $P_G(x)$ the characteristic polynomial of the adjacency matrix of G.

Suppose that $\{v_1, v_2, \ldots, v_n\}$ is the vertex-set of G and that $\{w_1, w_2, \ldots, w_m\}$ is the vertex-set of H. Without loss of generality we may assume that v_1 and w_1 are to be identified as a single vertex u. The rows of the adjacency matrix of the new graph can be ordered by $u, v_2, v_3, \ldots, v_n, w_2, w_3, \ldots, w_m$. Then the adjacency matrix has $\mathbf{A}(G - \{v_1\})$ and $\mathbf{A}(H - \{w_1\})$ as blocks on the diagonal.

Let $\mathbf{B} = x\mathbf{I} - \mathbf{A}(G \cdot H)$. We evaluate

$$\det \mathbf{B} = \sum_{\sigma} (-1)^{\operatorname{sgn}\sigma} b_{1,\sigma(1)} b_{2,\sigma(2)} \ldots b_{m+n-1,\sigma(m+n-1)},$$

by summing over all permutations σ of the vertices of $G \cdot H$. Let S_1 be the set of permutations σ such that $\sigma(u) = u$ or $\sigma(u)$ is a vertex in G. Similarly, let S_2 be the set of permutations such that $\sigma(u) = u$ or $\sigma(u)$ is a vertex in H. Then

$$\sum_{\sigma\in S_1} (-1)^{\operatorname{sgn}\sigma} b_{1,\sigma(1)} b_{2,\sigma(2)} \ldots b_{m+n-1,\sigma(m+n-1)} = P_G(x) P_{H-w_1}(x).$$

Similarly,

$$\sum_{\sigma\in S_2} (-1)^{\operatorname{sgn}\sigma} b_{1,\sigma(1)} b_{2,\sigma(2)} \ldots b_{m+n-1,\sigma(m+n-1)} = P_{G-v_1}(x) P_H(x).$$

This covers all the permutations, but counts those for which $\sigma(u) = u$ twice: this latter case gives $x P_{G-v} P_{H-w}$. In short, we have the following theorem.

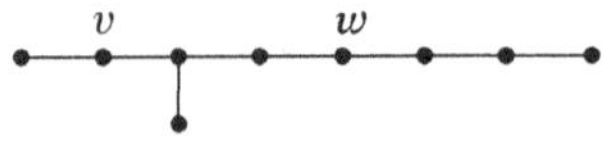

Fig. 10. The graph H

Theorem 9.1 *If $G \cdot H$ is formed by identifying the vertex v in G with the vertex w in H, then*

$$P_{G \cdot H}(x) = P_G(x)P_{H-w}(x) + P_{G-v}(x)P_H(x) - x\,P_{G-v}(x)P_{H-w}(x).$$

Schwenk [23] provides the example in Fig. 10: the graph H is to be used with roots v and w. One may compute that

$$P_{H-v}(x) = P_{H-w}(x) = x^2(x^2 - 2)(x^4 - 4x^2 + 2).$$

If we take any graph G with root u, then we may form $G \cdot H$ by identifying u with v, or u with w, and the two resulting graphs are cospectral. If G happens to be a tree, then the two resulting graphs are cospectral non-isomorphic trees. By iterating this process with the same root, we can get arbitrarily large families of cospectral trees.

Schwenk [24] has also shown that, as the number of vertices gets large, the probability that a tree contains a copy of H approaches 1; hence almost all trees have a cospectral mate. One may ask whether the same is true for graphs in general, but unfortunately almost nothing is known about the answer to this question.

References

1. N. Alon and V. D. Milman, λ_1-isoperimetric inequalities for graphs and superconductors, *J. Combin. Theory (B)* **38** (1985), 73–88.
2. N. Biggs, Intersection matrices for linear graphs, *Combinatorial Mathematics and its Applications* (ed. D. J. A. Welsh), Academic Press (1971), 15–23.
3. N. Bourbaki, *Groupes et Algèbres de Lie*, Masson, 1981.
4. A. E. Brouwer, A. M. Cohen and A. Neumaier, *Distance-Regular Graphs*, Springer-Verlag, 1989.
5. A. E. Brouwer and A. Neumaier, Graphs with spectral radius between 2 and $\sqrt{2+\sqrt{5}}$, *Linear Alg. Appl.* **114/115** (1989), 273–276.
6. P. J. Cameron, J. M. Goethals, J. J. Seidel and E. E. Shult, Line graphs, root systems, and elliptic geometry, *J. Algebra* **43** (1976), 305–327.
7. R. W. Carter, *Simple Groups of Lie Type*, Wiley, 1989.
8. D. Cvetković, On graphs whose second largest eigenvalue does not exceed 1, *Publ. Inst. Math. (Beograd)* **31** (1982), 15–20.
9. D. Cvetković, M. Doob and I. Gutman, On graphs whose spectral radius does not exceed $(2+\sqrt{5})^{1/2}$, *Ars Combinatorica* **14** (1982), 225–239.
10. D. Cvetković, M. Doob and H. Sachs, *Spectra of Graphs* (3rd ed.), Johann Ambrosius Barth, 1995.

11. D. Cvetković, M. Doob and S. Simić, Generalized line graphs, *J. Graph Theory* **5** (1981), 385–399.
12. D. Cvetković and S. Simić, On graphs whose second largest eigenvalue does not exceed $(\sqrt{5}-1)/2$, *Discrete Math.* **139** (1995), 213–227.
13. M. Doob and D. Cvetković, On spectral characterizations and embeddings of graphs, *Linear Alg. Appl.* **27** (1979), 17–26.
14. F. R. Gantmacher, *The Theory of Matrices*, Chelsea, 1959.
15. C. Godsil and G. Royle, *Algebraic Graph Theory*, Springer-Verlag, 2001.
16. W. Haemers, Which graphs are determined by their spectrum?, *Linear Alg. Appl.*
17. A. J. Hoffman and R. R. Singleton, On Moore graphs with diameter 2 and 3, *IBM J. Res. Develop.* **4** (1960), 497–504.
18. J. Li, *Subdominant Eigenvalues of Graphs*, PhD thesis, University of Manitoba, 1994.
19. A. Lubotsky, R. Phillips and P. Sarnak, Ramanujan graphs, *Combinatorica* **8** (1988), 261–277.
20. H. Minc and M. Marcus, *A Survey of Matrix Theory and Matrix Inequalities*, Prindle, Weber & Schmidt, 1964.
21. A. Mowshowitz, The characteristic polynomial of a graph, *J. Combin. Theory (B)* **12** (1972), 177–193.
22. R. C. Read and R. J. Wilson, *An Atlas of Graphs*, Oxford Univ. Press, 1998.
23. G. Sabidussi, Graph multiplication, *Math. Zeit.* **72** (1960), 446–457.
24. A. J. Schwenk, Almost all trees are cospectral, *New Directions in the Theory of Graphs* (ed. F. Harary), Academic Press (1973), 275–307.
25. A. J. Schwenk and R. J. Wilson, On the eigenvalues of a graph, *Selected Topics in Graph Theory* (eds. L. W. Beineke and R. J. Wilson), Academic Press (1979), 307–336.
26. S. Simić, Some notes on graphs whose second largest eigenvalue is less than $(\sqrt{5}-1)/2$, *Linear and Multilinear Algebra* **39** (1995), 59–71.
27. J. H. Smith, Some properties of the spectrum of a graph, *Combinatorial Structures and their Applications* (eds. R. Guy *et al.*), Gordon and Breach (1970), 403–406.

2

Graphs and matrices

RICHARD A. BRUALDI and BRYAN L. SHADER

Graphs and matrices enjoy a mutually beneficial relationship. On one level this relationship provides different viewpoints and conceptualizations of the same idea: but it is more than the simple translation of results in one framework into the other. Matrices, and the associated linear algebra, provide an important tool in graph theory, sometimes the only known tool. On the other hand, graphs provide a way to understand and take advantage of matrix structure. In this chapter we provide a glimpse of graphs and matrices working together.

1. Introduction

Matrices provide models for graphs (or models for the same phenomena modelled by graphs) that often illuminate their structure and allow the application of elementary but powerful linear algebraic techniques to their study. Conversely, associated with a matrix are graphs that capture in various ways the combinatorial structure of the matrix. Different graphs may reveal or emphasize different aspects of a matrix's structure. Taking this structure into account often leads to sharper theorems, improved inequalities and, sometimes, non-existence conclusions.

Let G be a graph (or a multigraph, general graph, digraph, weighted digraph, ...). The basic idea behind matrices associated with graphs is the following. Let the vertices of G be listed in some order – say, $v_1, v_2, \ldots, v_n$; the particular ordering is of no significance, but we have to choose one. We define a matrix $\mathbf{A} = (a_{ij})$ of order n, where the ij-entry is associated with the ordered pair of vertices (v_i, v_j). If G is a graph, then

$$a_{ij} = \begin{cases} 1, & \text{if } v_i \text{ and } v_j \text{ are joined by an edge} \\ 0, & \text{otherwise.} \end{cases}$$

As we saw in Chapter 1, the matrix $\mathbf{A}$ is the *adjacency matrix* of G and is a symmetric (0, 1)-matrix with zeros on the main diagonal.

It is an easy matter to modify the definition of $\mathbf{A}$ to take into account these different variations of the graph G. If G is a *multigraph*, then we let a_{ij} be the multiplicity of the edge joining v_i and v_j, and the adjacency matrix $\mathbf{A}$ is a non-negative integer matrix with zeros on the main diagonal. If loops are permitted, then $\mathbf{A}$ is a general non-negative integer matrix. If G is a *digraph* so that edges have a direction, then a_{ij} does not equal a_{ji} in general, and $\mathbf{A}$ is a general (0, 1)-matrix, or a general non-negative integer matrix if edges are allowed to have multiplicity greater than 1. Finally, if G is a *weighted graph* (or digraph) where edges have weights, then we let a_{ij} be the weight (positive, negative, zero or complex) of the edge from v_i to v_j.

Now we have closed the connection between graphs and matrices in that $\mathbf{A}$ can be an arbitrary real or complex matrix. Moreover, we can reverse our constructions: starting with a square matrix $\mathbf{A}$ we can define a graph of one of the types discussed above whose adjacency matrix is $\mathbf{A}$. So, in a sense, weighted graphs and matrices are interchangeable; for more on this theme, see [4].

As examples, consider the matrices

$$\mathbf{A} = \begin{pmatrix} 0 & 1 & 1 & 0 & 1 \\ 1 & 0 & 1 & 0 & 0 \\ 1 & 1 & 0 & 1 & 1 \\ 0 & 0 & 1 & 0 & 1 \\ 1 & 0 & 1 & 1 & 0 \end{pmatrix} \quad \text{and} \quad \mathbf{B} = \begin{pmatrix} 0 & 1 & 1 & 0 & 2 \\ 0 & 1 & 0 & 2 & 0 \\ 3 & 1 & 0 & 1 & 1 \\ 0 & 1 & 1 & 0 & 1 \\ 1 & 0 & 1 & 1 & 1 \end{pmatrix}.$$

The matrix $\mathbf{A}$ is a symmetric (0, 1)-matrix of order 5 with zeros on the main diagonal. Thus $\mathbf{A}$ is the adjacency matrix of a graph with 5 vertices. The matrix $\mathbf{B}$ is a non-negative integer matrix with maximum entry equal to 3, and is the adjacency matrix of a digraph each of whose edges has multiplicity at most 3 (or a weighted digraph with weights 0, 1, 2 and 3).

2. Some classical theorems

In this section we briefly mention some of the classic results that illustrate the powerful and beautiful relationship between graphs and matrices.

As we saw in Chapter 1, one of the earliest theorems to exploit the combinatorial structure of a matrix was the *Perron-Frobenius theorem*, concerning the spectral structure of a square non-negative matrix $\mathbf{A}$ (see [16] and [22]). Without any assumptions on $\mathbf{A}$, the following theorem holds.

Theorem 2.1 (Perron-Frobenius theorem) *Let* $\mathbf{A}$ *be a non-negative matrix of order* n*. Then* $\mathbf{A}$ *has a non-negative eigenvalue* ρ *such that every eigenvalue* λ *of* $\mathbf{A}$ *satisfies* $|\lambda| \leq \rho$*. Moreover,* $\mathbf{A}$ *has an entry-wise non-negative eigenvector* $\mathbf{x}$ *corresponding to the eigenvalue* ρ*.*

The *spectral radius* of a complex square matrix is the largest modulus of its eigenvalues. Thus, the number ρ in Theorem 2.1 is the spectral radius of the matrix $\mathbf{A}$.

Let G be the weighted digraph with adjacency matrix $\mathbf{A}$. One particular assumption on the digraph G leads to more detail concerning the eigenvalue structure of $\mathbf{A}$. Recall that a digraph is *strongly connected* if, for each ordered pair (v_i, v_j) of distinct vertices, there is a directed path in G from v_i to v_j.

Theorem 2.2 *Let* $\mathbf{A}$ *be a non-negative matrix of order* $n \geq 2$*, such that the digraph* G *corresponding to* $\mathbf{A}$ *is strongly connected. Then the spectral radius* ρ *of* $\mathbf{A}$ *is positive and has multiplicity* 1*; in addition,* $\mathbf{A}$ *has an entry-wise positive eigenvector* $\mathbf{x}$ *corresponding to the eigenvalue* ρ*. The eigenvalues* λ *of* $\mathbf{A}$ *with* $|\lambda| = \rho$ *are* $\rho e^{2\pi ij/k}$ $(j = 1, 2, \ldots, k)$*, where* k *is the greatest common divisor of the lengths of the cycles of* G*. The only non-negative eigenvectors of* $\mathbf{A}$ *are positive multiples of* $\mathbf{x}$*.*

Let $\mathbf{x} = (x_1, x_2, \ldots, x_n)^T$. The equation $\mathbf{Ax} = \rho\mathbf{x}$ implies that

$$\sum_{j=1}^{n} \sum_{i=1}^{n} a_{ij}\, x_j = \rho \sum_{j=1}^{n} x_j,$$

showing that ρ is a weighted average of the column sums $s_1, s_2, \ldots, s_n$ of $\mathbf{A}$. Hence

$$\min\{s_1, s_2, \ldots, s_n\} \leq \rho \leq \max\{s_1, s_2, \ldots, s_n\}.$$

In particular, this equation implies that if $\mathbf{A}$ is a (0, 1)-matrix corresponding to a digraph G, then the spectral radius of $\mathbf{A}$ lies between the minimum and maximum of the in-degrees of G. Since $\mathbf{A}$ and $\mathbf{A}^T$ have the same eigenvalues, a similar conclusion holds with 'in-degree' replaced by 'out-degree'.

A recent application of the digraph of a complex matrix $\mathbf{A} = (a_{ij})$ of order n to the localization of eigenvalues of $\mathbf{A}$ originates in *Geršgorin's theorem*. This theorem asserts that the eigenvalues of $\mathbf{A}$ lie in that part of the complex plane given by the union

$$\bigcup_{i=1}^{n} \{z \in \mathbf{C} : |z - a_{ii}| \le R_i\}$$

of n closed discs, where $R_i = \sum_{j \ne i} |a_{ij}|$ $(i = 1, 2, \ldots, n)$. If $\mathbf{A}$ is irreducible, then a boundary point of this region can be an eigenvalue of $\mathbf{A}$ only if it is a boundary point of each of the n discs. A better inclusion region takes into account the cycles of G (see [3] and [7]).

Theorem 2.3 *Let $\mathbf{A} = (a_{ij})$ be a complex matrix of order n with corresponding digraph G. The n eigenvalues of $\mathbf{A}$ lie in that part of the complex plane given by the union*

$$\bigcup_{\gamma} \left\{ z \in \mathbf{C} : \prod_{\gamma} |z - a_{ii}| \le \prod_{\gamma} R_i \right\}. \tag{1}$$

(Here $\bigcup_\gamma$ denotes the union over all cycles γ of G, and $\prod_\gamma$ denotes the product over those i for which v_i is a vertex of γ.) If G is strongly connected, then a boundary point of (1) *can be an eigenvalue of $\mathbf{A}$ only if it is a boundary point of each of the regions in its definition.*

Let G be a bipartite graph. Recall that the vertices of G can be partitioned into two sets X and Y such that each edge has one vertex in X and one vertex in Y; the pair $\{X, Y\}$ is a *bipartition* of the vertices of G. We choose a listing $x_1, x_2, \ldots, x_r$ for the vertices in X and a listing $y_1, y_2, \ldots, y_s$ for the vertices in Y, and concatenate these to get a listing $x_1, x_2, \ldots, x_r, y_1, y_2, \ldots, y_s$ of the vertices of G. Then the adjacency matrix $\mathbf{A}$ of G has the form

$$\mathbf{A} = \begin{pmatrix} \mathbf{O}_r & \mathbf{B} \\ \mathbf{B}^T & \mathbf{O}_s \end{pmatrix}$$

where $\mathbf{B}$ is an $r \times s$ matrix, and $\mathbf{O}_r$ and $\mathbf{O}_s$ are zero matrices of orders r and s, respectively. The matrix $\mathbf{B}$ is called the *biadjacency matrix* of G, and depends on the listing chosen for the vertices in X and Y. Using our previous observations, we deduce that bipartite graphs and (rectangular) (0, 1)-matrices are interchangeable, as are bipartite multigraphs and (rectangular) non-negative integral matrices, and weighted bipartite graphs and real or complex (rectangular) matrices. Adopting matrix terminology, we say that the bipartite graph is $r \times s$, and is *square* if $r = s$. If the bipartite graph G is connected, then the bipartition $\{X, Y\}$ is unique, and thus squareness is an intrinsic property of G. If G is not connected, then it has more than one bipartition, some of which may satisfy $r = s$.

A *matching* M in a graph G is a collection of independent edges. It is a *perfect matching* provided that each vertex of G belongs to some edge in M. Thus, a graph with an odd number of vertices cannot have a perfect matching. The graph G is *matching covered* provided that each edge belongs to at least one perfect matching. An $m \times n$ bipartite graph has a perfect matching only if it is square.

Let G be a square bipartite graph. To each edge of G we assign a weight, where the weights are algebraically independent numbers over the rational field $\mathbf{Q}$. We call the biadjacency matrix $\mathbf{A}$ of this weighted bipartite graph a *generic biadjacency matrix* over $\mathbf{Q}$. A theorem of Frobenius (see [16] and [17]) connects the structure of a bipartite graph and the determinant of its biadjacency matrix.

Theorem 2.4 *Let* $\mathbf{A}$ *be a generic biadjacency matrix corresponding to a square bipartite graph* G*. Then* G *is connected and matching covered if and only if* $\det \mathbf{A}$ *is an irreducible polynomial in the polynomial ring obtained from* $\mathbf{Q}$ *by adjoining the non-zero elements of* $\mathbf{A}$.

Another connection between bipartite graphs and matrices concerns scaling the rows and columns of a non-negative matrix so that all row and column sums are 1 (see [6] and [43]). Such a matrix is *doubly stochastic.*

Theorem 2.5 *Let* $\mathbf{A}$ *be a real matrix of order* n *corresponding to a square bipartite graph* G*. Then* G *is matching covered if and only if there exist positive diagonal matrices* $\mathbf{D}_1$ *and* $\mathbf{D}_2$ *such that* $\mathbf{D}_1\mathbf{A}\mathbf{D}_2$ *is doubly stochastic.*

With any multigraph G we can also associate a matrix that records the incidences between the vertices and edges. Let $v_1, v_2, \ldots, v_n$ be the vertices of G in some order, and let $e_1, e_2, \ldots, e_m$ be the edges in some order. We arbitrarily assign an orientation to each edge from one of its vertices (the *tail*) to the other (the *head*). The (oriented) *incidence matrix* of G, corresponding to the given orientation of G, is the $n \times m$ matrix $\mathbf{B} = (b_{ij})$ in which

$$b_{ij} = \begin{cases} 1, & \text{if } v_i \text{ is the tail of } e_j \\ -1, & \text{if } v_i \text{ is the head of } e_j \\ 0, & \text{otherwise.} \end{cases}$$

Each column of $\mathbf{B}$ contains exactly one 1 and exactly one -1. Note that $\mathbf{B}\mathbf{B}^T = \mathbf{D} - \mathbf{A}$, where $\mathbf{A}$ is the adjacency matrix of G and $\mathbf{D}$ is the diagonal matrix whose diagonal entries are the degrees of the vertices of G. The matrix $\mathbf{B}\mathbf{B}^T$ is the *Laplacian matrix* $\mathbf{L}$ of G. The following theorem relates the connectedness of a graph to the algebraic properties of its Laplacian matrix (see Chapter 4 and [7]).

Theorem 2.6 *Let G be a multigraph with n vertices. Then the rank of the Laplacian matrix* $\mathbf{L}$ *is at most* $n-1$, *with equality if and only if G is connected. If G is connected, then the number of spanning trees of G is the modulus of the determinant of any submatrix of* $\mathbf{L}$ *of order* $n-1$.

We conclude this section by remarking that Seidel [39] has introduced another matrix into the study of graphs. This is the $(0, 1, -1)$-matrix $\mathbf{S} = \mathbf{J}_n - \mathbf{I}_n - 2\mathbf{A}$, where $\mathbf{A}$ is the adjacency matrix of a graph G with n vertices and $\mathbf{J}_n$ is the all-1 matrix of order n. The matrix $\mathbf{S}$ can be obtained from $\mathbf{A}$ by replacing each 1 by -1 and each 0 not on the main diagonal by 1. This matrix has been useful in the study of strongly regular graphs (see Chapter 8).

3. Digraphs

We can view any graph G as a digraph, by regarding each edge joining two vertices as two arcs, one in each direction. If $\mathbf{A}$ is the adjacency matrix of G, then $\mathbf{A}$ is also the adjacency matrix of G viewed as a digraph. It is for this reason that we focus on digraphs in this section.

Let $\mathbf{A} = (a_{ij})$ be a matrix of order n, and let G be the weighted digraph corresponding to $\mathbf{A}$. The *weight* of a directed walk is the product of the weights of its arcs. A *directed* v_i-v_j *walk* is a directed walk in G from vertex v_i to vertex v_j.

From the definition of matrix multiplication we deduce that, for each positive integer k, the ij-entry of $\mathbf{A}^k$ is the sum of the weights of all the directed v_i-v_j walks of length k in G. If $\mathbf{A}$ is a $(0, 1)$-matrix, then the ij-entry of $\mathbf{A}$ is the number of directed v_i-v_j walks of length k. In particular, there is a directed v_i-v_j walk of length k if and only if the ij-entry of $\mathbf{A}^k$ is not 0. This last conclusion holds whenever $\mathbf{A}$ is a non-negative matrix, an assumption that prevents any cancellation in summing products of weights.

Let $\mathbf{A}$ be a non-negative matrix. Then $\mathbf{A}$ is *primitive* if there is a positive integer k for which each element of $\mathbf{A}^k$ is positive – that is, $\mathbf{A}^k$ is a *positive matrix.* Note that if $\mathbf{A}^k$ is positive, then so is $\mathbf{A}^l$ for each integer $l > k$. It follows from the above discussion that $\mathbf{A}$ is primitive if and only if there exists a positive integer k for which there are directed v_i-v_j walks of length k, for each i and j.

Let $\mathbf{P}$ be a permutation matrix of order n, corresponding to the permutation $i_1, i_2, \ldots, i_n$ of $\{1, 2, \ldots, n\}$. Thus $\mathbf{P}$ has 1s in positions (j, i_j) (for $1 \le j \le n$) and 0s elsewhere. The matrix $\mathbf{PAP}^T$ is obtained from $\mathbf{A}$ by permuting rows so that they are in the order $i_1, i_2, \ldots, i_n$, and then permuting columns so that they are also in the order $i_1, i_2, \ldots, i_n$: thus, the matrix $\mathbf{PAP}^T$ is obtained from $\mathbf{A}$ by *simultaneous permutations of its rows and columns.* Note that $\mathbf{A}$ and $\mathbf{PAP}^T$ represent the same digraph G, but with a different ordering of its vertices.

The matrix $\mathbf{A}$ is *reducible* if there exists a permutation matrix $\mathbf{P}$ such that

$$\mathbf{PAP}^T = \begin{pmatrix} \mathbf{A}_1 & \mathbf{O} \\ \mathbf{A}_{21} & \mathbf{A}_2 \end{pmatrix}, \qquad (2)$$

where $\mathbf{A}_1$ and $\mathbf{A}_2$ are matrices of positive orders n_1 and n_2. Assume that $\mathbf{A}$ is reducible and (2) holds. Then the vertex-set V of G can be partitioned into two non-empty subsets V_1 and V_2 of sizes n_1 and n_2, so that all arcs between V_1 and V_2 are directed from V_2 to V_1, and thus (V_2, V_1) is a *directed cut* of G. Conversely, a directed cut of G implies that $\mathbf{PAP}^T$ has the form given in (2) for some permutation matrix $\mathbf{P}$. Thus $\mathbf{A}$ is *reducible* if and only if the digraph G has a directed cut, and *irreducible* otherwise. Irreducibility of a matrix is equivalent to the strong connectivity of its digraph. We observe that a digraph is strongly connected if and only if it has no directed cut.

Defining $v \equiv w$ on the vertex-set V to mean that G has a directed v-w walk and a directed w-v walk, we obtain an equivalence relation that partitions V into equivalence classes $V_1, V_2, \ldots, V_k$. Each induced digraph $G[V_i]$ is a maximal strongly connected subdigraph of G, called a *strong component* of G. Let G^* be the digraph whose vertices are the sets $V_1, V_2, \ldots, V_k$, with an arc from V_i to V_j if and only if there is an arc in G from some vertex in V_i to some vertex in V_j $(i \neq j)$. Then G^* cannot have any closed directed walks. It follows that the sets $V_1, V_2, \ldots, V_k$ can be ordered as $V_{i_1}, V_{i_2}, \ldots, V_{i_k}$ in such a way that there is an arc from V_{i_p} to V_{i_q} in G^* only if $1 \leq q < p \leq k$. Ordering the vertices of V so that those in V_{i_1} come first, then those in $V_{i_2}, V_{i_3}, \ldots, V_{i_k}$, we deduce that there is a permutation matrix $\mathbf{P}$ for which

$$\mathbf{PAP}^T = \begin{pmatrix} \mathbf{A}_1 & \mathbf{0} & \cdots & \mathbf{0} \\ \mathbf{A}_{21} & \mathbf{A}_2 & \cdots & \mathbf{0} \\ \vdots & \vdots & \ddots & \vdots \\ \mathbf{A}_{k1} & \mathbf{A}_{k2} & \cdots & \mathbf{A}_k \end{pmatrix}, \qquad (3)$$

where $\mathbf{A}_1, \mathbf{A}_2, \ldots, \mathbf{A}_k$ are irreducible matrices. The matrices $\mathbf{A}_1, \mathbf{A}_2, \ldots, \mathbf{A}_k$, which correspond to the strong components of the digraph G, are uniquely determined up to simultaneous permutations of their rows and columns. The matrix in (3) is called the *Frobenius normal form* of $\mathbf{A}$, and $\mathbf{A}_1, \mathbf{A}_2, \ldots, \mathbf{A}_k$ are called the *irreducible components* of $\mathbf{A}$. In general, the Frobenius normal form of $\mathbf{A}$ is not unique. The uniqueness depends on the digraph G^* – that is, on the zero or non-zero character of the subdiagonal blocks $\mathbf{A}_{ij}$. If, for example, all the subdiagonal blocks are zero matrices, then the irreducible components can appear in any order along the diagonal of (3).

If G is a graph, then the matrix $\mathbf{A}$ is symmetric, and (3) reduces to

$$\mathbf{PAP}^T = \begin{pmatrix} \mathbf{A}_1 & \mathbf{0} & \cdots & \mathbf{0} \\ \mathbf{0} & \mathbf{A}_2 & \cdots & \mathbf{0} \\ \vdots & \vdots & \ddots & \vdots \\ \mathbf{0} & \mathbf{0} & \cdots & \mathbf{A}_k \end{pmatrix},$$

where $\mathbf{A}_1, \mathbf{A}_2, \ldots, \mathbf{A}_k$ are symmetric irreducible matrices that correspond to the connected components of G. Thus the symmetric matrix $\mathbf{A}$ is irreducible if and only if G is connected.

Now assume that $\mathbf{A}$ is an irreducible matrix of order n – that is, G is a strongly connected digraph. The *index of imprimitivity* of $\mathbf{A}$ and of G is the greatest common divisor of the lengths of the cycles of G, or equivalently, of the lengths of the closed directed walks in G. Since G is strongly connected, it must have a cycle if $n > 1$. The properties of strongly connected digraphs collected in the following lemma are straightforward to verify and lead to a normal form for irreducible matrices.

Lemma 3.1 *Let G be a strongly connected digraph with index of imprimitivity d. Then*

- *d is the greatest common divisor of the lengths of the closed directed walks containing any particular vertex;*
- *for each pair of vertices v and w, the lengths of all directed vw-walks are congruent modulo d;*
- *the vertex-set of G can be uniquely partitioned into d non-empty sets $U_1, U_2, \ldots, U_d$ in such a way that each arc of G is directed from some U_i to U_{i+1} $(i = 1, 2, \ldots, d; U_{d+1} = U_1)$, and the length of each directed walk from a vertex in U_i to a vertex in U_j is congruent to $j - i$ (modulo d), for $1 \le i, j \le d$.*

The sets $U_1, U_2, \ldots, U_d$ in Lemma 3.1 are called the *imprimitivity sets* of the strongly connected digraph G. If $|U_i| = n_i$ $(i = 1, 2, \ldots, d)$, then there is a permutation matrix $\mathbf{P}$ for which

$$\mathbf{PAP}^T = \begin{pmatrix} \mathbf{0}_{n_1} & \mathbf{A}_{12} & \mathbf{0} & \cdots & \mathbf{0} \\ \mathbf{0} & \mathbf{0}_{n_2} & \mathbf{A}_{23} & \cdots & \mathbf{0} \\ \vdots & \vdots & \vdots & \ddots & \mathbf{0} \\ \mathbf{0} & \mathbf{0} & \mathbf{0} & \cdots & \mathbf{A}_{d-1,d} \\ \mathbf{A}_{d1} & \mathbf{0} & \mathbf{0} & \cdots & \mathbf{0}_{n_d} \end{pmatrix}. \qquad (4)$$

The following lemma is due to Schur (see [7] for a proof).

Lemma 3.2 *A non-empty set of positive integers that is closed under addition contains all but a finite number of positive multiples of its greatest common divisor.*

We next show that primitive matrices are exactly those irreducible matrices with index of imprimitivity 1.

Theorem 3.3 *Let G be a strongly connected digraph with at least two vertices. Then the index of imprimitivity of G is* 1 *if and only if there is an integer N such that, for any two vertices v_i and v_j, there are directed v_i-v_j walks of each length $l \geq N$; equivalently, an irreducible non-negative matrix $\mathbf{A}$ is primitive if and only if its index of imprimitivity is* 1.

Proof Let d be the index of imprimitivity of G and assume that there is an integer N such that, for any two vertices v_i and v_j, there is a directed v_i-v_j walk of each length $l \geq N$. Then there are directed v_1-v_2 walks and directed v_2-v_1 walks of lengths N and $N + 1$. Hence G has closed directed walks of lengths $2N$ and $2N + 1$, and so $d = 1$.

Now assume that $d = 1$, and let $c_1, c_2, \ldots, c_p$ be the distinct lengths of the cycles of G. The strong connectivity of G implies that, for any two vertices v_i and v_j, there is a directed v_i-v_j walk that meets a cycle of each of the lengths $c_1, c_2, \ldots, c_p$. By Lemma 3.2, there are directed v_j-v_i walks of length l, for every sufficiently large integer l.

Now suppose that $\mathbf{A}$ is a non-negative irreducible matrix. Since the ij-entry of $\mathbf{A}^l$ is positive if and only if there is a directed v_i-v_j walk in the associated digraph, the theorem follows. ■

If G is a connected graph with symmetric adjacency matrix $\mathbf{A}$, then viewing G as a digraph, we see that G has a directed cycle of length 2; thus the index of imprimitivity of $\mathbf{A}$ is 1 or 2, and $\mathbf{A}$ is primitive if and only if G has a cycle of odd length – that is, G is not bipartite.

In our discussion of primitive non-negative matrices, we may restrict ourselves to (0, 1)-matrices. Let $\mathbf{A}$ be a primitive (0, 1)-matrix of order n. The *exponent* of $\mathbf{A}$, denoted by $\exp \mathbf{A}$, is the smallest positive integer t for which $\mathbf{A}^t$ is positive. In terms of primitive digraphs, the exponent is the smallest positive integer t for which there is a directed $v_i v_j$-walk of length t, for all i and j.

There has been considerable work done on determining exponents and related parameters. The following is a theorem of Dulmage and Mendelsohn [12].

Theorem 3.4 *Let $\mathbf{A}$ be a primitive* (0, 1)*-matrix of order n, and let l be the length of a shortest cycle in its digraph G. Then*

$$\exp \mathbf{A} \leq n + l(n - 2).$$

Proof Consider the matrix $\mathbf{A}^l$ and its digraph G^l. There is an arc in G^l from vertex v_i to vertex v_j exactly when there is a directed v_i-v_j walk of length l in G. Let L be the set of vertices of G^l that have loops. Then L has at least l elements, and the loops at vertices of L imply that there is a directed walk in G^l of length $n-1$ from each vertex in L to each vertex v_j. Hence in G there is a directed walk of length $l(n-1)$ from each vertex in L to each vertex v_j. For each vertex v_i there is a directed walk in G from v_i to some vertex in L of length $l_i \leq n-l$, and hence a directed v_i-v_j walk of length $n-l+l(n-1) = n+l(n-2)$. Hence $\mathbf{A}^{n+l(n-2)}$ is a positive matrix. ■

Since a primitive digraph with $n \geq 2$ vertices must have a cycle of length $l \leq n-1$, we get the following result of Wielandt [47].

Corollary 3.5 *The maximum exponent of a primitive matrix of order* $n\,(\geq 2)$ *is* $(n-1)^2+1$.

Proof It follows from Theorem 3.4 that the exponent of a primitive matrix of order n does not exceed $(n-1)^2+1$. The digraph obtained from a cycle of length n by inserting a chord between some pair of vertices at distance 2 on the cycle can be shown to have exponent $(n-1)^2+1$. ■

It can also be shown that, up to isomorphism, the digraph in the proof is the unique digraph with n vertices with exponent equal to $(n-1)^2+1$ (see, for example, [7]). The maximum exponent of a primitive symmetric (0, 1)-matrix $\mathbf{A}$ of order $n\,(\geq 2)$ equals $2n-2$, with equality if and only if the graph of $\mathbf{A}$ is a path with a loop at one end (see [42]).

The notion of primitivity has been generalized (see [15]). A *coloured digraph* is a digraph G in which each arc is labelled by one of a set $\{c_1, c_2, \ldots, c_t\}$ of colours. Given a directed walk α in G, we let $\vec{\alpha}$ denote the column vector whose ith coordinate is the number of arcs of α with colour c_i. Fix one vertex v, and define a relation on the vertices of G by $v_i \sim v_j$ whenever there exist directed walks α_i and α_j from v to v_i and v to v_j, respectively, for which $\vec{\alpha}_i = \vec{\alpha}_j$. It is easy to verify that, if G is strongly connected, this relation is an equivalence relation and is independent of v. The coloured digraph G is called *primitive* if there is exactly one equivalence class for $\sim$. Let $\gamma_1, \gamma_2, \ldots, \gamma_s$ be the cycles of G, let L be the subgroup of $\mathbf{Z}^t$ generated by $\{\vec{\gamma}_i : i = 1, 2, \ldots, s\}$, and let $\mathbf{M}$ be the $t \times s$ matrix whose ith column is $\vec{\gamma}_i$.

Theorem 3.6 *Let G be a strongly connected coloured digraph with n vertices and at least one arc of each of t colours. Then the following statements are equivalent:*
(1) The coloured digraph G is primitive.
(2) $L = \mathbf{Z}^t$.

(3) The greatest common divisor of the determinants of the $t \times t$ submatrices of $\mathbf{M}$ *is* 1.

(4) There exists a non-negative integer vector $\mathbf{h} = (h_1, h_2, \ldots, h_t)^T$ *such that, for any two vertices* v_i *and* v_j*, there is a directed* v_i-v_j *walk with* h_k *arcs of colour* c_k*, for* $k = 1, 2, \ldots, t$.

Proof The theorem follows from the truth of the implications (1) $\Rightarrow$ (2), (2) $\Longleftrightarrow$ (3), (2) $\Rightarrow$ (4), and (4) $\Rightarrow$ (1).

First assume that (1) holds. Let $\mathbf{e}_i$ be the $t \times 1$ vector whose ith entry is 1 and whose other entries are 0. The digraph D has an arc vw of colour c_i. Since D is primitive, there exist a directed v-w walk α and a directed w-v walk β for which $\vec{\alpha} = \vec{\beta}$. Since each closed directed walk of G can be decomposed into cycles, both $\vec{\alpha}$ and $\vec{\beta} + \mathbf{e}_i$ belong to L. Hence, $\mathbf{e}_i \in L$ for all i, and (2) holds.

The equivalence of (2) and (3) is a standard result concerning integer matrices, and can be found in [32].

To show that (2) $\Rightarrow$ (4), assume that (2) holds. Since G is strongly connected, there is a closed directed walk γ that passes through each vertex of D. Since $L = \mathbf{Z}^t$, there exist $s \times 1$ integer vectors $\mathbf{x}^{(j)}$ such that $\mathbf{M}\mathbf{x}^{(j)} = \mathbf{e}_j$ ($j = 1, 2, \ldots, t$).

Let $\mathbf{x}$ be a non-negative integer vector such that $\mathbf{x} \geq n\mathbf{x}^{(j)}$ (entry-wise) for $j = 1, 2, \ldots, t$, and let $\mathbf{h} = \vec{\gamma} + t\mathbf{M}\mathbf{x}$. We claim that, for any two vertices v and w, there exists a directed v-w walk α with $\vec{\alpha} = \mathbf{h}$. To see this, let p be a path from v to w with $\vec{\mathbf{p}} = (p_1, p_2, \ldots, p_t)^T$. Then $\vec{\mathbf{p}} = \mathbf{M}(\sum_{j=1}^{t} p_j \mathbf{x}^{(j)})$, and hence

$$\mathbf{h} = \vec{\gamma} + \vec{\mathbf{p}} + \mathbf{M}\left(\sum_{j=1}^{t}(\mathbf{x} - p_j\mathbf{x}^{(j)})\right).$$

Note that, for each i, the ith entry n_i of $\sum_{j=1}^{t}(\mathbf{x} - p_j\mathbf{x}^{(j)})$ is a non-negative integer. Thus, the directed walk α that starts at v, follows γ back to v, along the way goes n_i times around γ_i ($i = 1, 2, \ldots, t$), and then follows p to w, has $\vec{\alpha} = \mathbf{h}$. Hence (4) holds.

Clearly (4) implies (1). ■

The equivalence of (1) and (4) implies that if $t = 1$, so all arcs are coloured the same, then G is primitive as a coloured graph if and only if it is primitive as an uncoloured graph. The notion of exponents of coloured digraphs is introduced and studied in [41].

4. Biclique partitions of graphs

We next illustrate how elementary but powerful linear algebra can be used in the study of questions that concern the partitioning of the edges of a graph into certain types of subgraphs. A 'biclique partition' of a graph G is a partition of the edge-set of G into complete bipartite subgraphs. More formally, we make the following definitions. A *biclique* of a graph G is a subgraph whose edges form a complete bipartite subgraph. Given two disjoint subsets X and Y, $B(X, Y)$ denotes the biclique consisting of all edges joining a vertex in X and a vertex in Y. A *biclique partition* of G is a collection

$$B(X_1, Y_1), B(X_2, Y_2), \ldots, B(X_k, Y_k) \tag{5}$$

of bicliques whose edges partition the edges of G. Since a single edge can form a biclique, every graph has a biclique partition. The *biclique partition number* $\text{bp}(G)$ of G is the smallest number of bicliques that partition G.

Let $\mathbf{A}$ be the adjacency matrix of G. Then it is easy to verify that, for a biclique partition (5) of G,

$$\mathbf{A} = \mathbf{XY}^T + \mathbf{YX}^T = (\mathbf{X} \quad \mathbf{Y}) \begin{pmatrix} \mathbf{Y}^T \\ \mathbf{X}^T \end{pmatrix},$$

where $\mathbf{X}$ is the $n \times k$ matrix whose ij-entry is 1 if $v_i \in \mathbf{X}_j$, and 0 otherwise, and $\mathbf{Y}$ is defined analogously. Thus, a partition of G into k bicliques corresponds to a special type of factorization of $\mathbf{A}$ as the product of an $n \times 2k$ (0, 1)-matrix and a related $2k \times n$ (0, 1)-matrix.

The matrix $\mathbf{XY}^T$ is the adjacency matrix of the digraph obtained from G by orientating each edge in each $B(X_i, Y_i)$ from its vertex in X_i to its vertex in Y_i. Assume that G is the complete graph. A *tournament* is a digraph obtained by choosing an orientation for each edge of the complete graph G. Hence, partitions of the complete graph on n vertices into k bicliques correspond to factorizations $\mathbf{M} = \mathbf{XY}^T$ of the adjacency matrix $\mathbf{M}$ of a tournament as the product of an $n \times k$ (0, 1)-matrix and a $k \times n$ (0, 1)-matrix. The following lemma [11] implies that adjacency matrices of tournaments have special algebraic properties. Additional properties can be found in [30] and [40].

Lemma 4.1 *Let* $\mathbf{M}$ *be the adjacency matrix of a tournament on n vertices. Then the rank of* $\mathbf{M}$ *is at least* $n - 1$.

Proof Let $\mathbf{j}$ be the $n \times 1$ all-1 vector, and let $\mathbf{N}$ be the $(n + 1) \times n$ matrix obtained from $\mathbf{M}$ by appending $\mathbf{j}^T$ at the bottom. It suffices to prove that the columns of $\mathbf{N}$ are linearly independent.

Suppose that $\mathbf{Nx} = \mathbf{0}$. Then $\mathbf{Mx} = \mathbf{0}$ and $\mathbf{j}^T\mathbf{x} = 0$. Since $\mathbf{M}$ is the adjacency matrix of a tournament, $\mathbf{M} + \mathbf{M}^T = \mathbf{J}_n - \mathbf{I}_n$. Since $\mathbf{Mx} = \mathbf{0}$ and $\mathbf{x}^T\mathbf{j} = \mathbf{0}$, we have

$$0 = \mathbf{x}^T(\mathbf{M} + \mathbf{M}^T)\mathbf{x} = (\mathbf{x}^T\mathbf{j})^2 - \mathbf{x}^T\mathbf{x} = -\mathbf{x}^T\mathbf{x}.$$

Hence $\mathbf{x} = \mathbf{0}$, and so the columns of $\mathbf{N}$ are linearly independent. ■

Theorem 4.2 *The biclique partition number of K_n is $n - 1$.*

Proof A partition of K_n into k bicliques corresponds to a factorization of the adjacency matrix $\mathbf{M}$ of a tournament on n vertices as the product of an $n \times k$ matrix and a $k \times n$ matrix. The bound $\mathrm{bp}(K_n) \geq n - 1$ follows, since the rank of $\mathbf{M}$ is bounded above by k and below by $n - 1$. Since K_n can be partitioned into $n - 1$ stars, $\mathrm{bp}(K_n) = n - 1$. ■

This theorem, due to H. S. Witsenhausen, is known as the *Graham-Pollak theorem* and has several different proofs (see [20], [44], [34] and [9]). Interestingly, each of these proofs is linear-algebraic in nature. More generally, we have the following result (see [20]).

Theorem 4.3 *Let G be a graph with adjacency matrix $\mathbf{A}$. Then*

$$bp(G) \geq max\{n_-(\mathbf{A}), n_+(\mathbf{A})\},$$

where $n_-(\mathbf{A})$ and $n_+(\mathbf{A})$ are the number of negative and positive eigenvalues of $\mathbf{A}$.

Proof As argued in [21], the interlacing inequalities for eigenvalues of symmetric matrices (see Chapter 1 and [22]) imply that n_+ and n_- are subadditive functions – that is, for symmetric matrices $\mathbf{C}$ and $\mathbf{D}$ of the same order,

$$n_+(\mathbf{C} + \mathbf{D}) \leq n_+(\mathbf{C}) + n_+(\mathbf{D}) \text{ and } n_-(\mathbf{C} + \mathbf{D}) \leq n_-(\mathbf{C}) + n_-(\mathbf{D}).$$

Let $\mathbf{A} = \mathbf{XY}^T + \mathbf{YX}^T$, where $\mathbf{X}$ and $\mathbf{Y}$ are $n \times k$ matrices with jth columns $\vec{\mathbf{X}}_j$ and $\vec{\mathbf{Y}}_j$. Then

$$\mathbf{A} = \sum_{j=1}^{k} \left(\vec{\mathbf{X}}_j \vec{\mathbf{Y}}_j^T + \vec{\mathbf{Y}}_j \vec{\mathbf{X}}_j^T\right).$$

Since, for each j,

$$n_+\left(\vec{\mathbf{X}}_j \vec{\mathbf{Y}}_j{}^T + \vec{\mathbf{Y}}_j \vec{\mathbf{X}}_j{}^T\right) = n_-\left(\vec{\mathbf{X}}_j \vec{\mathbf{Y}}_j{}^T + \vec{\mathbf{Y}}_j \vec{\mathbf{X}}_j{}^T\right) = 1,$$

subadditivity implies that $n_+(\mathbf{A}) \leq k$ and $n_-(\mathbf{A}) \leq k$. The theorem now follows from the correspondence between partitions of G into k bicliques and factorizations of $\mathbf{A}$ as $\mathbf{A} = \mathbf{XY}^T + \mathbf{YX}^T$. ■

Note that $n_-(\mathbf{J}_n - \mathbf{I}_n) = n - 1$ and $n_+(\mathbf{J}_n - \mathbf{I}_n) = 1$, and hence Theorem 4.3 also shows that $\text{bp}(K_n) \geq n - 1$.

5. Bipartite graphs

As we observed in Section 1, bipartite graphs and rectangular matrices are interchangeable. The biadjacency matrix of an $r \times s$ bipartite graph with bipartition $\{X, Y\}$ is an $r \times s$ matrix. Properties of rectangular matrices can therefore be formulated in terms of bipartite graphs; conversely, the structure of bipartite graphs can be useful in investigations of rectangular matrices or non-symmetric square matrices.

Let G be an $r \times s$ bipartite graph with bipartition $\{X, Y\}$, and let $\mathbf{A}$ be its biadjacency matrix. The rows and columns of $\mathbf{A}$ are indexed by the elements of X and Y, respectively, and the edges of G correspond to 1s in $\mathbf{A}$. Given (possibly empty) subsets U of X and V of Y, the submatrix of $\mathbf{A}$ whose row indices are in U and whose column indices are in V is denoted by $\mathbf{A}[U, V]$; this submatrix is the biadjacency matrix of the induced subgraph $G[U, V]$ of G on the vertex-set $U \cup V$. It is a zero submatrix if and only if each edge of G contains a vertex of $X \backslash U$ or of $Y \backslash V$ – that is, if and only if $(X \backslash U) \cup (Y \backslash V)$ *covers* the edges of G.

A matching of G corresponds to a collection of 1s of $\mathbf{A}$ with no two 1s in the same row or column. The *matching number m(G)* is the largest number of edges in a matching of G. Since the edges of a matching are pairwise vertex-disjoint, the number of edges in a matching of G does not exceed the cardinality of each subset of vertices that covers the edges of G. Hence,

$$m(G) \leq \min\{r + s - (k + l)\}, \tag{6}$$

where the minimum is taken over all pairs (k, l) of non-negative integers for which $\mathbf{A}$ has a $k \times l$ zero submatrix. Here we consider the empty $0 \times s$ and $r \times 0$ submatrices as zero submatrices. König's theorem asserts that equality holds in (6).

Theorem 5.1 *Let G be an $r \times s$ bipartite graph with biadjacency matrix $\mathbf{A}$. Then $m(G)$ equals the minimum of $r + s - (k + l)$, taken over all pairs of integers (k, l) for which $\mathbf{A}$ has a $k \times l$ zero submatrix.*

Proof Let $d(\mathbf{A})$ denote the maximum sum of the dimensions of a zero submatrix of $\mathbf{A}$. It follows from the discussion preceding the theorem that

$$m(G) \leq (r + s) - d(\mathbf{A}).$$

To complete the proof, we show by induction on $r + s$ that G has a matching with $r + s - d(\mathbf{A})$ edges. This is clear if $r = 1, s = 1$, or $\mathbf{A} = \mathbf{O}$, so we assume that $r, s \geq 2$ and $\mathbf{A} \neq \mathbf{O}$.

First, suppose that $\mathbf{A}$ has a zero submatrix $\mathbf{A}[U, V]$ for which $|U| + |V| = d(\mathbf{A})$ and both U and V are non-empty. Each $p \times q$ zero submatrix of $\mathbf{A}[X\backslash U, Y]$ can be extended to a $(p + |U|) \times q$ zero submatrix of $\mathbf{A}$. Hence, by the definition of $d(\mathbf{A})$,

$$p + q \leq d(\mathbf{A}) - |U|.$$

It follows from the inductive hypothesis that $G[X\backslash U, Y]$ has a matching M_1 with $r - |U|$ edges. Similarly, $G[U, Y\backslash V]$ has a matching M_2 with $s - |V|$ edges. Hence, $M_1 \cup M_2$ is a matching of G with $r + s - |U| - |V| = r + s - d(\mathbf{A})$ edges.

Next, suppose that each zero submatrix $\mathbf{A}[U, V]$ of $\mathbf{A}$ with $|U| + |V| = d(\mathbf{A})$ has either $U = \emptyset$ or $V = \emptyset$. Without loss of generality we may assume that the top-left entry of $\mathbf{A}$ is 1. Let $\mathbf{B} = \mathbf{A}[X\backslash\{x_1\}, Y\backslash\{y_1\}]$. Since each zero submatrix of $\mathbf{A}$ whose dimensions sum to $d(\mathbf{A})$ is vacuous, the dimensions of each zero submatrix of B sum to at most $d(\mathbf{A}) - 1$. Hence, by induction, the induced subgraph $G[X\backslash\{x_1\}, Y\backslash\{y_1\}]$ has a matching M' of size $(r - 1) + (s - 1) - (d(\mathbf{A}) - 1) = r + s - d(\mathbf{A}) - 1$. Thus, M' along with the edge joining $x_1 y_1$ is a matching of G with $r + s - d(\mathbf{A})$ edges. ■

Now assume that G is an $r \times r$ bipartite graph – that is, G is square. A perfect matching of G corresponds to a permutation matrix $\mathbf{P}$ with $\mathbf{P} \leq \mathbf{A}$ (entry-wise). König's theorem implies that G has a perfect matching if and only if the dimensions of each zero submatrix of $\mathbf{A}$ sum to at most r.

Suppose further that G is a d-regular graph with $d > 0$. Then each row and column of $\mathbf{A}$ has precisely d 1s. Consider a zero submatrix $\mathbf{A}[U, V]$ of $\mathbf{A}$. Then the sum $(r - |V|)d$ of the elements of $\mathbf{A}$ in the columns indexed by $Y\backslash V$ is at least the sum $|U|d$ of the elements of $\mathbf{A}$ in the rows indexed by U. It follows that $|U| + |V| \leq r$, and hence that G has a perfect matching M. Removing the edges of M results in a $(d - 1)$-regular bipartite graph. Hence, by induction, the edges of the d-regular bipartite graph G can be partitioned into d perfect matchings: such a partitioning is called a *1-factorization* of G. In matrix terms, a square matrix (0, 1)-matrix $\mathbf{A}$ with constant row and column sums is the sum of permutation matrices.

Theorem 5.1 also implies the following fundamental result of Birkhoff about doubly stochastic matrices.

Theorem 5.2 *The set of all doubly-stochastic matrices of order n is the convex hull of the set of permutation matrices.*

Proof The set of doubly stochastic matrices (regarded as a subset of $\mathbf{R}^{n^2}$) is closed under convex combinations, and contains the permutation matrices of order n.

Let $\mathbf{S} = (s_{ij})$ be an $n \times n$ doubly-stochastic matrix, and let G be the $n \times n$ bipartite graph with an edge joining x_i and y_j if and only if $s_{ij} \neq 0$. We prove that $\mathbf{S}$ is a convex combination of permutation matrices, by induction on the number of non-zero entries of $\mathbf{S}$. Theorem 5.1 implies G has a perfect matching

$$\mathbf{M} = \{x_1 y_{i_1}, x_2 y_{i_2}, \ldots, x_m y_{i_m}\}.$$

Let $\mathbf{P}$ be the permutation matrix corresponding to $\mathbf{M}$, and let $\epsilon = \min\{s_{1i_1}, s_{2i_2}, \ldots, s_{mi_m}\}$. If $\epsilon = 1$, then $\mathbf{M} = \mathbf{P}$. Otherwise, $(\mathbf{S} - \epsilon\mathbf{P})/(1 - \epsilon)$ is a doubly-stochastic matrix with fewer non-zero entries than $\mathbf{S}$, and the theorem now follows by induction. ■

Theorem 2.4 refers to bipartite graphs that are connected and matching-covered, and we now give a matrix interpretation of these requirements. Let $\mathbf{A}$ be an $r \times s$ matrix and let G be the associated bipartite graph obtained by ignoring the weights. Then G is disconnected if and only if there exist permutation matrices $\mathbf{P}$ and $\mathbf{Q}$ such that $\mathbf{PAQ}$ has the form

$$\begin{pmatrix} \mathbf{A}_1 & \mathbf{0} \\ \mathbf{0} & \mathbf{A}_2 \end{pmatrix},$$

for some matrices $\mathbf{A}_1$ and $\mathbf{A}_2$. An $r \times r$ matrix is *fully indecomposable* if the sum of the dimensions of each of its non-empty zero submatrices is at most $r - 1$.

Assume that G is connected. Applying Theorem 5.1 to the subgraph obtained by deleting a vertex in each part of the bipartition of G, we deduce that a connected square bipartite graph is matching-covered if and only if its biadjacency matrix is fully indecomposable.

Let G be a bipartite graph with a perfect matching M, and let $\mathbf{A}$ be its biadjacency matrix. Using the Frobenius normal form for matrices and digraphs, we can derive a normal form for G and $\mathbf{A}$. Since G has a perfect matching M, there exists a permutation matrix $\mathbf{S}$ with $\mathbf{S} \leq \mathbf{A}$ (entry-wise). Each diagonal entry of $\mathbf{AS}^T$ is 1. Let $\mathbf{P}$ be a permutation matrix such that

$$\mathbf{P}(\mathbf{AS}^T)\mathbf{P}^T = \begin{pmatrix} \mathbf{A}_1 & \mathbf{0} & \cdots & \mathbf{0} \\ \mathbf{A}_{21} & \mathbf{A}_2 & \cdots & \mathbf{0} \\ \vdots & \vdots & \ddots & \vdots \\ \mathbf{A}_{k1} & \mathbf{A}_{k2} & \cdots & \mathbf{A}_k \end{pmatrix} \qquad (7)$$

is the Frobenius normal form of $\mathbf{AS}^T$, where $\mathbf{A}_1, \mathbf{A}_2, \ldots, \mathbf{A}_k$ are irreducible. Since the diagonal entries of each $\mathbf{A}_i$ are 1, Theorem 5.1 implies that each A_i

is fully indecomposable. Thus there exist permutation matrices **P** and **Q** such that **PAQ** has the form given in (7). The matrices $\mathbf{A}_1, \mathbf{A}_2, \ldots, \mathbf{A}_k$ are uniquely determined by **A** up to (not necessarily simultaneous) permutations of their rows and columns, and are called the *fully indecomposable components* of **A**. The matrix **A** is fully indecomposable if and only if it has only one fully indecomposable component.

6. Permanents

Let G be a square bipartite graph with adjacency matrix $\mathbf{A} = (a_{ij})$ of order n. As already observed, Theorem 5.1 implies that G has a perfect matching if and only if **A** does not have a zero matrix, the sum of whose dimensions is $n + 1$. Since the perfect matchings of G are in one-to-one correspondence with the permutation matrices **P** for which $\mathbf{P} \leq \mathbf{A}$ (entry-wise), the number of perfect matchings of G equals the number of permutations $i_1, i_2, \ldots, i_n$ of $\{1, 2, \ldots, n\}$ for which $a_{1i_1}a_{2i_2}\ldots a_{ni_n} = 1$.

More generally, if $\mathbf{A} = (a_{ij})$ is a real matrix of order n, the *permanent* of **A** is defined by

$$\operatorname{per} \mathbf{A} = \sum_{i_1, i_2, \ldots, i_n} a_{1i_1}a_{2i_2}\ldots a_{ni_n},$$

where the summation extends over all permutations $i_1, i_2, \ldots, i_n$ of $\{1, 2, \ldots, n\}$. It can be shown that per $\mathbf{A} =$ per $\mathbf{PAQ}$, for all permutation matrices **P** and **Q**.

If G is a bipartite graph, then the permanent of **A** equals the number of perfect matchings of G. If G is a weighted bipartite graph, we define the *weight of a perfect matching M* of G to be the product of the weights of the edges of M. Then per **A** is the sum of the weights of the perfect matchings of G.

The permanent resembles the determinant, but unlike the determinant, which can be computed in polynomial time using Gaussian elimination, there is no known way to compute the permanent efficiently. In fact, Valiant (see [45] and [46]) has shown that computing the permanent of a (0, 1)-matrix – that is, computing the number of perfect matchings of a bipartite graph – is a #P-complete problem. Thus computing the permanent is computationally equivalent to many other difficult counting problems.

The permanent also has meaning for digraphs. A *cycle cover* of a digraph is a collection of directed vertex-disjoint cycles containing all the vertices. Let $\mathbf{A} = (a_{ij})$ be a (0, 1)-matrix, and let G be the corresponding digraph. The fact that each permutation of $\{1, 2, \ldots, n\}$ can be written uniquely as a product of disjoint permutation cycles implies that the permanent of **A** counts the number of cycle covers of G.

We now turn to bounds on the permanent. Let $\mathbf{A} = (a_{ij})$ be a non-negative matrix of order n, and let $r_i = \sum_{j=1}^n a_{ij}$ and $s_j = \sum_{i=1}^n a_{ij}$ be the row and column sums of $\mathbf{A}$. Then an elementary bound is

$$\text{per}\,\mathbf{A} \le \min\left\{\prod_{i=1}^n r_i, \prod_{i=1}^n s_i\right\}. \tag{8}$$

Now assume that permutations have been applied to $\mathbf{A}$, so that

$$r_1 \le r_2 \le \cdots \le r_n \quad \text{and} \quad s_1 \le s_2 \le \cdots \le s_n; \tag{9}$$

then a theorem of Ostrand [33] gives the lower bound

$$\text{per}\,\mathbf{A} \ge \max\left\{\prod_{i=1}^n \max\{1, r_i - i + 1\}, \prod_{i=1}^n \max\{1, s_i - i + 1\}\right\},$$

provided that per $\mathbf{A} \ne 0$. If $t = \min\{r_1, r_2, \ldots, r_n, s_1, s_2, \ldots, s_n\}$, we deduce that per $\mathbf{A} \ge t!$ or per $\mathbf{A} = 0$.

Again, assuming that (9) holds, a theorem of Jurkat and Ryser [23] gives

$$\text{per}\,\mathbf{A} \le \prod_{i=1}^n \min\{r_i, s_i\},$$

thereby improving the bound (8). Following a conjecture of Minc, Brégman [2] proved, and later Schrijver [38] proved more elegantly, that

$$\text{per}\,\mathbf{A} \le \min\left\{\prod_{i=1}^n (r_i!)^{1/r_i}, \prod_{i=1}^n (s_i!)^{1/s_i}\right\}, \tag{10}$$

Now assume that G is k-regular, so that $\mathbf{A}$ has exactly k 1s in each row and column. Then (10) implies that

$$\text{per}\,\mathbf{A} \le (k!)^{n/k}.$$

Suppose that k is a divisor of n, and that $\mathbf{J}_k$ is the all-1 matrix of order k. Then the matrix

$$\mathbf{A} = \begin{pmatrix} \mathbf{J}_k & \mathbf{0} & \cdots & \mathbf{0} \\ \mathbf{0} & \mathbf{J}_k & \cdots & \mathbf{0} \\ \vdots & \vdots & \ddots & \vdots \\ \mathbf{0} & \mathbf{0} & \cdots & \mathbf{J}_k \end{pmatrix}$$

has permanent equal to $(k!)^{n/k}$. This implies the truth of a conjecture of Ryser.

Theorem 6.1 *Let n be a positive integer, and, let k be a divisor of n. Then the maximum number of* 1*-factors of a k-regular $n \times n$ bipartite graph is* $(k!)^{n/k}$.

Let $\beta(n, k)$ be the maximum number of perfect matchings in a k-regular $n \times n$ bipartite graph. Then the value $\beta(n, k)$ is given by Theorem 6.1, whenever k divides n. Also $\beta(n, n-1)$ equals the nth derangement number. In addition, $\beta(n, 2) = 2^{\lfloor n/2 \rfloor}$, with equality if and only if G has $n/2$ connected components isomorphic to $K_{2,2}$ if n is even, and $\lfloor n/2 \rfloor - 1$ components isomorphic to $K_{2,2}$ and one isomorphic to a 6-cycle if n is odd. If $n \geq 8$, the $(n-2)$-regular $n \times n$ bipartite graphs achieving $\beta(n, n-2)$ are the bipartite complements of the graphs achieving $\beta(n, 2)$. If $k = 3$ and n is not divisible by 3, then we have

$$\beta(3m+1, 3) = 6^{m-1}9 \ (m \geq 1), \quad \beta(5, 3) = 13,$$
$$\text{and} \quad \beta(3m+2, 3) = 6^{m-2}9^2 \ (m \geq 2).$$

We also have $\beta(4m+1, 4) = 24^{m-1}44$, for $m \geq 1$. For references and further discussion of the above results, see [7]. Finally, McKay and Wanless [31] have shown that if $n = kd$, with $d \geq 5$, then those bipartite graphs achieving $\beta(n, n-k)$ are the bipartite complements of those achieving $\beta(n, k)$.

In a series of papers culminating with [37], Schrijver obtained an exponential lower bound for the number of perfect matchings of k-regular bipartite multigraphs – that is, for the permanent of a non-negative integral matrix of order n with all row and column sums equal to k. Let $\lambda^*(n, k)$ be the minimum number of perfect matchings of a k-regular $n \times n$ bipartite graph, and let

$$\theta_k = \liminf_{n \to \infty} \lambda^*(n, k)^{1/n}.$$

The number θ_k gives the best exponential lower bound θ_k^n for the number of perfect matchings of k-regular bipartite graphs. Schrijver proved that $\theta_k = (k-1)^{k-1}/k^{k-2}$, from which the following result can be deduced.

Theorem 6.2 *Every k regular $n \times n$ bipartite multigraph has at least*

$$\left(\frac{(k-1)^{k-1}}{k^{k-2}}\right)^n$$

perfect matchings.

By Theorem 6.2,

$$\text{per } \mathbf{A} \geq \left(\frac{(k-1)^{k-1}}{k^{k-2}}\right)^n,$$

for each non-negative integral matrix $\mathbf{A}$ of order n with all row and column sums equal to k. For such a matrix $\mathbf{A}$, the matrix $(1/k)\mathbf{A}$ is doubly stochastic. Van der Waerden conjectured, and Egoryčhev [13] and Falikman [14] independently proved, a sharp lower bound for the minimum permanent of doubly stochastic matrices.

Theorem 6.3 *If* **A** *is a doubly stochastic matrix of order n, then*

$$\operatorname{per} \mathbf{A} \geq \frac{n!}{n^n},$$

with equality if and only if **A** *is* $(1/n)\mathbf{J}_n$.

In general, the lower bound of Theorem 6.3 for the number of perfect matchings in a regular bipartite graph is weaker than that of Theorem 6.2.

Since the definition of per **A** is similar to that of det **A**, and since there are good algorithms for computing the determinant, it is natural to consider whether there is a simple transformation that converts **A** to another matrix $\mathbf{A}'$ for which per $\mathbf{A} = \det \mathbf{A}'$. This possibility is considered in the next section.

7. Converting the permanent into the determinant

A *signed digraph* is a weighted digraph G^s with weights ± 1. Thus, its adjacency matrix is a $(0, 1, -1)$-matrix $\mathbf{A}^s = (a_{ij})$. Like the permanent of a $(0, 1)$-matrix, the determinant of $\mathbf{A}^s$ can be expressed in terms of cycle covers.

The *weight* $\operatorname{wt}(\gamma)$ of a cycle γ of G^s is the product of the weights of its arcs, and the *weight* of a cycle cover of G^s is the product of the weights of its cycles. Each cycle cover τ of G^s determines a permutation of $\{1, 2, \ldots, n\}$ whose permutation cycles correspond to digraph cycles, but not every permutation of $\{1, 2, \ldots, n\}$ corresponds to a cycle cover of G. The *sign* of a permutation τ is $(-1)^{n-k}$, where k is the number of cycles, and we denote this by $(-1)^\tau$. The *signed weight* $s(\tau)$ of a cycle cover τ is the product of the weight of τ and its sign. Using the correspondence between cycle covers of G^s and non-zero terms in the determinant expansion of $\mathbf{A}^s$, we have the formula

$$\det \mathbf{A}^s = \sum_{\sigma \in S_n} (-1)^\sigma \prod_{i=1}^{n} a_{i\sigma_i} = \sum_{\tau \in C(G^s)} s(\tau),$$

where $C(G^s)$ is the set of cycle covers of G^s.

Now let G be the digraph obtained from G^s by ignoring the signs of its arcs, and let **A** be the adjacency matrix of G. Since the cycle covers of G^s are the same as those of G, the triangle inequality implies that

$$\operatorname{per} \mathbf{A} = \sum_{\tau \in C(G)} 1 \geq \Big| \sum_{\tau \in C(G^s)} s(\tau) \Big| = |\det \mathbf{A}^s|, \tag{11}$$

with equality if and only if the cycle covers of G^s all have the same signed weight. This suggests a possible way of computing the permanent of a $(0, 1)$-matrix **A**: replace certain 1s of **A** by -1s so as to obtain a matrix $\mathbf{A}^s$ with the property that all cycle covers of G^s have the same signed weight, and then calculate $|\det \mathbf{A}^s|$.

An assignment of this type may not be possible. For example, consider $\mathbf{J}_3$, the all-1 matrix of order 3. Then per $\mathbf{J}_3 = 6$, and by Hadamard's inequality for determinants, the absolute value of the determinant of each 3×3 matrix of ± 1s is at most $3^{3/2}$, which is less than 6. It is thus impossible to use this technique to convert the computation of per $\mathbf{J}_3$ into the computation of the determinant of a $\mathbf{J}_3^s$.

In the case that $\mathbf{A}$ is a (0, 1)-matrix, per $\mathbf{A} \neq 0$ can be efficiently checked using König's theorem. If per $\mathbf{A} \neq 0$ and such an assignment is possible, we say that $\mathbf{A}$ is *convertible*. More precisely, a square (0, 1)-matrix $\mathbf{A}$ is *convertible* if per $\mathbf{A} \neq 0$ and it is possible to assign minus signs to some non-zero elements of $\mathbf{A}$ to obtain a matrix $\mathbf{A}^s$ such that per $\mathbf{A} = |\det \mathbf{A}^s|$; such a matrix $\mathbf{A}^s$ is a *conversion of* $\mathbf{A}$. Note that the conversion $\mathbf{A}^s$ applies not only to the permanent of $\mathbf{A}$, but to the permanent of any matrix obtained from $\mathbf{A}$ by replacing its 1s by arbitrary numbers. More generally, we call an assignment of minus signs to some of the elements of $\mathbf{A}$ a *signing* of $\mathbf{A}$.

The matrix $\mathbf{J}_3$ has no conversion, but the matrix

$$\mathbf{A}^s = \begin{pmatrix} 1 & -1 \\ 1 & 1 \end{pmatrix}$$

is a conversion of $\mathbf{J}_2$. We say that a bipartite graph G is *convertible* if its biadjacency matrix is convertible.

To test whether a signing $\mathbf{A}^s$ is a conversion of $\mathbf{A}$, we need not examine all cycle covers of $\mathbf{A}$. The following basic theorem of Bassett, Maybee and Quirk [1] asserts that if $\mathbf{I}_n \leq \mathbf{A}$ (entry-wise) (which implies that there is a loop at each vertex of G^s), then one need check only the weights of the cycles of G^s. We note that if per $\mathbf{A} \neq 0$, then there exists a permutation matrix $\mathbf{P}$ such that $\mathbf{I}_n \leq \mathbf{PA}$. Clearly, $\mathbf{A}^s$ is a conversion of $\mathbf{A}$ if and only if $\mathbf{PA}^s$ is a conversion of $\mathbf{PA}$. Also, if $\mathbf{D}$ is a diagonal matrix each of whose diagonal entries is 1 or -1, then $\mathbf{A}^s\mathbf{D}$ is a conversion of $\mathbf{AD}$ if and only if $\mathbf{A}^s$ is a conversion of $\mathbf{A}$. Thus there is no loss of generality in assuming that $\mathbf{I}_n \leq \mathbf{A}$, and that each element on the main diagonal of $\mathbf{A}^s$ is -1.

Theorem 7.1 *Let* $\mathbf{A}$ *be a* (0, 1)*-matrix with* $\mathbf{I}_n \leq \mathbf{A}$, *and let* $\mathbf{A}^s$ *be a signing of* $\mathbf{A}$ *with every diagonal entry* -1. *Then* $\mathbf{A}^s$ *is a conversion of* $\mathbf{A}$ *if and only if each cycle in the signed digraph of* $\mathbf{A}^s$ *has weight* -1.

Proof Let G be the digraph of $\mathbf{A}$, and let G^s be the signed digraph of $\mathbf{A}^s$.

First suppose that $\mathbf{A}^s$ is a conversion of $\mathbf{A}$. Then equality holds in (11), and so each cycle cover of $\mathbf{A}^s$ has the same signed weight. The cycle cover consisting of n loops has signed weight $(-1)^n$. If γ is an l-cycle of G^s, then the cycle cover of G^s consisting of γ and $n-l$ loops has weight $(-1)^{n-1}\text{wt}(\gamma)$, and so $\text{wt}(\gamma) = -1$.

Conversely, suppose that each cycle has weight -1. Let τ be a cycle cover consisting of l cycles. Since each cycle has weight -1, the signed weight of τ is

$(-1)^{n-1}(-1)^l = (-1)^n$. All cycle covers have the same signed weight $(-1)^n$, and by (11), $\mathbf{A}^s$ is a conversion of $\mathbf{A}$. ■

Given a signed digraph G^s, we consider the unsigned digraph G' obtained from G^s by ignoring the weights and replacing each arc ij with sign $+1$ by a path $i \to v_{ij} \to j$ of length 2, by inserting a new vertex v_{ij}. It is easy to verify that G^s has no cycles of positive weight if and only if G' has no cycles of even length. Thus, the problem of recognizing whether $\mathbf{A}^s$ is a conversion of $\mathbf{A}$ is equivalent to that of recognizing whether a digraph G' has a cycle of even length. This was a major unsolved problem until recently. We describe the solution to this even-cycle problem below.

Theorem 7.1 yields another proof that $\mathbf{J}_3$ is not convertible. If G^s is a signed digraph corresponding to a signing $\mathbf{A}^s$ of $\mathbf{J}_3$ with diagonal entries -1, then each arc of G that is not a loop is contained in exactly two cycles and G^s has five non-trivial cycles, and so not all cycles of G^s can have weight -1.

As another application of Theorem 7.1, we prove a result of Kasteleyn that every connected matching-covered planar bipartite graph is convertible. If $\vec{G}$ is an orientation of a plane bipartite graph G, and if α is a cycle of G, then a *forward edge* of α is an edge whose orientation agrees with the clockwise orientation of α. The number of forward edges of α is denoted by $n(\alpha)$.

Corollary 7.2 *Let G be a 2-edge-connected matching-covered plane bipartite graph. If there exists an orientation $\vec{G}$ of G for which each cycle G that bounds a face has an odd number of forward edges, then G is convertible.*

Proof Suppose that $\vec{G}$ is an orientation of G for which each cycle that bounds a face has an odd number of forward edges. Let $M = \{x_1y_1, x_2y_2, \ldots, x_ny_n\}$ be a perfect matching of G. By reversing the orientation of each edge with one vertex in a given subset of vertices, we may assume that each edge of x_iy_i of M is oriented in $\vec{G}$ from x_i to y_i. Let $\mathbf{A}^s = (a_{ij})$ be the signing of the biadjacency matrix of G with $a_{ij} = 0$ if x_i is not joined to y_j in G, $a_{ij} = -1$ if $x_i \to y_j$ in $\vec{G}$, and $a_{ij} = 1$ if $y_j \to x_i$ in $\vec{G}$. Let G^s be the signed digraph of $\mathbf{A}^s$, and let $x_1, x_2, \cdots, x_k, x_1$ be a directed cycle γ of G^s. By Theorem 7.1, it suffices to show that the weight of γ is -1. Note that

$$x_1y_1x_2y_2 \ldots x_ky_kx_1 \tag{12}$$

is a cycle α in the bipartite graph G. Let v, e and f denote the number of vertices, edges and faces, respectively, in the region R bounded by α, and let $\alpha_1, \alpha_2, \ldots, \alpha_f$ be the cycles in G that bound the faces in R. Each edge in R, but not in α, is a forward edge in one face of R and a backward edge in another, and each edge in α belongs to exactly one face bounded by R. This, and the assumption that each

$n(\alpha_i)$ is odd, imply that

$$n(\alpha) + e - 2k = \sum_{i=1}^{f} n(\alpha_i) \equiv f (\text{mod } 2).$$

Hence, $n(\alpha) \equiv e + f (\text{mod } 2)$. By Euler's theorem for planar graphs, $v - e + f = 1$. Since G has a perfect matching containing half of the edges of α, v is even. We conclude that $n(\alpha)$ is odd.

In (12), the vertices of α are either in clockwise or anticlockwise order. In the former case, each edge of α belonging to M is a forward edge and there are an odd number of backward edges lying in α but not in M. Since these edges correspond to edges of γ of weight -1, $\text{wt}(\gamma) = -1$. In the latter case, each edge of α belonging to M is a backward edge, and (13) implies that there are an odd number of forward edges lying in α but not in M. Since these edges correspond to edges of γ of weight -1, $\text{wt}(\gamma) = -1$. ■

The existence of such an orientation $\vec{G}$ is easily established by induction on the number of edges. More generally, the biadjacency matrix of a planar bipartite graph with a perfect matching is convertible (see [24]). Kasteleyn used this result to determine the number n_{pq} of ways of tiling a $p \times q$ chessboard (with p or q even) with dominoes. It is easy to see that n_{pq} is the number of perfect matchings of the planar bipartite graph G_{pq} whose vertices correspond to the squares of the $p \times q$ chessboard and whose edges correspond to adjacent squares. As outlined in [8], one can use Corollary 7.2 to find a conversion of the biadjacency matrix of G_{pq}, and then to calculate the eigenvalues of the conversion to attain the closed formula:

$$n_{pq} = 2^{pq/2} \prod_{k=1}^{p} \prod_{l=1}^{q} \left(\cos^2 \left(\frac{\pi k}{p+1} \right) + \cos^2 \left(\frac{\pi l}{q+1} \right) \right)^{1/4} \approx e^{0.29pq}.$$

This technique, and more sophisticated algebraic techniques, have been used to give closed formulas for tilings of other objects (see [28]).

The complete bipartite graph $K_{3,3}$ is non-planar, and since its biadjacency matrix is $\mathbf{J}_3$ it is also not convertible. More generally, Little [29] obtained the following characterization of convertible matrices. An *even subdivision* of a bipartite graph G is a graph H obtained by replacing the edges of G by internally disjoint paths, each with an even number of vertices.

Theorem 7.3 *Let G be a bipartite graph with a perfect matching. Then G is convertible if and only if G does not contain a spanning subgraph H in which one component is an even subdivision of $K_{3,3}$ and other components (if any) are single edges.*

The *Heawood graph,* which has biadjacency matrix $\mathbf{I}_7 + \mathbf{C}_7 + \mathbf{C}_7^3$, where $\mathbf{C}_7$ is the permutation matrix with 1s in positions $(1, 2), (2, 3), \ldots, (6, 7), (7, 1)$, is convertible but not planar. In fact,

$$\text{per}\left(\mathbf{I}_7 + \mathbf{C}_7 + \mathbf{C}_7^3\right) = \det\left(\mathbf{I}_7 + \mathbf{C}_7 + \mathbf{C}_7^3\right) = 24.$$

In 1999, Robertson, Seymour and Thomas [36] gave a different characterization of convertible graphs, and used it to develop a polynomial-time algorithm for recognizing convertible matrices, thereby solving the even-cycle problem. Their characterization asserts that every convertible bipartite graph can be constructed by piecing together planar bipartite graphs and Heawood graphs in a particular way.

Let G_1 and G_2 be connected matching-covered bipartite graphs, and let x_1y_1 and x_2y_2 be edges in G_1 and G_2, respectively. The *1-join* of G_1 and G_2 is the bipartite graph obtained from G_1 and G_2 by identifying x_1 with x_2 and y_1 with y_2, and adjoining all edges of the form xy, where xy_1 is an edge of G_1 and x_2y is an edge of G_2. A *brace* is a connected bipartite graph G for which each pair of disjoint edges of G is contained in a perfect matching of G. Now let G_1 and G_2 be braces, and let $w_1x_1y_1z_1w_1$ and $w_2x_2y_2z_2w_2$ be cycles in G_1 and G_2, respectively. The *2-join* of G_1 and G_2 is the bipartite graph obtained by identifying w_1 with w_2, x_1 with x_2, y_1 with y_2, and z_1 with z_2. It is not difficult to show that every 1-join or 2-join of convertible bipartite graphs is convertible (see [8]). Also, if G is convertible and if G' is a bipartite graph which has a perfect matching and is obtained from G by deleting some edges, then G' is convertible.

Since a matrix is convertible if and only if each of its fully indecomposable components is convertible, we can restrict ourselves to characterizing convertible matching-covered bipartite graphs. We now state the characterization of convertible bipartite graphs given in [36].

Theorem 7.4 *Let G be a connected matching-covered bipartite graph.*

(a) If G is not a brace, then it is convertible if and only if it can be obtained from the 1*-join of convertible graphs by removing a (possibly empty) subset of edges.*

(b) If G is a brace, then it is convertible if and only if it is either a planar bipartite graph or the Heawood graph, or is a spanning subgraph of a 2*-join of planar bipartite braces.*

8. Chordal graphs and perfect Gaussian elimination

A graph G is *chordal* if each cycle γ of length greater than 3 has a *chord* – that is, an edge joining two non-consecutive vertices of γ. In a chordal graph, no induced subgraph is a cycle of length greater than 3. Complete graphs, in particular cycles of length 3, are chordal, as are all trees. Chordal graphs are useful data structures in solving sparse symmetric systems of linear equations.

A *simplicial vertex* of a graph G is a vertex v whose neighbours are all joined to each other. Thus, v is a simplicial vertex of G if and only if the neighbours of v induce a complete graph. An ordering $v_1, v_2, \ldots, v_n$ of the vertices of G is a *simplicial* (or *perfect*) *elimination ordering* if v_i is a simplicial vertex of the subgraph induced by $\{v_i, v_{i+1}, \ldots, v_n\}$, for $i = 1, 2, \ldots, n-1$. Any ordering of the vertices of a complete graph is a simplicial elimination ordering. Any ordering of the vertices of a tree obtained by successively deleting end-vertices is a simplicial elimination ordering. A cycle of length 4 or more has no simplicial elimination ordering.

Dirac [11] derived two characterizations of chordal graphs. The first is the existence of a simplicial elimination ordering. The other is in terms of *vertex separators* of a pair of non-adjacent vertices v and w – that is, sets of vertices S, not containing v or w, for which the subgraph induced on the complement of S is disconnected, with v and w in different components. If no proper subset of S separates v and w, then S is a *minimal vertex separator* of v and w. For non-adjacent vertices v and w, the complement of $\{v, w\}$ is a vertex separator, and hence there is a minimal vertex separator of v and w.

Theorem 8.1 *Let G be a graph. The following statements are equivalent.*
(1) For each minimal vertex separator S of G, the induced subgraph $G[S]$ is a complete graph.
(2) G is a chordal graph.
(3) G has a simplicial elimination ordering.

Proof (1) $\Rightarrow$ (2) Assume that (1) holds, and consider a cycle $\gamma = vxwa_1 \ldots a_k v$ of length greater than 3, so that $k \geq 1$. If v and w are adjacent, then γ has a chord. Otherwise, let S be a minimal vertex separator of v and w. Then S contains x and at least one of $a_1, a_2, \ldots, a_k$, and so γ has a chord. Thus G is chordal.

(2) $\Rightarrow$ (1) Assume that G is chordal. Let S be a minimal vertex separator of vertices v and w, and let G_v and G_w be the connected components of $G - S$ containing v and w, respectively. The minimality assumption implies that each vertex in S is adjacent to some vertex in G_v and some vertex in G_w. If S has only one vertex, then $G[S]$ is a complete graph. Otherwise, let x and y be distinct vertices in S. Then there exist an x-y path all of whose internal vertices are in G_v, and an y-x path all of whose internal vertices are in G_w. We may choose such paths to have smallest length. Hence, there is a cycle γ of length greater than 3, all of whose vertices (except for x and y) belong to G_v or G_w. Since G is chordal, the cycle γ has a chord. But our assumptions imply that the only possible chord is an edge joining x and y. Thus $G[S]$ is a complete graph, and (1) holds.

(3) $\Rightarrow$ (2) Let γ be a cycle of length greater than 3. The vertex v of γ with the smallest index in a simplicial elimination scheme has the property that the two

vertices adjacent to v on the cycle are adjacent to each other, and hence determine a chord of γ. It follows that G is chordal.

(2) $\Rightarrow$ (3) To prove that a chordal graph has a simplicial elimination ordering, it suffices to show that every chordal graph has a simplicial vertex. Any vertex of a complete graph is a simplicial vertex. We show by induction that *each chordal graph G which is not complete has at least two non-adjacent simplicial vertices.* Let v and w be non-adjacent vertices of G, and let S be a vertex separator of v and w. Let $G_v = G[U]$ and $G_w = G[W]$. By what we have proved, $G[S]$ is a complete graph. If the chordal graph $G[S \cup U]$ is not complete, then it has two non-adjacent simplicial vertices, by the induction assumption, at least one of which must be in U, since $G[S]$ is complete. If $G[S \cup U]$ is complete, then any vertex in U is a simplicial vertex of G. Thus U contains a simplicial vertex of G, and similarly so does W, and these two vertices are non-adjacent. ■

Let $\mathbf{A} = (a_{ij})$ be a symmetric invertible matrix of order n, all of whose diagonal entries are non-zero. Gaussian elimination to solve a symmetric system of linear equations $\mathbf{Ax} = \mathbf{b}$ proceeds by successively pivoting on a non-zero diagonal element (using elementary row operations and the corresponding column operations, to 'zero out' the non-zero elements in the row and column of the pivot element) until $\mathbf{A}$ is reduced to a diagonal matrix. If $\mathbf{A}$ is a large sparse matrix, and if the graph G whose edges correspond to the non-zero off-diagonal elements of $\mathbf{A}$ is chordal, then by following a simplicial elimination ordering of G when choosing pivots for Gaussian elimination on $\mathbf{A}$, a zero element of $\mathbf{A}$ remains zero throughout the elimination. This is usually described by saying that 'no fill-in occurs,' implying, in particular, that sparseness is preserved. We are also assuming that the elements on the main diagonal remain non-zero throughout Gaussian elimination. This implies that a data structure for $\mathbf{A}$ can be constructed by using only the positions of the non-zero elements of $\mathbf{A}$, reducing storage requirements substantially.

Bipartite graphs provide a model for Gaussian elimination for non-symmetric matrices also. We briefly discuss the bipartite analogue of chordal graphs and simplicial elimination orderings. A bipartite graph with a cycle of length greater than 3 can never be chordal. A graph G is *chordal-bipartite* if it is bipartite and if each cycle of length greater than 4 has a chord. Complete bipartite graphs – in particular, cycles of length 4 – are chordal-bipartite. Trees are both chordal and chordal-bipartite.

Let G be a bipartite graph with bipartition $\{X, Y\}$. An edge e joining v in X and w in Y is *bisimplicial* if the set Y_v of neighbours of v, together with the set X_w of neighbours of w, induces a complete bipartite graph. Let $M = (e_1, e_2, \ldots, e_k)$ be a sequence of edges of G that form a matching, let $e_i = p_i q_i (i = 1, 2, \ldots, k)$, and let $P_i = \{p_1, p_2, \ldots, p_{i-1}\}$ and $Q_i = \{q_1, q_2, \ldots, q_{i-1}\}$, for $i = 1, 2, \ldots, k + 1$.

Then M is a *bisimplicial elimination ordering* for G if e_i is a bisimplicial edge of the subgraph $G - (P_i \cup Q_i)$ $(i = 1, 2, \ldots, k)$ and the induced subgraph $G - (P_{k+1} \cup Q_{k+1})$ has no edges. In performing Gaussian elimination on a non-symmetric linear system $\mathbf{Ax} = \mathbf{b}$, a bisimplicial elimination ordering provides a sequence of pivots on the elements of $\mathbf{A}$ corresponding to the edges $e_1, e_2, \ldots, e_k$, for which no fill-in results. Golumbic and Goss [19] proved the following theorem.

Theorem 8.2 *Every chordal bipartite graph has a bisimplicial elimination ordering.*

The converse of Theorem 8.2 is false, as can be seen by the graph consisting of a 6-cycle and one end-edge incident to each vertex of the cycle. For more information on chordal bipartite graphs, see [18] and [19].

9. Ranking players in tournaments

Recall that, as its name suggests, a *tournament* models the results of a round-robin tournament on a set of n players, in which each player plays each of the others exactly once and there are no ties. The vertices correspond to the players, and there is an arc from v to w if v beats w.

Given the results of a round-robin tournament G, it is natural to try to rank the players. If the vertices of G can be ordered as $(v_{i_1}, \ldots, v_{i_n})$, where $v_{i_j} \to v_{i_k}$ if and only if $j \leq k$, then G is a *transitive tournament* and there is no ambiguity in the ranking: player v_{i_1} is the strongest, v_{i_2} is the second strongest, and so on. More generally, suppose that G is not strongly connected, so that its adjacency matrix $\mathbf{A}$ is reducible. Then G contains a directed cut (U, W), and since G is a tournament, $v_i \to v_j$ for all $v_i \in U$ and $v_j \in W$. Any ranking of the players must rank those players in U as stronger than those in W. Hence, in ranking the results of a round-robin tournament, we can restrict our attention to ranking the results of the subtournaments corresponding to the irreducible components of $\mathbf{A}$.

Suppose now that G is strongly connected. It is well known that there then exists a Hamiltonian cycle in G. The existence of such a cycle causes any ranking to have inconsistencies, since any player v can argue that he is at least as good as any other player w by considering the path from v to w. Yet it is still desirable to have some ranking of the players.

A first attempt is to rank the players according to the number of games they win. The *score* of a vertex v_i is its out-degree, and the *score vector* $\mathbf{s}$ of G is the vector whose ith entry is the score of v_i. Thus $\mathbf{Aj} = \mathbf{s}$, where $\mathbf{j}$ is the $n \times 1$ all-1 vector. This ranking scheme has two drawbacks. First, by the pigeon-hole principle, each strongly connected tournament has at least two players with the same score, and so there are always ties in the ranking. Second, this ranking scheme assumes that all wins are valued equally, but a win against a 'strong' player should arguably be valued more than a win against a 'weak' player. Several 'power ranking' methods have been proposed that take into account the strength of each player's opponents.

We discuss two such ranking methods. The first is known as the *Kendall-Wei method,* and it can be motivated as follows. The vector $\mathbf{s} = \mathbf{Aj}$ records the scores of the players, and so its ith entry is a measure of the strength of player v_i. The ith entry $\sum_{i:i\to j} S_j$ of $\mathbf{As}$ is the sum of the strengths (according to $\mathbf{s}$) of the players that v_i beats. Thus, $\mathbf{As}$ accounts for the strengths of the opponents that player v_i beats. Now the entries of $\mathbf{As}$ can be viewed as a measure of the strength of the players, and we can consider $\mathbf{A}^2\mathbf{s} = \mathbf{A}(\mathbf{As})$, the vector whose ith entry is the sum of the strengths (according to $\mathbf{As}$) of the players that i beats. This can be repeated to obtain a sequence $\mathbf{r}_1 = \mathbf{s}$, $\mathbf{r}_2 = \mathbf{As}$, $\mathbf{r}_3 = \mathbf{Ar}_2, \ldots$ of strength vectors. We normalize this sequence so that the sum of the entries of each vector is 1 to obtain a sequence

$$\mathbf{r}'_1 = \mathbf{s}/\mathbf{j}^T\mathbf{s}, \quad \mathbf{r}'_2 = \mathbf{r}_2/\mathbf{j}^T\mathbf{r}_2, \ldots .$$

It is not difficult to show that the limits $\rho = \lim_{n\to\infty} \mathbf{j}^T\mathbf{r}'_n$ and $\mathbf{x} = \lim_{n\to\infty} \mathbf{r}'_n$ exist and satisfy $A\mathbf{x} = \rho\mathbf{x}$. We conclude from Theorem 2.2 that ρ is the spectral radius of $\mathbf{A}$ and that $\mathbf{x}$ is the unique non-negative eigenvector of $\mathbf{A}$ whose entries sum to 1. We are led to the Kendall-Wei method: rank the players of G according to the entries of $\mathbf{x}$, with stronger players corresponding to larger entries of $\mathbf{x}$.

The *variance* of the vector $\mathbf{x}$ is defined by

$$\text{var}\,(\mathbf{x}) = \sum_{i<j} (x_i - x_j)^2,$$

and provides a measure of how evenly matched the players are, according to the Kendall-Wei method; var($\mathbf{x}$) is larger when there is a greater disparity among the strengths of the players. In particular, var($\mathbf{x}$) = 0 if and only if all the players of G have equal strength. The following result relates var($\mathbf{x}$) and ρ (see [30]).

Theorem 9.1 Let *G be a strongly connected tournament with adjacency matrix* $\mathbf{A}$. *Let ρ be the spectral radius of* $\mathbf{A}$, *and let* $\mathbf{x}$ *be the positive eigenvector of* $\mathbf{A}$ *corresponding to ρ with* $\mathbf{j}^T\mathbf{x} = 1$. *Then:*

(1) var$(\mathbf{x}) = (n - 1 - 2\rho)\mathbf{x}^T\mathbf{x}$;

(2) $\rho \leq (n-1)/2$, *with equality if and only if n is odd and G is regular.*

Proof We have

$$\begin{aligned}\sum_{i<j}(x_i - x_j)^2 &= (n-1)\sum_{i=1}^{n} x_i^2 - 2\sum_{i<j} x_i x_j \\ &= (n-1)\mathbf{x}^T\mathbf{x} - \mathbf{x}^T(\mathbf{J}_n - \mathbf{I}_n)\mathbf{x} \\ &= (n-1)\mathbf{x}^T\mathbf{x} - \mathbf{x}^T(\mathbf{A} + \mathbf{A}^T)\mathbf{x} \\ &= (n-1)\mathbf{x}^T\mathbf{x} - 2\rho\mathbf{x}^T\mathbf{x} \\ &= (n-1-2\rho)\mathbf{x}^T\mathbf{x}.\end{aligned}$$

Hence (1) holds, and $\rho \le (n-1)/2$ with equality if and only if $\mathrm{var}(\mathbf{x}) = 0$. Since $\mathrm{var}(\mathbf{x}) = 0$ if and only if $\mathbf{x} = (1/n)\mathbf{j}$, and since $\mathbf{j}$ is an eigenvector of $\mathbf{A}$ if and only if $\mathbf{A}$ has constant row sums, (2) holds. ■

In the light of Theorem 9.1, it is natural to ask which strongly connected tournaments on n vertices have the smallest and largest spectral radii – in other words, which strongly connected tournaments are the least or most evenly matched, according to the Kendall-Wei scheme? Theorem 9.1 implies that, for n odd, the largest spectral radius is $(n-1)/2$, and the tournaments achieving this are the regular tournaments. Brualdi and Li [5] conjectured that, for $n = 2k$, the largest spectral radius occurs when the adjacency matrix has the form

$$\begin{pmatrix} \mathbf{B} & \mathbf{B}^T \\ \mathbf{B}^T + \mathbf{I} & \mathbf{B} \end{pmatrix},$$

where $\mathbf{B}$ is the adjacency matrix of a transitive tournament with k vertices. In [27] Kirkland shows that, for sufficiently large k, a tournament of order $2k$ with largest spectral radius must have half of its players with score k, and the other half with score $k-1$. Brualdi and Li [5] conjectured that the minimum spectral radius is achieved by the tournament with $v_i \to v_j$ for i and j with $i > j+1$, and $v_i \to v_{i+1}$ for $i = 1, 2, \ldots, n-1$. This conjecture was also proved by Kirkland [26].

We can also use the positive eigenvector $\mathbf{y}$ of $\mathbf{A}^T$, with $\mathbf{j}^T\mathbf{y} = 1$, to rank the players of a tournament. This corresponds to taking into account the weaknesses of those players defeating a given player; smaller entries of $\mathbf{y}$ correspond to stronger players. A ranking scheme of Ramanujacharyula [35] ranks the players according to the strength-to-weakness ratio x_i/y_i, with larger values corresponding to better players. More about this ranking scheme can be found in [25].

The method Pagerank™, used by the search engine Google to rank web-pages, is similar to the Kendall-Wei ranking scheme. Unlike many search engines that work by searching for web-pages having many words in common with a given query, Google focuses on links between pages and searches for 'high-quality' pages that are related to the query. The quality of a page is determined by how many relevant pages refer to it, the quality of those referring pages, how many relevant pages the page refers to, and the quality of those pages. More precisely, Google first uses text-matching techniques to generate a set V of web-pages that are relevant to the query. Then Google forms a weighted digraph G whose vertices correspond to the elements of V, and where the weight of the edge vw is the number of times that web-page v links to web-page w.

Let $\mathbf{A}$ be the adjacency matrix of G. Then the ith entry of $\mathbf{x}^{(1)} = \mathbf{A}^T\mathbf{j}$ measures how often a relevant site refers to site i. Large entries of $\mathbf{x}^{(1)}$ are potential 'experts' for the query. Similarly, the ith entry of $\mathbf{y}^{(1)} = \mathbf{Aj}$ measures the number of relevant sites referred to by i, and large entries of $\mathbf{y}^{(1)}$ correspond to potential 'information sources' for the query. But not all links on the web are of equal value. Pagerank™ takes into account the quality of the sites linked to and from a site. This is done by considering the sequences and $\mathbf{x}^{(i)} = \mathbf{A}^T\mathbf{y}^{(i-1)}$ and $\mathbf{y}^{(i)} = \mathbf{Ax}^{(i-1)}$ for $i \geq 2$. Thus, for example, the ith entry of $\mathbf{x}^{(2)} = \mathbf{A}^T\mathbf{y}^{(1)}$ is the sum of the weights (according to $\mathbf{y}^{(1)}$) of the sites that refer to page i. It is easy to see that $\mathbf{x}^{(2i)} = (\mathbf{AA}^T)^i\mathbf{j}$ and $\mathbf{y}^{(2i)} = (\mathbf{A}^T\mathbf{A})^i\mathbf{j}$ $(i = 1, 2, \ldots)$. Thus, if $\mathbf{A}^T\mathbf{A}$ is irreducible, then the normalized sequences

$$\mathbf{x}^{(2i)}/\mathbf{j}^T\mathbf{x}^{(2i)} \quad \text{and} \quad \mathbf{y}^{(2i)}/\mathbf{j}^T\mathbf{y}^{(2i)}$$

converge to positive eigenvectors $\mathbf{u}$ and $\mathbf{v}$ of $\mathbf{A}^T\mathbf{A}$ and $\mathbf{AA}^T$, respectively. Pagerank™ uses the ith entry of $\mathbf{u} + \mathbf{v}$ as a measure of the relevant expertise of site i, and then lists the sites in non-increasing order based on the values of the entries of $\mathbf{u}$. Hence, Google rests mathematically on the Perron-Frobenius theorem, with which we began our discussion.

References

1. L. Bassett, J. Maybee and J. Quirk, Qualitative economics and the scope of the correspondence principle, *Econometrica* **36** (1968), 544–563.
2. L. M. Brégman, Certain properties of non-negative matrices and their permanents, *Dokl. Akad. Nauk SSSR* **211** (1973), 27–30 (= *Soviet Math. Dokl.* **14** (1973), 945–949).
3. R. A. Brualdi, Matrices, eigenvalues, and directed graphs, *Linear and Multilinear Algebra* **11** (1982), 143–165.
4. R. A. Brualdi, The symbiotic relationship of combinatorics and matrix theory, *Linear Alg. Appl.* **162–164** (1992), 65–105.
5. R. A. Brualdi and Q. Li, Problem 31, *Discrete Math.* **43** (1983), 329–330.
6. R. A. Brualdi, S. V. Parter and H. Schneider, The diagonal equivalence of a non-negative matrix to a stochastic matrix, *J. Math. Anal. Appl.* **16** (1966), 31–50.
7. R. A. Brualdi and H. J. Ryser, *Combinatorial Matrix Theory,* Encyclopedia of Mathematics and its Applications **39**, Cambridge Univ. Press, 1991.

8. R. A. Brualdi and B. L. Shader, *Matrices of Sign Solvable Linear Systems,* Cambridge Tracts in Mathematics **116**, Cambridge Univ. Press, 1995.
9. F. R. K. Chung, R. L. Graham and P. M. Winkler, On the addressing problem for directed graphs, *Graphs Combin.* **1** (1985), 41–50.
10. D. de Caen, The ranks of tournament matrices, *Amer. Math. Monthly* **98** (1991), 829–831.
11. G. A. Dirac, On rigid circuit graphs, *Abh. Math. Sem. Univ. Hamburg* **25** (1961), 71–76.
12. A. L. Dulmage and N. S. Mendelsohn. Gaps in the exponent set of primitive matrices, *Illinois J. Math.* **8** (1964), 642–656.
13. G. P. Egoryčev, A solution of van der Waerden's permanent problem, *Dokl. Akad. Nauk SSSR* **258** (1981), 1041–1044 (= *Soviet Math. Dokl.* **23** (1981), 619–622).
14. D. I. Falikman, A proof of van der Waerden's conjecture on the permanent of a doubly stochastic matrix, *Mat. Zametki* **29** (1981), 931–938 (= *Math. Notes* **29** (1981), 475–479).
15. E. Fornasini and M. Valcher, Primitivity of positive matrix pairs: algebraic characterization, graph theoretic description and 2D systems interpretation, *SIAM J. Matrix Anal. Appl.* **19** (1998), 71–88.
16. G. Frobenius, Über Matrizen aus nicht negativen Elementen, *Sitzungsber. Preuss. Akad. Wiss. Berlin* (1912), 456–477.
17. G. Frobenius, Über zerlegbare Determinanten, *Sitzungsber. Preuss. Akad. Wiss. Berlin* (1917), 274–277.
18. M. C. Golumbic, *Algorithmic Graph Theory and Perfect Graphs,* Academic Press, 1980.
19. M. C. Golumbic and C. F. Goss, Perfect elimination and chordal bipartite graphs, *J. Graph Theory* **2** (1978), 155–163.
20. R. L. Graham and H. O. Pollak, On the addressing problem for loop switching, *Bell System Tech. J.* **50** (1971), 2495–2519.
21. D. Gregory, B. L. Shader and V. Watts, Biclique decompositions and Hermitian rank, *Linear Alg. Appl.* **292** (1999), 267–280.
22. R. A. Horn and C. R. Johnson, *Matrix Analysis,* Cambridge Univ. Press, 1985.
23. W. B. Jurkat and H. J. Ryser, Term ranks and permanents of non-negative matrices, *J. Algebra* **5** (1967), 342–357.
24. P. W. Kasteleyn, Dimer statistics and phase transitions, *J. Math. Phys.* **4** (1963), 287–293.
25. S. Kirkland, Spectral radii of tournament matrices whose graphs are related by an arc reversal, *Linear Alg. Appl.* **217** (1995), 179–202.
26. S. Kirkland, On the minimum Perron value of an irreducible tournament matrix, *Linear Alg. Appl.* **244** (1996), 277–304.
27. S. Kirkland, A note on Perron vectors for almost regular tournament matrices, *Linear Alg. Appl.* **266** (1997), 43–47.
28. G. Kuperberg, Symmetries of plane partitions and the permanent-determinant method, *J. Combin. Theory (A)* **68** (1994), 115–151.
29. C. H. C. Little, A characterization of convertible (0, 1)-matrices, *J. Combin. Theory (B)* **18** (1975), 187–208.
30. J. S. Maybee and N. J. Pullman, Tournament matrices and their generalizations I, *Linear and Multilinear Algebra* **28** (1990), 57–70.
31. B. D. McKay and I. M. Wanless. Maximising the permanent of (0, 1)-matrices and the number of extensions of latin rectangles, *Elec. J. Combin.* **5** (1998), R11.

32. M. Newman, *Integral Matrices,* Academic Press, 1972.
33. P. A. Ostrand, Systems of distinct representatives II. *J. Math. Anal. Appl.* **32** (1970), 1–4.
34. G. W. Peck, A new proof of a theorem of Graham and Pollak, *Discrete Math.* **49** (1984), 327–328.
35. C. Ramanujacharyula, Analysis of preferential experiments, *Psychometrika* **29** (1964), 257–261.
36. N. Robertson, P. D. Seymour and R. Thomas, Permanents, Pfaffian orientations, and even directed circuits, *Ann. of Math.* **150** (1999), 929–975.
37. A. Schrijver, A short proof of Minc's conjecture, *J. Combin. Theory (A)* **25** (1978), 80–83.
38. A. Schrijver, Counting 1-factors in regular bipartite graphs, *J. Combin. Theory (B)* **72** (1998), 122–135.
39. J. J. Seidel, A survey of two-graphs. *Teorie Combinatoire* **I** (ed. B. Segre), Accademia Nazionale dei Lincei (1976), 481–511.
40. B. L. Shader, On tournament matrices, *Linear Alg. Appl.* **162–164** (1992), 335–368.
41. B. L. Shader and S. Suwilo, Exponents of non-negative matrix pairs, *Linear Alg. Appl.*, **363** (2003), 275–293.
42. J. Y. Shao, The exponent set of symmetric primitive matrices, *Scientia Sinica (A)* **XXX**, 348–358.
43. R. Sinkhorn and P. Knopp, Concerning non-negative matrices and doubly stochastic matrices, *Pacific J. Math.* **21** (1967), 343–348.
44. H. Tverberg, On the decomposition of K_n into complete bipartite graphs, *J. Graph Theory* **6** (1982), 493–494.
45. L. G. Valiant, Completeness classes in algebra, *Proc. 11th ACM Symp. Theory of Computing* (1979), 249–261.
46. L. G. Valiant, The complexity of computing the permanent, *Theoret. Comp. Sci.* **8** (1979), 189–201.
47. H. Wielandt, Unzerlegbare, nicht negative Matrizen, *Math. Zeit.* **52** (1950), 642–645.
48. T. H. Wei, *The Algebraic Foundations of Ranking Theory,* PhD thesis, Cambridge, 1952.

3

Spectral graph theory

DRAGOŠ CVETKOVIĆ and PETER ROWLINSON

In this chapter we are concerned with the spectrum of the adjacency matrix of a finite graph. We present a selection of recent results related to graph angles, star sets, star partitions, star complements, and graphs with least eigenvalue -2*. In addition, we consider a spectral reconstruction problem, graphs that are cospectral or almost cospectral, and graphs for which all eigenvalues are integers.*

1. Introduction

The matrices commonly used to specify a finite graph include the Laplacian matrix (see Chapter 4), the Seidel matrix (see Section 2 below), and the (0, 1)-adjacency matrix. In this chapter we are concerned with the spectra of adjacency matrices, a topic that features prominently in Chapters 1 and 2 and in the monographs [12], [13] and [21]. The subject has continued to attract the attention of researchers, and it is impossible to review in one short chapter all of the important results obtained in the last few years. Accordingly, it was necessary to be selective, and the choice made here reflects the authors' own interests.

Some basic facts about graph spectra can be found in Chapter 1 and [46, Ch. 11]. Here we extend the discussion of eigenvalues to related algebraic invariants, such as the angles between eigenspaces and coordinate axes treated in Section 2. These lead naturally to the notion of a *star partition*, which provides a one-to-one correspondence (in general, not unique) between the eigenvalues and the vertices of a graph (see Section 3). In such a correspondence the vertices associated with a common eigenvalue μ are said to form a *star set* for μ, while the subgraph induced by the remaining vertices is called a *star complement* for μ. In Section 4 we discuss star sets and a technique for constructing the graphs with a prescribed star complement. This technique makes it possible not only to characterize a number of graphs, but also to complete the description of all graphs with least eigenvalue -2. The problem of describing the exceptional graphs that arise here was one of some 25 years' standing, and it is discussed in Section 5.

Another long-standing problem concerns the reconstructibility of the characteristic polynomial of a graph from the characteristic polynomials of the vertex-deleted subgraphs. This has recently been solved for trees, and a proof of the reconstructibility in this case is given in Section 6. On the other hand, a conjecture from 1983 on 'almost cospectral' graphs has recently been disproved. The context is a construction known as a non-complete extended p-sum of graphs, discussed in Section 7. This construction also provides one means of generating graphs for which every eigenvalue is an integer: these 'integral graphs' are discussed in Section 8.

The use of computers is an intrinsic feature of research in discrete mathematics and, in particular, in the theory of graph spectra. The chapter includes computational results related to minimal cospectral graphs with the same angles, and to integral graphs with ten or fewer vertices.

2. Angles

Let G be a graph with vertex-set $V(G) = \{1, 2, \ldots, n\}$ and let $\mathbf{A}$ be the adjacency matrix of G. Thus, $\mathbf{A} = (a_{ij})$, where $a_{ij} = 1$ if i and j are adjacent and $a_{ij} = 0$ otherwise. The eigenvalues of $\mathbf{A}$ are graph invariants, because a relabelling of the vertices results in a matrix similar to $\mathbf{A}$. Accordingly, we may define $\phi_G(x)$, the *characteristic polynomial* of G, as $\det(x\mathbf{I} - \mathbf{A})$. We take the distinct eigenvalues of $\mathbf{A}$ to be $\mu_1, \mu_2, \ldots, \mu_m$, where $\mu_1 > \mu_2 > \cdots > \mu_m$. These eigenvalues, together with their respective multiplicities $k_1, k_2, \ldots, k_m$, constitute the *spectrum* of G.

Since in general a graph is not determined by its spectrum (see Chapter 1), we seek further algebraic invariants which might distinguish cospectral graphs. As an orthogonally diagonalizable matrix, $\mathbf{A}$ is determined by n linearly independent eigenvectors and the corresponding eigenvalues $\lambda_1 \geq \lambda_2 \geq \cdots \geq \lambda_n$. Accordingly, it is natural to turn our attention to eigenspaces, but we must bear in mind that the coordinates of vectors are permuted when the vertices are relabelled. This presents no difficulty when all the coordinates coincide, and so, abusing terminology, we can define the *main angles* of G as the cosines of the angles between the all-1 vector $\mathbf{j}$ and the eigenspaces of $\mathbf{A}$. For $i = 1, 2, \ldots, m$, let $\mathbf{P}_i$ be the $n \times n$ matrix which represents the orthogonal projection of $\mathbf{R}^n$ onto the eigenspace $\mathcal{E}(\mu_i)$ with respect to the standard basis $\{\mathbf{e}_1, \mathbf{e}_2, \ldots, \mathbf{e}_n\}$ of $\mathbf{R}^n$. Then the main angles of G are the numbers $\beta_1, \beta_2, \ldots, \beta_m$, where $\beta_i = \|\mathbf{P}_i\mathbf{j}\|/\sqrt{n}$. Note that $(\beta_1, \beta_2, \ldots, \beta_m) = (1, 0, \ldots, 0)$ if and only if $\mathbf{j} \in \mathcal{E}(\mu_1)$ – that is, if and only if G is regular. In general, we say that μ_i is a *main eigenvalue* if $\mathbf{j} \notin \mathcal{E}(\mu_i)^\perp$.

Given the spectrum of G, a knowledge of the invariant $(\beta_1, \beta_2, \ldots, \beta_m)$ is equivalent (see [39]) to a knowledge of either of two other spectra that appear in the literature. One is the *Seidel spectrum*, which is the spectrum of the matrix $\mathbf{J} - \mathbf{I} - 2\mathbf{A}$ (where $\mathbf{J}$ is the all-1 matrix), and the other is the *derived spectrum* of G, as defined by Neumaier [37].

We also observe that, given the spectrum of G, a knowledge of the invariant $(\beta_1, \beta_2, \ldots, \beta_m)$ is equivalent to a knowledge of the spectrum of the complement $\overline{G}$, or of the *cone* over G (obtained from G by adding a vertex adjacent to every vertex of G). For the observation concerning $\overline{G}$, note first that the spectral decomposition of $\mathbf{A}$ is

$$A = \mu_1\mathbf{P}_1 + \mu_2\mathbf{P}_2 + \cdots + \mu_m\mathbf{P}_m,$$

where $\mathbf{P}_i^2 = \mathbf{P}_i = \mathbf{P}_i^T\,(i = 1, 2, \ldots, m)$ and $\mathbf{P}_i\mathbf{P}_j = \mathbf{0}$ (for $i \neq j$). Now the characteristic polynomial of $\overline{G}$ is given by:

$$\begin{aligned}\phi_{\overline{G}}(x) &= \det((x+1)\mathbf{I} + \mathbf{A} - \mathbf{J})\\ &= \det((x+1)\mathbf{I} + \mathbf{A}) - \mathbf{j}^T\operatorname{adj}((x+1)\mathbf{I} + \mathbf{A})\mathbf{j}\\ &= (-1)^n\phi_G(-x-1)\left(1 - \mathbf{j}^T((x+1)\mathbf{I} + \mathbf{A})^{-1}\mathbf{j}\right)\\ &= (-1)^n\phi_G(-x-1)\left(1 - n\sum_{i=1}^{m}\frac{\beta_i^2}{x+1+\mu_i}\right).\end{aligned} \tag{1}$$

We place the second observation in a more general context. For $\Delta \subseteq V(G)$, let G_Δ denote the graph obtained from G by adding a vertex adjacent to the vertices in Δ, and let $\mathbf{r}$ be the characteristic vector of Δ – that is, $\mathbf{r} = \sum_{j\in\Delta}\mathbf{e}_j$. Then G_Δ has characteristic polynomial

$$\begin{aligned}\begin{vmatrix} x & -\mathbf{r}^T\\ -\mathbf{r} & x\mathbf{I} - \mathbf{A}\end{vmatrix} &= x\det(x\mathbf{I} - \mathbf{A}) - \mathbf{r}^T\operatorname{adj}(x\mathbf{I} - \mathbf{A})\mathbf{r}\\ &= \phi_G(x)\left(x - \sum_{i=1}^{m}\frac{\|\mathbf{P}_i\mathbf{r}\|^2}{x - \mu_i}\right).\end{aligned}$$

By [46, Cor. 3.3] we have $\phi_{G_{\{j\}}}(x) = x\phi_G(x) - \phi_{G-j}(x)$, and so

$$\phi_{G-j}(x) = \phi_G(x) \sum_{i=1}^{m} \frac{\alpha_{ij}^2}{x - \mu_i}, \tag{2}$$

where $\alpha_{ij} = \|\mathbf{P}_i \mathbf{e}_j\|$. The numbers α_{ij} are called the *angles* of G; they are actually the cosines of the angles between the coordinate axes and the eigenspaces. We may label the vertices of G so that the columns of the $m \times n$ matrix (α_{ij}) are ordered lexicographically; then (α_{ij}) is a graph invariant, called the *angle matrix* of G. Since the diagonal entries of $\mathbf{A}^k$ are $\sum_{i=1}^m \mu_i^k \alpha_{ij}^2$ $(j = 1, 2, \ldots, n)$, the columns of the angle matrix are all the same if and only if, for each $k \in \mathbf{N}$, the number of j-j walks of length k is independent of the vertex j; such a graph is said to be *walk-regular* (see [30]).

It follows from (2) that, given the spectrum of G, a knowledge of the angles of G is equivalent to a knowledge of the spectra of the vertex-deleted subgraphs of G. (This is the context in which to view Theorem 7.3 of [46]; in particular, if $\{\mathbf{x}_1, \mathbf{x}_2, \ldots, \mathbf{x}_{k_i}\}$ is an orthonormal basis for the eigenspace $\mathcal{E}(\mu_i)$, then $\sum_{h=l}^{k_i} (\mathbf{e}_1 \cdot \mathbf{x}_h)^2 = \alpha_{i1}^2$.)

The angles of G satisfy the following relations (see [21, Ch. 4]):

$$\sum_{j=1}^{n} \alpha_{ij}^2 = k_i \quad \text{and} \quad \sum_{i=1}^{m} \alpha_{ij}^2 = 1.$$

Since the number of j-j walks of length k in G is $\sum_{i=1}^m \mu_i^k \alpha_{ij}^2$ (see [46, Thm. 2.4]), the spectrum and angles of G determine the vertex-degrees in G. It is also the case that the spectrum and angles determine the numbers of 4-cycles and 5-cycles. On the other hand, the following example shows that a graph may not be determined by its angles, main angles and spectrum.

Example 1 The two graphs depicted in Fig. 1 are non-isomorphic, but they are both 4-regular and have the same eigenvalues, the same angles and the same main angles. The ten vertices are labelled so that the angle sequences $(\alpha_{1j}, \alpha_{2j}, \ldots, \alpha_{mj})$

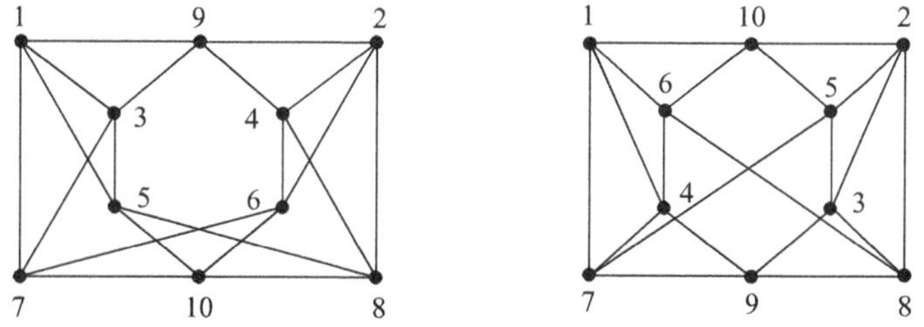

Fig. 1.

coincide for $j = 1, 2, \ldots, 10$; equivalently, for each j, the graphs obtained by deleting the vertex j are cospectral.

It has been shown by a computer search (see [15]) that graphs with fewer than 10 vertices are characterized by their eigenvalues and angles. However, there are 58 pairs of cospectral graphs on 10 vertices with the property that the graphs within each pair have the same angles. Moreover, they also have the same main angles (a fact for which we do not have an explanation), and no multiple eigenvalue is a main eigenvalue. By (1), the characteristic polynomial of a complementary graph $\overline{G}$ is determined by the characteristic polynomial and the main angles of G, and so the graphs from 29 of the 58 pairs are the complements of those from the other 29 pairs.

A construction from [21, pp. 113–114] shows that there is an infinite set of cospectral trees with the same angles. The trees in the smallest example given there have order 35, but an exhaustive computer search has revealed that there is just one example among trees with up to 20 vertices (see [15]). The trees from this pair have 19 vertices, and it is surprising that there are no examples with 20 vertices. The trees in question are displayed in Fig. 2 as T_1 and T_2. The subtree T identified by the heavy lines is well known in constructions of cospectral graphs, mainly because the graphs $T - 4$ and $T - 7$ are cospectral.

The vertices in T_1 and T_2 are labelled so that $T_1 - i$ is cospectral with $T_2 - i$, for $i = 1, 2, \ldots, 19$. Note that $T_1 - 5$ and $T_2 - 5$ both have two components with 10 and 8 vertices, which however are not cospectral. In $T_1 - 5$ the components have the following spectra (where non-integer eigenvalues are given to three places of decimals):

$$\pm 2.074, \pm 1.414, \pm 1.414, \pm 0.835, 0, 0 \quad \text{and} \quad \pm 2.222, \pm 1.240, \pm 0.726, 0, 0.$$

On the other hand, in $T_2 - 5$ the two components have the spectra

$$\pm 2.222, \pm 1.414, \pm 1.240, \pm 0.726, 0, 0 \quad \text{and} \quad \pm 2.074, \pm 1.414, \pm 0.835, 0, 0.$$

If we delete vertex 6, the components even have different numbers of vertices, yet $T_1 - 6$ and $T_2 - 6$ are still cospectral.

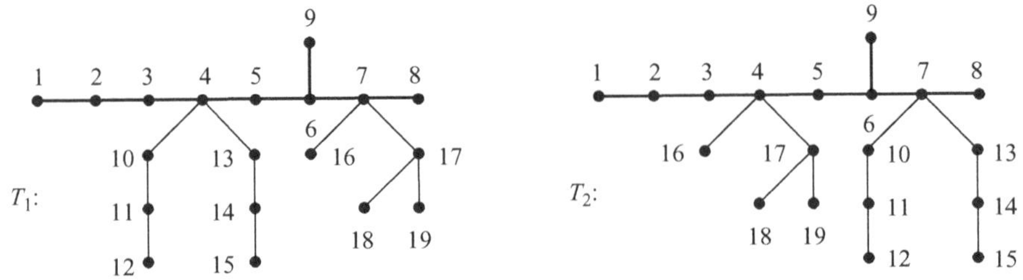

Fig. 2.

If we try to generalize this example, we encounter difficulties. Suppose that we form the graph H_1 by attaching any two rooted graphs K and L at vertices 4 and 7 of T, and then form H_2 by interchanging L and K. The formula for the characteristic polynomial of a graph obtained by coalescing rooted graphs in this way (see [21, p. 159]) shows that the following pairs are cospectral: H_1 and H_2, $H_1 - i$ and $H_2 - i$ for $i = 4, 7$ or any vertex i in K or L. However, for other values of i, the pairs are not cospectral, except for the special graphs $K = T_1$ and $L = T_2$ shown in Fig. 2.

An exhaustive search for cospectral graphs on 10 vertices (see [36]) shows that there exists a set $\mathcal{S}$ of 21 cospectral graphs with 10 vertices and 20 edges. The complements of these graphs are also cospectral and have 25 edges. Computations show also that, in both cases, the graphs are distinguished by their angles (see [10]). We reproduce here some data concerning the graphs in $\mathcal{S}$:

eigenvalues:

$$4.380, 1.686, 1.162, 0.542, 0, 0, -1.295, -1.526, -2.286, -2.663;$$

coefficients of the characteristic polynomial:

$$1, 0, -20, -18, 84, 76, -119, -72, 56, 0, 0;$$

main angles:

$$0.956, 0.025, 0.066, 0.151, 0.207, 0.044, 0.109, 0.019, 0.032.$$

These huge sets of cospectral graphs should perhaps be exploited in experiments to order graphs by their angles, for the following reason. Experience shows that it is appropriate to order graphs by their eigenvalues or spectral moments; then cospectral graphs remain to be ordered, and it is natural to use angles for this purpose because they determine the vertex-degrees.

Although graphs cannot in general be characterized by eigenvalues and angles, for certain classes of graphs (for example, trees, unicyclic graphs, bicyclic graphs, tree-like cubic graphs) it is feasible to construct all the graphs in a given class with prescribed eigenvalues and angles. Details may be found in [21, Ch. 5].

We conclude this section by noting another consequence of (2). Since $\mathbf{P}_i \neq \mathbf{0}$ we can always find a vertex j such that $\mathbf{P}_i\mathbf{e}_j \neq \mathbf{0}$, and then the multiplicity of μ_i as an eigenvalue of $G - j$ is $k_i - 1$. Repeated application of this argument shows that, for each $i \in \{1, 2, \ldots, m\}$, there exists a set Y_i of vertices such that

$$|Y_i| = k_i \quad \text{and} \quad \phi_{G-Y_i}(\mu_i) \neq 0. \tag{3}$$

We shall see in the next section that there always exists a partition of the vertex-set $V(G) = \{Y_1, Y_2, \ldots, Y_m\}$ such that (3) holds for each i. Such a partition is called a *polynomial partition* for G.

3. Star sets and star partitions

Since the column space of $\mathbf{P}_i$ is $\mathcal{E}(\mu_i)$, we know that, for each $i = 1, 2, \ldots, m$, there exists a set X_i of k_i vertices such that the vectors $\mathbf{P}_i\mathbf{e}_j (j \in X_i)$ form a basis for $\mathcal{E}(\mu_i)$. Such a set is called a *star set* for μ_i in G. (This terminology reflects the fact that the vectors $\mathbf{P}_i\mathbf{e}_j (j = 1, \ldots, n)$ form a *eutactic star*, as defined by Seidel [47].) A *star partition* for G is a partition

$$V(G) = \{X_1, X_2, \ldots, X_m\} \tag{4}$$

such that X_i is a star set for μ_i. In this situation, if $\mathcal{B}_i = \{\mathbf{P}_i\mathbf{e}_j : j \in X_i\}$, then $\mathcal{B}_1 \cup \mathcal{B}_2 \cup \ldots \cup \mathcal{B}_m$ is a basis for $\mathbf{R}^n$ and is called a *star basis* associated with G. Note that G is determined by its spectrum and a star basis. The following result [21, Thm. 7.2.9] shows that the star partitions for G are precisely the polynomial partitions for G.

Theorem 3.1 *Let G be a graph and X be a subset of $V(G)$. Then for any eigenvalue μ_i of G with multiplicity k_i, the following statements are equivalent:*
(1) $\{\mathbf{P}_i\mathbf{e}_j : j \in X\}$ is a basis of $\mathcal{E}(\mu_i)$.
(2) $\mathbf{R}^n = \mathcal{E}(\mu_i) \oplus \mathcal{V}$, where $\mathcal{V} = \langle \mathbf{e}_j : j \notin X \rangle$.
(3) $|X| = k_i$, and μ_i is not an eigenvalue of $G - X$.

We now prove the fundamental existence theorem.

Theorem 3.2 *Every graph has a star partition.*

Proof Let $\{\mathbf{x}_1, \mathbf{x}_2, \ldots, \mathbf{x}_n\}$ be a basis of $\mathbf{R}^n$, obtained by stringing together arbitrary fixed bases of $\mathcal{E}(\mu_1), \mathcal{E}(\mu_2), \ldots, \mathcal{E}(\mu_m)$; say, $\mathcal{E}(\mu_i)$ has basis $\{\mathbf{x}_h : h \in R_i\}$, where $\{R_1, R_2, \ldots, R_m\}$ is a fixed partition of $\{1, 2, \ldots, n\}$. Let $\mathbf{T}$ be the transition matrix from the basis $\{\mathbf{x}_1, \mathbf{x}_2, \ldots, \mathbf{x}_n\}$ to the basis $\{\mathbf{e}_1, \mathbf{e}_2, \ldots, \mathbf{e}_n\}$. Thus $\mathbf{T} = (t_{hj})$, where

$$\mathbf{e}_j = \sum_{h=1}^{n} t_{hj}\mathbf{x}_h \quad (j = 1, 2, \ldots, n).$$

On projecting orthogonally onto $\mathcal{E}(\mu_i)$, we have

$$\mathbf{P}_i\mathbf{e}_j = \sum_{h \in R_i} t_{hj}\mathbf{x}_h. \tag{5}$$

We say that a partition $\{C_1, C_2, \ldots, C_m\}$ of $\{1, 2, \ldots, n\}$ is *feasible* if $|C_i| = k_i$, for $i = 1, 2, \ldots, m$. For such a partition let $\mathbf{T}_i$ be the $k_i \times k_i$ submatrix of $\mathbf{T}$ whose rows are indexed by R_i and whose columns are indexed by C_i. The corresponding multiple Laplacian development of det **T** has the form

$$\det \mathbf{T} = \sum \left\{ \pm \prod_{i=1}^{m} \det \mathbf{T}_i \right\},$$

where the sum is taken over all $n!/(k_1!\,k_2!\cdots k_m!)$ feasible partitions. Since **T** is invertible, some term $\Pi_{i=1}^{m}\det \mathbf{T}_i$ is non-zero – say, that determined by the partition $\{X_1, X_2, \ldots, X_m\}$. This partition is a star partition because, in view of (5), the invertibility of $\mathbf{T}_i$ guarantees that each $\mathbf{x}_h(h \in R_i)$ is a linear combination of the vectors $\mathbf{P}_i\mathbf{e}_j(j \in X_i)$. ■

This existence theorem may be strengthened as follows: if X is any star set for an eigenvalue μ_i, then there exists a star partition (4) with $X_i = X$ (see [21, Thm. 7.4.5]). In any case, we can always label the n vertices of G with the n eigenvalues of G in such a way (in general, not unique) that the vertices labelled μ_i constitute a star set for μ_i. This is how the vertices are labelled in the examples shown in Fig. 3. The second example there is one of 750 star partitions of the Petersen graph; these fall into ten isomorphism classes determined by the automorphism group of the graph (see [21, Sec. 7.7]).

An important consequence of Theorem 3.2 is that we can associate with any graph a star basis which is canonical, in the sense that two graphs are isomorphic if and only if they have the same spectrum and the same canonical basis. To see why this is possible, note first that a graph has only finitely many star partitions and hence determines only finitely many star bases; then, as a canonical star basis, we can take one that is extremal in some lexicographical ordering of bases. This crude approach is grossly inefficient, involving as it does all permutations of coordinates, and significant improvements in complexity have been achieved by the use of a recursive procedure for ordering vertices and star bases simultaneously (see [21, Ch. 8]). We remark here only that one can always find a star partition in

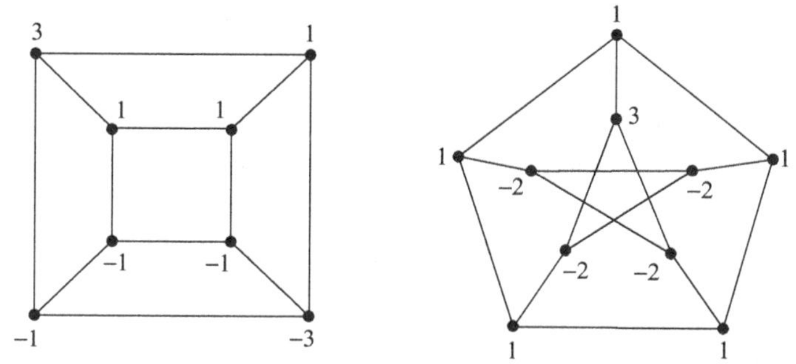

Fig. 3.

polynomial time – this is a consequence of Edmonds' matroid intersection theorem [28]. If one could find a canonical star basis in polynomial time, then we would have a polynomial algorithm for determining whether two graphs are isomorphic. This graph isomorphism problem was the original motivation for introducing star partitions, but as we shall see in the next section, it transpires that individual star sets are important in their own right.

4. Star complements

Let X be a star set for the eigenvalue μ of the graph G, and let $H = G - X$. We call H a *star complement* for μ in G; it is the subgraph of G induced by the complement $\overline{X}$ of X in $V(G)$. A star complement for μ is called a *μ-basic subgraph* in [29]. Here we discuss the influence of a star complement on the structure of a graph (see [21, Ch. 7]).

Theorem 4.1 *Let X be a star set for μ in G.*
(a) If $\mu \neq 0$, then $\overline{X}$ is a dominating set in G.
(b) If $\mu \neq -1$ or 0, then $\overline{X}$ is a location-dominating set in G – that is, the $\overline{X}$-neighbourhoods of vertices in X are distinct and non-empty.

In addition to the notation above, let $|X| = |V(G)| - t$, so that $t = |V(H)| = \operatorname{codim} \mathcal{E}(\mu)$. It follows from Theorem 4.1 that if $\mu \neq -1$ or 0 then $|X| < 2^t$, and hence $|V(G)| < t + 2^t$, a bound that can be improved to $t + \frac{1}{2}(t-1)(t+4)$ when $t > 1$ (see [39]). This bound of the form $\frac{1}{2}t^2 + O(t)$ is asymptotically best possible as $t \to \infty$, because in the line graph $L(K_t)$ the eigenspace of -2 has codimension t. When $\mu = 0$, non-adjacent vertices in X with the same $\overline{X}$-neighbourhood are called *duplicate vertices*; and when $\mu = -1$, adjacent vertices in X with the same $\overline{X}$-neighbourhood are called *coduplicate vertices* (see [29]).

Since duplicate or coduplicate vertices correspond to repeated rows of $\mu\mathbf{I} - \mathbf{A}$, it is clear that $|V(G)|$ cannot be bounded in terms of t when $\mu = -1$ or 0. However, these exceptional values do not normally obstruct our arguments, because we can specify a graph to within duplicate or coduplicate vertices. A graph without such vertices is called a *core graph*. It follows from a theorem of Kotlov and Lovász [35] that if $\mu = -1$ or 0 and G is a core graph, then $|V(G)| = O(2^{t/2})$.

In view of the foregoing remarks, there are only finitely many graphs (finitely many core graphs when $\mu = -1$ or 0) with a star complement of prescribed order, or equivalently, with an eigenspace of prescribed codimension. (The graphs with an eigenspace $\mathcal{E}(\mu)$ ($\mu \neq -1, 0$) of codimension at most 5 are determined in [43].) There are relatively few graphs with a prescribed star complement, and this is the basis for characterizing graphs by star complements, as illustrated below.

The next result shows that any graph is determined uniquely by

(*i*) an eigenvalue μ;

(*ii*) a star complement H for μ;

(*iii*) the embedding of H in G.

Given (*ii*) and (*iii*), we know all the edges between vertices in $\overline{X}$ and all the edges between X and $\overline{X}$; then knowledge of μ enables us to find all the edges between vertices of X. This establishes the role of a single eigenvalue in determining the structure of a graph.

Theorem 4.2 (Reconstruction theorem) *Let X be a star set for μ in the graph G. If $G - \overline{X}$ and $G - X$ have adjacency matrices $\mathbf{A}_X$ and $\mathbf{C}$, respectively, then G has an adjacency matrix of the form*

$$\begin{pmatrix} \mathbf{A}_X & \mathbf{B}^T \\ \mathbf{B} & \mathbf{C} \end{pmatrix},$$

where

$$\mu\mathbf{I} - \mathbf{A}_X = \mathbf{B}^T(\mu\mathbf{I} - \mathbf{C})^{-1}\mathbf{B}. \tag{6}$$

Proof Clearly the adjacency matrix of G has the form $\mathbf{A} = \begin{pmatrix} \mathbf{A}_X & \mathbf{B}^T \\ \mathbf{B} & \mathbf{C} \end{pmatrix}$ for some matrix $\mathbf{B}$. We have

$$\mu\mathbf{I} - \mathbf{A} = \begin{pmatrix} \mu\mathbf{I} - \mathbf{A}_X & -\mathbf{B}^T \\ -\mathbf{B} & \mu\mathbf{I} - \mathbf{C} \end{pmatrix},$$

where $\mu\mathbf{I} - \mathbf{C}$ is invertible. In particular, if $|X| = k$ then the matrix $(-\mathbf{B} \mid \mu\mathbf{I} - \mathbf{C})$ has rank $n - k$. But $\mu\mathbf{I} - \mathbf{A}$ has rank $n - k$, and so the rows of $(-\mathbf{B} \mid \mu\mathbf{I} - \mathbf{C})$ form a basis for the row space of $\mu\mathbf{I} - \mathbf{A}$. Hence there exists a $k \times (n - k)$ matrix $\mathbf{L}$ such that

$$(\mu\mathbf{I} - \mathbf{A}_X \mid -\mathbf{B}^T) = \mathbf{L}(-\mathbf{B} \mid \mu\mathbf{I} - \mathbf{C}).$$

Now $\mu\mathbf{I} - \mathbf{A}_X = -\mathbf{L}\mathbf{B}$ and $-\mathbf{B}^T = \mathbf{L}(\mu\mathbf{I} - \mathbf{C})$, and the result then follows by eliminating $\mathbf{L}$. ■

The converse of Theorem 4.2 is also true: if (6) holds and $\mathbf{A}_X$ has size $k \times k$, then the null space of $\mu\mathbf{I} - \mathbf{A}$ consists of the vectors $\begin{pmatrix} \mathbf{x} \\ (\mu\mathbf{I} - \mathbf{C})^{-1}\mathbf{B}\mathbf{x} \end{pmatrix}$, where $\mathbf{x} \in \mathbf{R}^k$.

We now consider a means of constructing the graphs having H as a star complement for μ, by adding to H a suitable set X of vertices (the *star complement technique*). Such a graph is specified by the H-neighbourhoods of vertices in X, or equivalently by the columns $\mathbf{b}_u (u \in X)$ of the matrix $\mathbf{B}$ from (6). In view of

Theorem 4.2 and its converse, we may add the set X if and only if we can add each pair of vertices in X. Explicitly, by equating entries in (5), we see that X is a star set for μ if and only if

$$\mathbf{b}_u^T(\mu\mathbf{I} - \mathbf{C})^{-1}\mathbf{b}_u = \mu, \quad \text{for all } u \in X, \tag{7}$$

$$\text{and} \quad \mathbf{b}_u^T(\mu\mathbf{I} - \mathbf{C})^{-1}\mathbf{b}_v \in \{-1, 0\}, \quad \text{for all distinct } u, v \in X. \tag{8}$$

To describe all the graphs with H as a star complement for μ, it suffices to determine those graphs for which X is maximal, since any graph with H as a star complement for μ is an induced subgraph of such a graph.

Example 2 We illustrate the star complement technique in the simple case when $\mu = 1$ and H is the 5-cycle 123451. To check (7) and (8), we simply sum the entries in an appropriate submatrix of $(\mu\mathbf{I} - \mathbf{C})^{-1}$, which in this case is the circulant matrix with first row $(1, 0, -1, -1, 0)$. We find from (7) that we may add a single vertex u to H if its H-neighbourhood $N(u)$ consists of either (a) a single vertex $\delta(u)$, or (b) three consecutive vertices of the 5-cycle. By (8), two distinct vertices u and v may be added in just four ways as follows (where $u \sim v$ means that the vertices u and v are adjacent):

- both are of type (a), $\delta(u) \nsim \delta(v)$, $u \sim v$;
- both are of type (a), $\delta(u) \sim \delta(v)$, $u \nsim v$;
- both are of type (b), $|N(u) \cap N(v)| = 2$, $u \sim v$;
- u is of type (a), v is of type (b), $\delta(u) \sim v$, but $\delta(u)$ is not the middle vertex of $N(v)$, $u \nsim v$.

It follows that at most two vertices of type (b) may be added, and that the maximal graphs with H as a star complement for the eigenvalue 1 are those shown in Fig. 4 (with 2, 1, 0 vertices of type (b), respectively): in the figure, the vertices of H are shown in black.

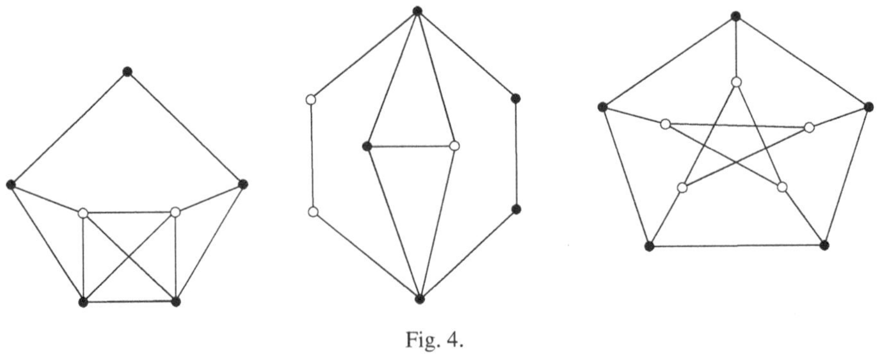

Fig. 4.

The general problem here is to find all of the graphs having a given graph as a star complement, or equivalently, to find all of the solutions $\mathbf{A}_X$, $\mathbf{B}$, μ of (6) for a given matrix $\mathbf{C}$. The 'restricted problem' is to find the solutions $\mathbf{A}_X$, $\mathbf{B}$ of (6) for a given matrix $\mathbf{C}$ and a given eigenvalue μ (see Example 2). We give the results of six such investigations; in some cases we can characterize certain graphs as those that are maximal extensions of a prescribed star complement.

(1) [41, Thm. 3.6] If $H \cong K_{1,5}$ and $\mu \neq -1$, then $\mu = 1$ and G is an induced subgraph of the Clebsch graph (see [7, p. 35]). If $H \cong K_{1,5}$ and $\mu = -1$, then the core subgraph of G is an induced subgraph of one of two graphs with 15 and 16 vertices.
(2) [34, Thm. 3.1] If $H \cong K_{2,5}$, $\mu \neq -1$ and $|X| > 1$, then $\mu = 1$ and G is an induced subgraph of the Schläfli graph (see [7, p. 32]). All of the regular graphs with $K_{2,5}$ as a star complement have been determined.
(3) [42, Thm. 2.2 and Cor. 2.3] If G is r-regular, $H \cong K_{1,r}$ and $0 < r \neq \mu$, then $r = \mu(\mu^2 + 3\mu + 1)$ and $|V(G)| = (\mu^2 + 3\mu)^2$. Furthermore, if $\mu = 1$, then G is the Clebsch graph; if $\mu = 2$, then G is the Higman-Sims graph (see [7, p. 107]).
(4) [33, Thm. 7.3] If $H \cong 6K_1 \cup K_{1,16}$, $\mu = 2$ and $|X|$ is maximal, then G is the McLaughlin graph (see [3, p. 373]).
(5) [2, Thm. 2.4] If $H \cong C_t$ (t odd, $t \geq 5$) and $\mu = -2$, then G is an induced subgraph of the line graph $L(K_t)$.
(6) [22, Thm. 2.1] If $H \cong \overline{K_{1,t-3} \cup 2K_1}$ ($t \geq 4, t \neq 8$) and $\mu = -2$, then G is an induced subgraph of a generalized line graph $L(K_{t-u}; u, 0, 0, \ldots, 0)$, for $0 \leq u \leq t - 3$ (see [7, p. 52]).

In (5) and (6), μ is equal to the least eigenvalue -2 of G. The star complements for -2 that can arise in this situation are discussed in [22] and [23]. A connected graph with least eigenvalue $\lambda_n \geq -2$ which is not a generalized line graph is called an *exceptional graph*, and the problem of determining all the exceptional graphs has only recently been solved (see [17]). The solution, described in the next section, uses the star complement technique to find those exceptional graphs that are maximal, in the sense that any exceptional graph is an induced subgraph of such a graph.

5. Exceptional graphs

If the graph G with adjacency matrix $\mathbf{A}$ has least eigenvalue $\lambda_n \geq -2$, then $\mathbf{A} + 2\mathbf{I}$ is the *Gram matrix* of a set S of n vectors with the property that the angle between any two of them is 60° or 90°. It is well known that if G is exceptional, then G is representable in E_8 (see [3, Ch. 3] or [7, Ch. 3]): this means that S can be taken to be a subset of the root system E_8 defined as follows (see also Chapter 1, Table 2). If $\{\mathbf{e}_1, \mathbf{e}_2, \ldots, \mathbf{e}_8\}$ is an orthonormal basis for $\mathbf{R}^8$, then E_8 consists of the 112 vectors of the form $\pm\mathbf{e}_i \pm \mathbf{e}_j (i \neq j)$, together with the 128 vectors of the form $\frac{1}{2}\sum_{i=1}^{8} \epsilon_i \mathbf{e}_i$, where $\epsilon_i \pm 1$ and $\Pi_{i=1}^{8} \epsilon_i = 1$. The root system E_7 consists of the vectors in E_8 orthogonal to a fixed vector in E_8, while E_6 consists of the vectors in E_8 orthogonal to a pair of vectors at 60°.

The connected graphs with least eigenvalue $\lambda_n > -2$ were determined by Doob and Cvetković [27] in 1979; accordingly, we consider only graphs with least eigenvalue -2. It was proved in [23] that such a graph is exceptional if and only if it has an exceptional star complement for -2. Such a star complement has least eigenvalue greater than -2 (by interlacing), and hence is one of 573 graphs of the following types (see [27]):

type I: one of 20 graphs on 6 vertices representable in E_6;
type II: one of 110 graphs on 7 vertices representable in E_7, but not E_6;
type III: one of 443 graphs on 8 vertices representable in E_8, but not E_7.

The graphs of type III are one-vertex extensions of graphs of type II, which are in turn one-vertex extensions of graphs of type I. Vectors in E_8 which represent the additional vertices are referred to below as *extension vectors*. The 443 graphs of type III are described in [5]. The 110 graphs of type II are identified in [17] by means of the list of 7-vertex graphs in [12]. The 20 graphs of type I are identified in [20]: they belong to the family $\mathcal{F}$ of 31 minimal forbidden subgraphs that characterize generalized line graphs, the other eleven having -2 as their least eigenvalue (see [14]). Accordingly, a graph is exceptional if and only if its least eigenvalue is greater than or equal to -2 and it contains a graph of type I as an induced subgraph (see [23, Prop. 3.1]). Since $\mathcal{F}$ was determined in [27] independently of root systems, the star complement technique may be used to determine the exceptional graphs without recourse to root systems.

Let G be an exceptional graph with adjacency matrix $\mathbf{A}$, let H be a star complement for the eigenvalue -2, and let $t = 6, 7$ or 8. If H is representable in E_t, then so is G, since $\mathbf{A} + 2\mathbf{I}$ is the Gram matrix of vectors which span a t-dimensional space (see [6]). If G is a maximal exceptional graph, then H is of type III, since otherwise G is representable in E_7, and then within E_8 we could add an extension vector to obtain a larger exceptional graph.

We say that a graph is *H-maximal* if it is maximal with respect to the property of having H as a star complement for the eigenvalue -2. In order to describe the exceptional graphs representable in E_6, it suffices to find the H-maximal graphs for star complements H of type I; for those representable in E_7, but not E_6, we take H to be of type II; while if H is of type III, the H-maximal graphs are precisely the maximal exceptional graphs. Ten H-maximal graphs arise when H is of type I, and they are described in [17, Ex. 5]. When H is of type II there are

39 H-maximal graphs, and some details are given in [23, Sec. 3]. For H of type III, Lepović used a computer to determine the maximal star sets that can be added to H (see [29, Algorithm 2.4], and [17]). He found that there are 473 maximal exceptional graphs in all. Moreover, the results reveal how these graphs can be constructed independently of a computer search (see [18] and [24]).

To describe some of these graphs, recall from [7, p. 59] that two graphs with the same vertex-set V are *switching-equivalent* if, for some partition of $V(G)$ into two sets P and Q, the non-edges and edges between P and Q are interchanged; for example, the two graphs of Fig. 1 are switching-equivalent, with $P = \{1, 2, 7, 8\}$ and $Q = \{3, 4, 5, 6, 9, 10\}$.

First we give the distribution of maximal exceptional graphs over the number of vertices:

number of vertices:	22	28	29	30	31	32	33	34	36
number of graphs:	1	1	432	25	7	3	1	2	1

The 432 graphs on 29 vertices include 430 cones over graphs switching-equivalent to the line graph $L(K_8)$. Of the other 43 maximal exceptional graphs, 37 are extensions of such cones, with maximum degree 28, while the remaining six graphs have maximum degree less than 28. None of the maximal exceptional graphs is regular, but the list includes graphs that are of interest for a variety of reasons. For instance, several graphs having only integer eigenvalues arise; such graphs are discussed in Section 8. The examples here include some that have only three distinct eigenvalues but are not strongly regular. The two graphs on 34 vertices are cospectral, but can be distinguished by their angles. All but a few of the graphs have the graph $K_{1,2} \cup 5K_1$ as one of several star complements for -2; in particular, this provides a means of constructing the maximal exceptional graphs that are not 29-vertex cones.

6. Reconstructing the characteristic polynomial

Here we consider one of four reconstruction problems discussed by Schwenk [45] (see also [12, Section 3.5] and [46, Section 12]). For any graph G with vertex-set $\{1, 2, \ldots, n\}$, let $\mathcal{P}(G)$ be the multiset consisting of the characteristic polynomials $\phi_{G-i}(x)$ of the vertex-deleted subgraphs $G - i$ $(i = 1, 2, \ldots, n)$.

Problem 1 *Is it true that, for $n > 2$, the characteristic polynomial $\phi_G(x)$ of a graph G is determined uniquely by $\mathcal{P}(G)$?*

For $n > 2$, it is known that $\phi_G(x)$ is determined by the vertex-deleted subgraphs themselves: this result is due to Tutte [53]. Problem 1 was posed by the first author at the 18th International Scientific Colloquium in Ilmenau in 1973, and the first results were obtained by Gutman and Cvetković [31] (see also [13, p. 267] and [12, pp. 68–70]). No examples of non-unique reconstruction of the characteristic polynomial are known. Some relations between Problem 1 and the Ulam reconstruction problem for graphs are described in [21, Sec. 5.4].

Since $\phi'_G(x) = \sum_{i=1}^{n} \phi_{G-i}(x)$ (see [13, p. 60]), we can readily determine the characteristic polynomial $\phi_G(x)$, except for the constant term. If we know just one eigenvalue of G, then the constant term is determined. In particular, this is the case if we know a multiple root λ of some polynomial $\phi_{G-i}(x)$, for then (by interlacing) λ is an eigenvalue of G.

In [31], Problem 1 was solved affirmatively for regular graphs and for a broad class of bipartite graphs including trees without a 1-factor. Here we indicate how the result was extended to the remaining trees in [11] and [16]. We denote by $\sigma(G)$ the set of distinct eigenvalues of the graph G.

Theorem 6.1 *Let H be a graph of order n (≥ 3) with exactly two connected components. If these components have different orders, then the characteristic polynomial of H is determined uniquely by $\mathcal{P}(H)$.*

Proof Suppose, by way of contradiction, that there exists at least one graph $G \neq H$ such that $\phi_G(x) = \phi_H(x) + a$ ($a \neq 0$) and $\phi_{G-i}(x) = \phi_{H-i}(x)$, for $i = 1, 2, \ldots, n$. Let H_1 and H_2 be the two components of H, with H_j of order n_j and $n_1 > n_2$. Clearly,

$$\sigma(H - i) = \sigma(H_1) \cup \sigma(H_2 - i) \quad (i \in V(H_2)). \tag{9}$$

Since $H - i$ has no repeated eigenvalues, the same is true of H_1, and we let $\sigma(H_1) = \{\lambda_1^*, \lambda_2^*, \ldots, \lambda_{n_1}^*\}$, where $\lambda_1^* > \lambda_2^* > \cdots > \lambda_{n_1}^*$.

Let v be a fixed vertex of H_2. Since $|\sigma(H_2 - v)| < n_1 - 1$, there exists at least one index $i = i_0$ ($1 \leq i_0 \leq n_1 - 1$) such that no eigenvalue of $H_2 - v$ lies in the open interval $(\lambda_{i_0+1}^*, \lambda_{i_0}^*)$. We deduce that

$$\left(\lambda_{i_0+1}^*, \lambda_{i_0}^*\right) \cap \sigma(H - v) = \emptyset. \tag{10}$$

Since $\phi_{G-v}(x) = \phi_{H-v}(x)$, we know from (9) that $\lambda_{i_0}^*$ and $\lambda_{i_0+1}^*$ lie in $\sigma(G - v)$. By the interlacing theorem, there exists at least one eigenvalue α of G in the interval $(\lambda_{i_0+1}^*, \lambda_{i_0}^*)$. Since $\phi_G(\alpha) = 0$ and $\phi_G(\lambda_{i_0+1}^*) = \phi_G(\lambda_{i_0}^*) = a$, there exist at least two eigenvalues α, β of G in $(\lambda_{i_0+1}^*, \lambda_{i_0}^*)$.

Finally, using the interlacing theorem again, we see that $G - v$ has at least one eigenvalue $\gamma \in [\alpha, \beta] \subseteq (\lambda_{i_0+1}^*, \lambda_{i_0}^*)$. Since $\sigma(G - v) = \sigma(H - v)$, this is a contradiction to (10). ■

In similar vein we have also the following result.

Theorem 6.2 *If H is a disconnected graph with at least three components, then the characteristic polynomial of H is determined uniquely by $\mathcal{P}(H)$.*

Proof Suppose that H has components $H_1, H_2, \ldots, H_k$ $(k > 2)$. We may assume that all eigenvalues of H are simple, and hence that $\lambda_1(H_1) > \lambda_1(H_2) > \cdots > \lambda_1(H_k)$. It follows that

$$(\lambda_1(H_1), \lambda_1(H_2)) \cap \sigma(H_k - v) = \emptyset \quad (v \in V(H_k)),$$

and the proof now follows as before. ■

In proving the next result, we make use of the fact (see [13, p. 37]) that the characteristic polynomial of a tree with n vertices has constant term $(-1)^{n/2}$ or 0, according as it does or does not have a 1-factor. We write $e(G)$ for the number of edges of the graph G.

Theorem 6.3 *If G is a tree, then its characteristic polynomial is determined uniquely by $\mathcal{P}(G)$.*

Proof Suppose, by way of contradiction, that there exists a tree G whose characteristic polynomial is not uniquely determined by $\mathcal{P}(G)$. In view of our introductory remarks, we know that such a tree G has a 1-factor, and so the number of vertices is even – say $n = 2k$, where $k > 1$. Now there exists at least one graph H such that $\phi_H(x) = \phi_G(x) + a$ $(a \neq 0)$ and $\phi_{H-i}(x) = P_{G-i}(x)$ $(i = 1, 2, \ldots, n)$. Such a graph H is not connected, since otherwise it is a tree, the number of edges being known from the coefficient of x^{n-2} in $\phi_H(x)$. Since $a \neq 0$, this tree does not have a 1-factor, and so $\phi_H(0) = 0$. By [46, Cor. 3.3], if u is the neighbour of an end-vertex of H, then 0 is a multiple eigenvalue of $H - u$. Then G has 0 as an eigenvalue, giving a contradiction.

By Theorems 6.1 and 6.2, H has exactly two connected components H_1 and H_2, each with exactly k vertices. Since $e(H) = e(G) = 2k - 1$, we know that one component, say H_1, is a unicyclic graph and the other component H_2 is a tree. Let $\lambda_1^* > \lambda_2^* > \cdots > \lambda_k^*$ be the eigenvalues of H_2. If there are a vertex v of H_1 and an index i_0 such that $(\lambda_{i_0+1}^*, \lambda_{i_0}^*) \cap \sigma(H_1 - v) = \emptyset$, then the proof proceeds as in Theorem 5.1. Otherwise, for any vertex v of H_1, the eigenvalues $\gamma_1, \gamma_2, \ldots, \gamma_{k-1}$ of $H_1 - v$ interlace those of H_2 – that is,

$$\gamma_i \in (\lambda_{i+1}^*, \lambda_i^*) \quad (i = 1, 2, \ldots, k - 1). \tag{11}$$

Now, because it has no multiple eigenvalues, the unicyclic graph H_1 is not a cycle, and so we may choose v to be an end-vertex. Then $H_1 - v$ is unicyclic, and we deduce from (11) the contradiction

$$2(k-1) = e(H_1 - v) = \sum_{i=1}^{k-1} \gamma_i^2 < \sum_{i=1}^{k} (\lambda_i^*)^2 = 2e(H_2) = 2(k-1).$$

This completes the proof. ■

7. Non-complete extended p-sums of graphs

Next we consider a very general graph operation called the *non-complete extended p-sum* of graphs, which we abbreviate as *NEPS* (see [19]); these graphs were discussed briefly in Chapter 1.

Let $\mathcal{B}$ be a set of non-zero binary n-tuples in $\{0, 1\}^n$. The NEPS with basis $\mathcal{B}$ of graphs $G_1, G_2, \ldots, G_n$ is the graph with vertex-set $V(G_1) \times V(G_2) \times \cdots \times V(G_n)$, where the vertices $(x_1, x_2, \ldots, x_n)$ and $(y_1, y_2, \ldots, y_n)$ are adjacent if and only if there exists an n-tuple $(\beta_1, \beta_2, \ldots, \beta_n) \in \mathcal{B}$ such that $x_i = y_i$ when $\beta_i = 0$, and x_i is adjacent to y_i in G_i when $\beta_i = 1$.

We identify some special cases where a graph G is the NEPS with basis $\mathcal{B}$ of two graphs G_1 and G_2:

- if $\mathcal{B} = \{(0, 1), (1, 0)\}$, then G is the *sum* $G_1 + G_2$ of G_1 and G_2;
- if $\mathcal{B} = \{(1, 1)\}$, then G is the *product* $G_1 \Box G_2$ of G_1 and G_2;
- if $\mathcal{B} = \{(0, 1), (1, 0), (1, 1)\}$, then G is the *strong product* $G_1 * G_2$ of G_1 and G_2.

(A variety of terms and notations for these particular constructions can be found in the literature.)

The *p-sum* of n graphs is a NEPS whose basis consists of all n-tuples with exactly p entries equal to 1. The 2-sum of three graphs G_1, G_2 and G_3 is denoted by $D(G_1, G_2, G_3)$.

The foregoing definitions, together with some basic results on NEPS and early references, can be found in [13]; for a more recent review of results, see [25]. The proofs of the next two theorems are given in [13, Sec. 2.5].

Theorem 7.1 *Let $\mathbf{A}_1, \mathbf{A}_2, \ldots, \mathbf{A}_n$ be adjacency matrices of graphs $G_1, G_2, \ldots, G_n$, respectively. The NEPS G with basis $\mathcal{B}$ of graphs $G_1, G_2, \ldots, G_n$ has as adjacency matrix the matrix $\mathbf{A}$ given by*

$$\mathbf{A} = \sum_{\beta \in \mathcal{B}} \mathbf{A}_1^{\beta_1} \otimes \mathbf{A}_2^{\beta_2} \ldots \otimes \mathbf{A}_n^{\beta_n}. \tag{12}$$

Here $\mathbf{A}_k^0$ is the identity matrix of the same size as $\mathbf{A}_k$, $\mathbf{A}_k^1 = \mathbf{A}_k$, and $\otimes$ denotes the Kronecker product of matrices.

One consequence of Theorem 7.1 is the following result.

Theorem 7.2 *If* $p_i = |V(G_i)|$ *and if* $\lambda_{i1}, \lambda_{i2}, \ldots, \lambda_{ip_i}$ *is the spectrum of* G_i $(i = 1, 2, \ldots, n)$*, then the spectrum of the NEPS of* $G_1, G_2, \ldots, G_n$ *with basis* $\mathcal{B}$ *consists of all possible values* $\Lambda_{i_1,i_2,\ldots,i_n}$*, where*

$$\Lambda_{i_1,i_2,\ldots,i_n} = \sum_{\beta \in \mathcal{B}} \lambda_{1i_1}^{\beta_1} \lambda_{2i_2}^{\beta_2} \ldots \lambda_{ni_n}^{\beta_n} \quad (i_k = 1, 2, \ldots, p_k; k = 1, 2, \ldots, n). \tag{13}$$

Together with expressions (12) and (13) we consider the function

$$f(x_1, x_2, \ldots, x_n) = \sum_{\beta \in \mathcal{B}} x_1^{\beta_1} x_2^{\beta_2} \ldots x_n^{\beta_n}, \tag{14}$$

which will be of interest later.

First we quote the following theorem which lists some of the classes of graphs identified in [25] as being closed under the NEPS operation. An *even graph* is one in which each vertex has even degree, a *singular graph* is one with 0 as an eigenvalue, and a *transitive graph* is one whose automorphism group acts transitively on the vertices.

Theorem 7.3 *The following classes of graphs are closed under the NEPS operation:*
(1) regular graphs;
(2) even graphs;
(3) singular graphs;
(4) integral graphs;
(5) transitive graphs;
(6) walk-regular graphs.

Here, (1) and (2) follow from the fact that the vertex-degrees in a NEPS can be expressed in terms of the vertex-degrees of the graphs on which the operation is performed, while (3) and (4) follow from Theorem 6.2. (5) follows from the observation that the automorphism group of a NEPS contains the direct product of the automorphism groups of the constituent graphs, while (6) follows from the characterization of walk-regular graphs in terms of angles (see Section 2 and [25, Remark 3.5]).

It is a well-known fact that the product of two connected bipartite graphs G_1 and G_2 is disconnected, with two components. Similar situations arise more generally in a NEPS, and graph eigenvalues are a good means of investigating the conditions under which a NEPS is connected or bipartite (see [13, Sec. 7.4]).

Two graphs are *almost cospectral* if their non-zero eigenvalues (and their multiplicities) coincide. In [12] it was conjectured that, if a NEPS of bipartite graphs is disconnected, then its components are almost cospectral. This conjecture is true for the product of graphs, and we shall show (following [9]) that it is also true for the 2-sum of three graphs.

Connectedness properties of the NEPS, from Theorems 7.1 and 7.2, are related to the function (14), as described in [13]; for a recent treatment, see [49]. In the case of the 2-sum of three graphs, the function (14) is $x_1x_2 + x_1x_3 + x_2x_3$, which is even with respect to the variables x_1, x_2, x_3, in the sense that it remains unchanged when these variables are simultaneously changed in sign. By [13, Thm. 7.12] we conclude that the 2-sum of three bipartite graphs is a disconnected graph with two components. For example, $D(K_2, K_2, K_2) = 2K_4$; geometrically, this says that if the vertices of $D(K_2, K_2, K_2)$ are represented by the vertices of a cube, then the 2-sum consists of two tetrahedra.

The proof of the following lemma is straightforward.

Lemma 7.4 *Let $x_1, x_2, \ldots, x_r$ and $y_1, y_2, \ldots, y_s$ be non-increasing sequences of non-zero real numbers. If*

$$\sum_{i=1}^{r} x_i^k = \sum_{j=1}^{s} y_j^k \quad \textit{for all } k \in \mathbf{N},$$

then $r = s$ and $x_i = y_i$ $(i = 1, 2, \ldots, r)$.

We can then prove the following theorem.

Theorem 7.5 *Two graphs are almost cospectral if and only if, for each $k \in \mathbf{N}$, they have the same number of closed walks of length k.*

Proof Assume first that G and H are two graphs that satisfy the condition concerning walks. If $\lambda_1, \lambda_2, \ldots, \lambda_m$ and $\mu_1, \mu_2, \ldots, \mu_n$ are the eigenvalues of G and H, respectively, then $\sum_{i=1}^{m} \lambda_i^k = \sum_{j=1}^{n} \mu_j^k$ for all $k \in \mathbf{N}$. By Lemma 7.4, G and H are almost cospectral. The reverse implication is immediate. ∎

We can use Theorem 7.5 to prove in a new way that if G and H are connected bipartite graphs, then the two components A and B of the product $G \times H$ are almost cospectral. Let $(x_1, y_1), (x_2, y_2), \ldots, (x_k, y_k)$ be a closed walk of length k in A. Then $(x_1, y_2), (x_2, y_3), \ldots, (x_k, y_1)$ is a closed walk of length k in B. For each $k \geq 2$, this mapping is a bijection between the sets of closed walks of length k in the graphs A and B. Since no graph has any closed walks of length 1, we deduce from Theorem 7.5 that the components A and B are almost cospectral. We apply the same idea to the 2-sum of three graphs.

Let $a_1, a_2, \ldots, a_n$ be a finite sequence of real numbers. We count the groups of mutually equal members successively following one another in this sequence, by defining the function ψ recursively as follows: $\psi(1) = 1$, and for $i = 2, 3, \ldots, n$,

$$\psi(i+1) = \begin{cases} \psi(i), & \text{if } a_{i+1} = a_i \\ \psi(i) + 1, & \text{otherwise.} \end{cases}$$

If $a_n = a_1$ and $\psi(n) = k$, we redefine those values of $\psi(i)$ equal to k to be 1 instead. Let s be the maximum value of ψ and define $b_{\psi(i)} = a_i$ $(i = 1, 2, \ldots, n)$. The *quasi-cyclic shift* of the sequence $a_1, a_2, \ldots, a_n$ is the sequence obtained from $a_1, a_2, \ldots, a_n$ by replacing each a_i with $b_{\psi(i)+1}$ (where $b_{s+1} = b_1$).

Theorem 7.6 *The two components of the 2-sum of three bipartite graphs are almost cospectral.*

Proof Let H_1 and H_2 be the two components, and let

$$(x_1, y_1, z_1), (x_2, y_2, z_2), \ldots, (x_k, y_k, z_k) \tag{15}$$

be a closed walk of length k (≥ 2) in H_1. We construct from (15) a closed walk of the same length in H_2, and this will establish the required one-one correspondence.

The sequence $z_1, z_2, \ldots, z_k$ has the properties that z_i and z_{i+1} are either equal or adjacent $(i = 1, 2, \ldots, k-1)$ and $z_k = z_1$.

If not all of the vertices $z_1, z_2, \ldots, z_k$ are the same, then the sequence $z_1, z_2, \ldots, z_k$ determines a closed walk of length at most k in the corresponding graph. Let $w_1, w_2, \ldots, w_k$ be the quasi-cyclic shift of the sequence $z_1, z_2, \ldots, z_k$. It is easy to see that the sequence

$$(x_1, y_1, w_1), (x_2, y_2, w_2), \ldots, (x_k, y_k, w_k)$$

is a closed walk of length k in H_2.

On the other hand, if $z_1 = z_2 = \cdots = z_k = z$, then

$$(x_1, y_2, z), (x_2, y_3, z), \ldots, (x_k, y_1, z)$$

is a closed walk in H_2 corresponding to (15).

Thus, for each k, the number of closed walks of length k is the same in H_1 and H_2. By Theorem 7.5, the two components are almost cospectral. ■

It was conjectured that the components of a NEPS of bipartite graphs are always almost cospectral, but this is not true in general (see [51]). A counter-example is provided by the graph $(P_3 \times P_3) + K_2$, and modified conjectures are discussed in [51].

8. Integral graphs

A graph is *integral* if all its eigenvalues are integers. The quest for integral graphs was initiated by Harary and Schwenk [32]. The thirteen connected cubic integral graphs were obtained by Cvetković and Bussemaker (see [8] and [4]), and independently by Schwenk [44]. Cvetković [8] proved that the set of connected regular integral graphs of any fixed degree is finite.

Radosavljević and Simić [48] determined the thirteen connected non-regular non-bipartite integral graphs with maximum degree 4. The corresponding problem for bipartite graphs is not yet solved; see [48] for some details. Recently, Stevanović [52] determined all 24 connected 4-regular integral graphs without 3 or -3 in the spectrum.

The search for integral graphs becomes easier if we restrict ourselves to the product of graphs. If G is connected, non-bipartite, r-regular and integral, then the product $G \times K_2$ is connected, bipartite, r-regular and integral, since the eigenvalues of K_2 are 1 and -1. Therefore, in determining r-regular integral graphs we need consider only bipartite graphs, and later extract non-bipartite graphs G from the decompositions of bipartite graphs of the form $G \times K_2$. On the other hand, if G is bipartite, then $G \times K_2 = 2G$ and we cannot obtain new graphs by iterating the product with K_2.

A system of Diophantine equations for the spectrum of a connected 4-regular bipartite integral graph has been obtained in [26]. The equations derive from an interpretation of the first six spectral moments in terms of the numbers of short cycles. The system has a finite number of solutions, and according to [26] the graphs in question have at most 5040 vertices. The non-existence of graphs with some of the spectra obtained in [26] was established in [50] using graph angles. A list of the 65 known 4-regular connected integral graphs is given in [26].

There are exactly 150 connected integral graphs with up to 10 vertices (see [1]), and we reproduce here some of the details. The numbers i_n of connected integral graphs with n vertices are given for $n = 1, 2, \ldots, 10$ in the following table:

n:	1	2	3	4	5	6	7	8	9	10
i_n:	1	1	1	2	3	6	7	22	24	83

The connected integral graphs with up to five vertices are easily identified from the table of spectra of connected graphs with up to five vertices, given in [13]: they are K_1, K_2, K_3, K_4, C_4, K_5, $\overline{2K_1 \cup K_3}$ and $K_{1,4}$.

The six connected integral graphs on six vertices have identification numbers 1, 9, 51, 52, 106 and 109 in the table of connected graphs with six vertices given in [20]. They are K_6, the octahedron $\overline{3K_2}$, $C_3 + K_2$, $K_{3,3}$, C_6, and the unique tree with degree sequence (3, 3, 1, 1, 1, 1).

There are seven connected integral graphs on seven vertices, with identification numbers 4, 191, 235, 300, 710, 841 and 853 in the table of connected graphs on seven vertices given in [12]. The connected integral graphs on eight, nine and ten vertices are specified in [1] by their adjacency matrices, and also by pictures in the case of eight or nine vertices. Their spectra are provided in all cases.

There are no cospectral (non-isomorphic) connected integral graphs on fewer than eight vertices. There is just one triplet of connected integral graphs on eight vertices; there are three pairs on nine vertices; and ten pairs, one triplet, two quadruplets and one quintuplet on ten vertices. None of the graphs in these sets is regular, and in all cases the cospectral graphs can be distinguished by angles. We have already noted that the smallest cospectral graphs with the same angles have ten vertices (see [15]), and we find that there are no integral graphs among the 58 pairs of such cospectral graphs.

There are integral cospectral graphs with fewer than eight vertices, but then at least one of them is disconnected. These cases include the smallest pair of cospectral graphs, $K_{1,4}$ and $C_4 \cup K_1$, which have the common spectrum $(2, 0, 0, 0, -2)$. Also, the subdivision graph $S(K_{1,3})$ has as a cospectral mate the disconnected graph $C_6 \cup K_1$, with spectrum $(2, 1, 1, 0, -1, -1, -2)$.

Some cospectral integral graphs have the same main angles as well. This happens in several pairs: in one pair on nine vertices and in six pairs on ten vertices, as well as in a triplet on ten vertices. Note that, in each of the two aforementioned cospectral quadruplets, the main angles are not the same for all four graphs. The complements of cospectral graphs sharing the same main angles are cospectral. The complements of graphs from one cospectral quadruplet form two pairs of cospectral graphs, but not a cospectral quadruplet. Similarly, the complements of graphs from the other quadruplet do not provide a cospectral quadruplet; in this case, we obtain only a cospectral triplet. However, among the complements of graphs in all these sets of cospectral graphs, only the complements of two graphs on nine vertices are integral. Both complements are disconnected: the first has three components, two of them being isolated vertices, while the second has two components, one an isolated vertex. Cospectral integral graphs with cospectral integral complements were studied for the first time in [38], where an example with eleven vertices is given. It is interesting that we have the same situation with isolated vertices as above.

Several other graphs in our lists have integral complements; some of these complements are disconnected, but we can readily identify the pairs of connected complementary integral graphs. They include one pair on eight vertices, one on nine vertices, and four on ten vertices. The smallest such pair consists of the 6-vertex graphs C_6 and $C_3 + K_2$. All of these graphs are regular. Self-complementary graphs also feature in our lists: examples include a non-regular graph of order 8 and two regular graphs of order 9.

References

1. K. Balinska, D. Cvetković, M. Lepović and S. Simić, There are exactly 150 connected integral graphs up to 10 vertices, *Univ. Beograd, Publ. Elektrotehn. Fak., Ser. Mat.* **10** (1999), 95–105.
2. F. K. Bell, Characterizing line graphs by star complements, *Linear Alg. Appl.* **296** (1999), 15–25.
3. A. E. Brouwer, A. M. Cohen and A. Neumaier, *Distance-Regular Graphs,* Springer-Verlag, 1989.
4. F. C. Bussemaker and D. Cvetković, There are exactly 13 connected, cubic, integral graphs, *Univ. Beograd, Publ. Elektrotehn. Fak. Ser. Mat. Fiz.* **544–576** (1976), 43–48.
5. F. C. Bussemaker and A. Neumaier, Exceptional graphs with smallest eigenvalue -2 and related problems, *Math. of Computation* **59** (1992), 583–608.
6. P. J. Cameron, J. M. Goethals, J. J. Seidel and E. E. Shult, Line graphs, root systems, and elliptic geometry, *J. Algebra* **43** (1976), 305–327.
7. P. J. Cameron and J. H. van Lint, *Designs, Graphs, Codes and their Links,* Cambridge Univ. Press, 1991.
8. D. Cvetković, Cubic integral graphs, *Univ. Beograd, Publ. Elektrotehn. Fak. Ser. Mat. Fiz.* **498–541** (1975), 107–113.
9. D. Cvetković, On the 2-sum of three graphs: variations on the graph product disconnectedness theme, *Bull. Acad. Serbe Sci. Arts, Cl. Sci. Math. Natur., Sci. Math.* **118** (**24**) (1999), 107–117.
10. D. Cvetković, Characterizing properties of some graph invariants related to electron charges in the Hückel molecular orbital theory, *Proc. DIMACS Workshop on Discrete Mathematical Chemistry, March 1998, DIMACS Ser. Discrete Math. Theoret. Comp. Sci.* **51** (2000), 79–84.
11. D. Cvetković, On the reconstruction of the characteristic polynomial of a graph, *Discrete Math.* **212** (2000), 45–52.
12. D. Cvetković, M. Doob, I. Gutman and A. Torgašev, *Recent Results in the Theory of Graph Spectra,* North-Holland, 1988.
13. D. Cvetković, M. Doob and H. Sachs, *Spectra of Graphs* (3rd ed.), Johann Ambrosius Barth Verlag, 1995.
14. D. Cvetković, M. Doob and S. Simić, Generalized line graphs, *J. Graph Theory* **5** (1981), 385–399.
15. D. Cvetković and M. Lepović, Cospectral graphs with the same angles and with a minimal number of vertices, *Univ. Beograd, Publ. Elektrotehn. Fak., Ser. Mat.* **8** (1997), 88–102.
16. D. Cvetković and M. Lepović, Seeking counterexamples to the reconstruction conjecture for characteristic polynomials of graphs, and a positive result, *Bull. Acad. Serbe Sci. Arts, Cl. Sci. Math. Natur., Sci. Math.* **116** (**23**) (1998), 91–100.
17. D. Cvetković, M. Lepović, P. Rowlinson and S. Simić, A database of star complements of graphs, *Univ. Beograd, Publ. Elektrotehn. Fak., Ser. Mat.* **9** (1998), 103–112.
18. D. Cvetković, M. Lepović, P. Rowlinson and S. K. Simić, The maximal exceptional graphs, *J. Combin. Theory (B)* **86** (2002), 347–363.
19. D. Cvetković and R. Lučić, A new generalization of the p-sum of graphs, *Univ. Beograd, Publ. Elektrotehn. Fak., Ser. Mat. Fiz.* **302–319** (1970), 67–71.

20. D. Cvetković and M. Petrić, A table of connected graphs on six vertices, *Discrete Math.* **50** (1984), 37–49.
21. D. Cvetković, P. Rowlinson and S. K. Simić, *Eigenspaces of Graphs*, Cambridge Univ. Press, 1997.
22. D. Cvetković, P. Rowlinson and S. K. Simić, Some characterizations of graphs by star complements, *Linear Alg. Appl.* **301** (1999), 81–97.
23. D. Cvetković, P. Rowlinson and S. K. Simić, Graphs with least eigenvalue −2: the star complement technique, *J. Alg. Combin.* **14** (2001), 5–16.
24. D. Cvetković, P. Rowlinson and S. K. Simić, Constructions of the maximal exceptional graphs with largest degree less than 28, *Bull. Acad. Serbe Sci. Arts, Cl. Sci. Math. Natur., Sci. Math.* **122 (26)** (2001), 115–131.
25. D. Cvetković and S. Simić, Non-complete extended p-sum of graphs, graph angles and star partitions, *Publ. Inst. Math. (Beograd)* **53(67)** (1993), 4–16.
26. D. Cvetković, S. Simić and D. Stevanović, 4-regular integral graphs, *Univ. Beograd, Publ. Elektrotehn. Fak., Ser. Mat.* **9** (1998), 89–102.
27. M. Doob and D. Cvetković, On spectral characterizations and embeddings of graphs, *Linear Alg. Appl.* **27** (1979), 17–26.
28. J. Edmonds, Matroid Intersection, *Discrete Optimization 1* (ed. P. L. Hammer *et al.*) Annals of Discrete Mathematics **4**, North-Holland (1979), 39–49.
29. M. N. Ellingham, Basic subgraphs and graph spectra, *Austral. J. Combin.* **8** (1993), 247–265.
30. C. D. Godsil and B. D. McKay, Feasibility conditions for the existence of walk-regular graphs, *Linear Algebra Appl.* **30** (1980), 51–61.
31. I. Gutman and D. Cvetković, The reconstruction problem for characteristic polynomials of graphs, *Univ. Beograd, Publ. Elektrotehn. Fak. Ser. Mat. Fiz.* **498–541** (1975), 45–48.
32. F. Harary and A. J. Schwenk, Which graphs have integral spectra?, *Graphs and Combinatorics, Proc. Conf. Graph Theory and Combinatorics, George Washington Univ., June 1973* (eds. R. Bari and F. Harary), Springer-Verlag (1974), 45–51.
33. P. S. Jackson, *Star sets and related aspects of algebraic graph theory,* PhD thesis, University of Stirling, 1999.
34. P. S. Jackson and P. Rowlinson, On graphs with complete bipartite star complements, *Linear Alg. Appl.* **298** (1999), 9–20.
35. A. Kotlov and L. Lovász, The rank and size of graphs, *J. Graph Theory* **23** (1996), 185–189.
36. M. Lepović, Some statistical data on graphs with 10 vertices, *Univ. Beograd, Publ. Elektrotehn. Fak., Ser. Mat.* **9** (1998), 79–88.
37. A. Neumaier, Derived eigenvalues of symmetric matrices, with applications to distance geometry, *Linear Alg. Appl.* **134** (1990), 107–120.
38. Z. Radosavljević and S. Simić, Computer aided search for all graphs such that graph and its complement have its spectrum bounded from below by −2, *Ars Combin.* **24A** (1987), 21–27.
39. P. Rowlinson, The characteristic polynomials of modified graphs, *Discr. Appl. Math.* **67** (1996), 209–219.
40. P. Rowlinson, On graphs with multiple eigenvalues, *Linear Alg. Appl.* **283** (1998), 75–85.

41. P. Rowlinson, Star sets and star complements in finite graphs: a spectral construction technique, *Proc. DIMACS Workshop on Discrete Mathematical Chemistry, March 1998, DIMACS Ser. Discrete Math. Theoret. Comp. Sci.* **51** (2000), 323–332.
42. P. Rowlinson, Star sets in regular graphs, *J. Combin. Math. Combin. Computing* **34** (2000), 3–22.
43. P. Rowlinson and F. K. Bell, Graph eigenspaces of small codimension, *Discrete Math.* **220** (2000), 271–278.
44. A. J. Schwenk, Exactly thirteen connected cubic graphs have integral spectra, *Theory and Applications of Graphs, Proc. Internat. Conf. Western Michigan Univ., Kalamazoo, Mich., May 1976* (eds. Y. Alavi and D. Lick), Springer-Verlag (1978), 516–533.
45. A. J. Schwenk, Spectral reconstruction problems, *Ann. New York Acad. Sci.* **328** (1979), 183–189.
46. A. J. Schwenk and R. J. Wilson, On the eigenvalues of a graph, *Selected Topics in Graph Theory* (eds. L. W. Beineke and R. J. Wilson), Academic Press (1978), 307–336.
47. J. J. Seidel, Eutactic stars, *Combinatorics* (eds. A. Hajnal and V. T. Sós), North-Holland (1978), 983–989.
48. S. Simić and Z. Radosavljević, The nonregular, nonbipartite, integral graphs with maximum degree four, *J. Combin. Inf. Syst. Sci.* **20** (1995), 9–26.
49. D. Stevanović, When is NEPS of graphs connected?, *Linear Alg. Appl.* **301** (1999), 137–143.
50. D. Stevanović: Nonexistence of some 4-regular integral graphs, *Univ. Beograd, Publ. Elektrotehn. Fak., Ser. Mat.* **10** (1999), 81–86.
51. D. Stevanović, When can the components of NEPS of connected bipartite graphs be almost cospectral?, *Linear Alg. Appl.* **311** (2000), 35–44.
52. D. Stevanović, 4-regular integral graphs avoiding ± 3 in the spectrum, *Univ. Beograd, Publ. Elektrotehn. Fak., Ser. Mat.* **14** (2003).
53. W. T. Tutte, All the king's horses: a guide to reconstruction, *Graph Theory and Related Topics* (eds. J. A. Bondy and U. S. R. Murty), Academic Press (1979), 15–33.

4
Graph Laplacians

BOJAN MOHAR

Eigenvalues of Laplace matrices of graphs are related to various combinatorial properties of graphs. They can be used to extract information about some parameters that are hard to compute or estimate, most notably those related to expansion and vertex partitions. The smallest and largest eigenvalue can be expressed as solutions to a quadratic optimization problem. It turns out that the right generalized setting for this is semi-definite programming, where duality theory leads to powerful applications, one of which is given in Section 5. A more general setting of weighted graphs is presented, which brings us also to the study of simple random walks whose transition matrix can be expressed via the related Laplacian.

1. Introduction

Eigenvalues of graphs have been extensively studied since the early developments of graph theory in the 1960s (see, for example, Chapter 1, [4], [11], [12] or [23]). Most of the early works considered eigenvalues of adjacency matrices of graphs. New developments in the 1980s made it clear that eigenvalues and eigenvectors of the related Laplace matrices of graphs enter the theory in several applications more naturally than eigenvalues of adjacency matrices (see [8], [37] and [40]), and therefore deserve their own study.

Many properties of eigenvalues of graphs can be expressed in terms of simple graph invariants, such as vertex degrees or the number of short cycles. Such results are most interesting from the point of view of linear algebra. However, from the viewpoint of applications in combinatorics, it is more interesting to study relations between eigenvalues and 'hard' graph invariants. Such an approach was very successful in the study of distance-regular graphs; we refer to the excellent monographs [4] and [23]. One of the most fascinating aspects of such applications is that eigenvalues appear as a tool to prove results that appear to have nothing to do with the eigenvalues themselves.

Applications of eigenvalue methods in combinatorics, graph theory and combinatorial optimization have a long history. For example, eigenvalue bounds on the chromatic number were formulated by Wilf [48] and Hoffman [28] in the 1960s. Another early application, in the area of graph partition, is due to Fiedler [17] and Donath and Hoffman [14]. An important use of eigenvalues is Lovász's notion of the ϑ-function [31].

The next important result was the use of eigenvalues in the construction of superconcentrators and expanders by Alon and Milman [1]. Isoperimetric properties of graphs and their eigenvalues play a crucial role in the design of various randomized algorithms. These applications are based on the so-called 'rapidly mixing Markov chains' (see Section 8).

There is an increasing interest in the application of eigenvalues to combinatorial optimization problems. For example, an eigenvalue approach was used in the study of the quadratic assignment problem and general graph partition problems [43], the max-cut problem [13], and labelling problems ([25] and [29]). Spectral partitioning, which is based on eigenvectors of Laplace eigenvalues of graphs, has proved to be a successful heuristic approach in the design of partition algorithms ([25] and [27]), in parallel computation [45], clustering [24], ranking ([25] and [29]), and in graph drawing [19]. We refer to [40] for additional applications.

Laplace matrices of graphs are closely related to the Laplacian, the second order differential operator $\Delta(f) = -\text{div}(\text{grad}\, f)$. This relation yields an important bilateral link between the spectral geometry of Riemannian manifolds and graph theory, and makes it possible to use results about graphs in the study of Laplacians on manifolds and, conversely, to transfer results about Laplacians on manifolds to graphs. Each of these directions has given rise to exciting new discoveries (see Buser [7], Brooks [6] and Chung [8]). An interesting approach, using Laplace matrices and discrete Schrödinger operators to obtain certain minor monotone graph invariants, has been developed by Colin de Verdière. His remarkable results, which connect with several central areas of mathematics, are nicely presented in [9].

In this chapter, only basic information on the Laplace matrices of graphs is presented. For further and more specialized results, there are several books and survey papers, such as [4], [8], [9], [11], [33] and [40].

2. The Laplacian of a graph

Let $G = (V, E)$ be an undirected finite graph with n vertices and m edges. We also consider *weighted graphs* with a *weight function* that assigns a non-negative real weight a_{vw} to each pair v, w of vertices. We require that the weights satisfy the following properties:

- $a_{vw} > 0$ if $vw \in E$, and $a_{vw} = 0$ if $vw \notin E$;
- $a_{vw} = a_{wv}$, for all $v, w \in V$.

Unweighted graphs can be viewed as weighted graphs in which a_{vw} is the number of edges between v and w.

In the unweighted case, the degree $\deg v$ of a vertex $v \in V$ is the number of edges of G incident with v. In the weighted case, it is defined by

$$\deg v = \sum_{w \in V} a_{vw}.$$

Recall that the maximum and minimum vertex degrees in G are denoted by $\Delta(G)$ and $\delta(G)$, respectively.

Given a graph G, its *(weighted) adjacency matrix* $\mathbf{A}(G) = (a_{vw})$ is the $n \times n$ matrix, with rows and columns indexed by V, whose entries are the edge-weights. The *degree matrix* $\mathbf{D}(G) = \mathrm{diag}(\deg v : v \in V)$ *is the diagonal matrix indexed by* V *with the vertex-degrees on the diagonal. The difference*

$$\mathbf{L}(G) = \mathbf{D}(G) - \mathbf{A}(G) \tag{1}$$

is the Laplace matrix (or *Laplacian*) of G.

One of the first applications of Laplace matrices of graphs is in the statement of the well-known *matrix-tree theorem (see [4, Ch. 6]).*

Theorem 2.1 (Matrix-tree theorem) *Let G be a graph and let $\mathbf{L} = \mathbf{L}(G)$ be its Laplace matrix. If T is a spanning tree of G, let $a(T)$ be the product of the edge-weights of T. For any vertices v and w, the (weighted) number of spanning trees of G is*

$$\tau(G) = \sum \{a(T) : T \text{ a spanning tree of } G\} = |\det \mathbf{L}_w^v|,$$

where $\mathbf{L}_w^v$ is the submatrix of $\mathbf{L}$ obtained by deleting the column corresponding to v and the row corresponding to w.

The most common proofs of Theorem 2.1 use the related *oriented incidence matrix* $\mathbf{Q}$ of G, which is defined as follows. Orient arbitrarily each edge of G. Then the matrix $\mathbf{Q}$ is the $n \times m$ matrix with rows and columns indexed by V and E, whose (v, e)-entry is $-\sqrt{a_{vw}}$ if v is the initial vertex of $e = vw, \sqrt{a_{wv}}$ if v is the terminal vertex of $e = wv$, and 0 if v and e are not incident. If e is a loop, then the (v, e)-entry is 0, irrespective of its weight.

Theorem 2.2 *Let G be a graph, and let $\mathbf{Q}$ be an oriented incidence matrix with respect to some orientation of its edges. Then*

$$\mathbf{L}(G) = \mathbf{Q}\mathbf{Q}^T.$$

Theorem 2.2 shows that the product $\mathbf{Q}\mathbf{Q}^T$ is independent of the chosen orientation of the edges of G.

Let $\mathbf{R}^V = \{\mathbf{f}\colon V \to \mathbf{R}\}$ be the set of functions from V to $\mathbf{R}$. If $\mathbf{f} \in \mathbf{R}^V$ and $v \in V$, denote by $\mathbf{f}_v$ the value of $\mathbf{f}$ at the vertex v. With the usual operations of sum and scalar multiplication, $\mathbf{R}^V$ becomes a real vector space of dimension n, endowed with the inner product

$$\langle \mathbf{f}, \mathbf{g} \rangle = \sum_{v \in V} \mathbf{f}_v \mathbf{g}_v.$$

The corresponding norm in $\mathbf{R}^V$ is

$$\|\mathbf{f}\| = \langle \mathbf{f}, \mathbf{f} \rangle^{1/2} = \left(\sum_{v \in V} \mathbf{f}_v^2 \right)^{1/2}$$

The matrix $\mathbf{L}$ (and other matrices indexed by the vertices of G) acts on $\mathbf{R}^V$ as a linear operator. Its action is determined by the rule of matrix-vector multiplication, where $\mathbf{g} = \mathbf{L}\mathbf{f}$ is the function defined by the formula

$$\mathbf{g}_v = (\mathbf{L}\mathbf{f})_v = \sum_{w \in V} \mathbf{L}_{vw} \mathbf{f}_w = \sum_{w \in V} a_{vw}(\mathbf{f}_v - \mathbf{f}_w) \quad (v \in V).$$

There is a natural quadratic form associated with $\mathbf{L}$.

Theorem 2.3

$$\langle \mathbf{f}, \mathbf{L}\mathbf{f} \rangle = \sum_{vw \in E} a_{vw}(\mathbf{f}_v - \mathbf{f}_w)^2. \tag{2}$$

Proof Since $\mathbf{L} = \mathbf{Q}\mathbf{Q}^T$, the definition of $\mathbf{Q}$ implies that

$$\langle \mathbf{f}, \mathbf{L}\mathbf{f} \rangle = \langle \mathbf{f}, \mathbf{Q}\mathbf{Q}^T \mathbf{f} \rangle = \langle \mathbf{Q}^T \mathbf{f}, \mathbf{Q}^T \mathbf{f} \rangle = \sum_{e \in E} (\mathbf{Q}^T \mathbf{f})_e^2 = \sum_{vw \in E} a_{vw}(\mathbf{f}_v - \mathbf{f}_w)^2.$$

■

For graphs without isolated vertices, there is another common definition of a graph Laplacian. Since it is related to Markov chains and random walks (see Section 8), it is also known as the *transition Laplace matrix*,

$$\mathbf{L}'(G) = \mathbf{D}^{-1/2}\mathbf{L}(G)\mathbf{D}^{-1/2},$$

where $\mathbf{D}$ is the degree matrix. In particular, we have

$$\mathbf{L}'(G)_{vw} = \begin{cases} 1 - \dfrac{a_{vv}}{\deg(v)}, & \text{if } v = w \\ -\dfrac{a_{vw}}{\sqrt{\deg(v)\deg(w)}}, & \text{if } vw \in E(G) \\ 0, & \text{otherwise.} \end{cases} \tag{3}$$

Observe that $\mathbf{L}'(G)$ is symmetric, and that for a d-regular loopless graph,

$$\mathbf{L}(G) = d \cdot \mathbf{L}'(G).$$

Notice also that the Laplace matrix $\mathbf{L}(G)$ does not change if we add loops to G, while the transition Laplace matrix $\mathbf{L}'(G)$ is changed by the addition of loops. Transition Laplace matrices are extensively treated in [8].

3. Laplace eigenvalues

By definition, $\mathbf{L}(G)$ is a real symmetric matrix, and equation (2) (together with the non-negativity of edge-weights) implies that it is positive semi-definite. It therefore has n non-negative real eigenvalues $\lambda_i = \lambda_i(G)$, which in this chapter we take in non-decreasing order

$$0 \leq \lambda_1 \leq \lambda_2 \leq \cdots \leq \lambda_n.$$

These eigenvalues are called the *Laplace eigenvalues* of the graph G. It is easy to see that 0 is always an eigenvalue of $\mathbf{L}(G)$, and that $\mathbf{j} = (1, 1, \ldots, 1)^T$ is a corresponding eigenvector.

Theorem 3.1 *The multiplicity of* 0 *as an eigenvalue of* $\mathbf{L}(G)$ *is equal to the number of connected components of G.*

This result implies that $\lambda_1(G) = 0$ is a simple eigenvalue of $\mathbf{L}(G)$ if and only if the graph G is connected. It is also obvious, after writing $\mathbf{L}(G)$ in block diagonal form, that the Laplace eigenvalues of G are the union of the eigenvalues of its connected components.

Since the sum of all the eigenvalues equals the trace of the matrix, we have

$$\sum_{i=1}^{n} \lambda_i(G) = \sum_{v \in V} \deg(v) = 2|E(G)|, \tag{4}$$

where $|E(G)|$ should be replaced by the sum of all edge-weights in the case of weighted graphs.

If $\mathbf{M}$ is a matrix with real eigenvalues, we use the notation $\lambda_i(\mathbf{M})$ to denote the ith smallest eigenvalue of $\mathbf{M}$ (respecting the multiplicities). We use the symbol $\lambda_{\max}(\mathbf{M})$ to denote the maximum eigenvalue of $\mathbf{M}$, and sometimes write $\lambda_{\min}(\mathbf{M})$ instead of $\lambda_1(\mathbf{M})$.

Let G be a (weighted) *k-regular graph*. Then equation (1) implies that λ is an eigenvalue of $\mathbf{L}(G)$ if and only if $k - \lambda$ is an eigenvalue of the weighted adjacency matrix $\mathbf{A}(G)$. More precisely,

$$\lambda_i(\mathbf{L}(G)) = k - \lambda_{n-i+1}(\mathbf{A}(G)) \quad (i = 1, 2, \ldots, n).$$

This result enables us to use known results about the eigenvalues of the adjacency matrix of a regular graph in the study of its Laplace eigenvalues. For example, the eigenvalues of the adjacency matrix of the complete graph K_n are $n - 1$ and -1 (the latter with multiplicity $n - 1$), and therefore, $\lambda_1(K_n) = 0$, and $\lambda_i(K_n) = n$ for $2 \le i \le n$. Similarly, the Laplace eigenvalues of the n-cycle C_n are precisely the numbers

$$\nu_k = 2 - 2\cos\left(\frac{2k\pi}{n}\right) \quad (k = 1, 2, \ldots, n). \tag{5}$$

If G is a simple unweighted graph and $\overline{G}$ is its complement, then

$$\mathbf{L}(G) + \mathbf{L}(\overline{G}) = n\mathbf{I} - \mathbf{J}, \tag{6}$$

where $\mathbf{J}$ is the all-1 matrix. The same relation holds in the weighted case if we define the edge-weights $\overline{a}_{vw}$ of $\overline{G}$ to be $1 - a_{vw}$ ($v, w \in V$, $v \ne w$). In order to have non-negative weights in $\overline{G}$, we need $0 \le a_{vw} \le 1$ for $v, w \in V$.

Let $\mathbf{f}^1, \mathbf{f}^2, \ldots, \mathbf{f}^n$ be an orthogonal system of eigenvectors of $\mathbf{L}(G)$ such that $\mathbf{f}^1 = \mathbf{j}$ and $\mathbf{L}(G)\mathbf{f}^i = \lambda_i \mathbf{f}^i$, for $i = 1, 2, \ldots, n$. By (6), we get $\mathbf{L}(\overline{G})\mathbf{f}^1 = 0$ and $\mathbf{L}(\overline{G})\mathbf{f}^i = (n - \lambda_i)\mathbf{f}^i$, for $2 \le i \le n$. This proves the following result.

Theorem 3.2 *Suppose that the edge-weights of a graph G satisfy $0 \le a_{vw} \le 1$, and let $\overline{G}$ be the graph with edge-weights $\overline{a}_{vw} = 1 - a_{vw}$. Then $\lambda_1(\overline{G}) = 0$, and $\lambda_i(\overline{G}) = n - \lambda_{n-i+2}(G)$ for $2 \le i \le n$.*

For example, if $G = K_{r,s}$, then $\overline{G} = K_r \cup K_s$. So, $\overline{G}$ has eigenvalues 0 (twice), r ($r - 1$ times), and s ($s - 1$ times). Hence, the Laplace eigenvalues of $K_{r,s}$ are 0, s ($r - 1$ times), r ($s - 1$ times), and $r + s$.

Recall that the *Cartesian product* $G \,\square\, H$ of graphs G and H has vertex-set $V(G \,\square\, H) = V(G) \times V(H)$, where (v_1, v_2) is adjacent to (w_1, w_2) if and only if $v_1 = w_1$ and $v_2 w_2 \in E(H)$, or $v_2 = w_2$ and $v_1 w_1 \in E(G)$. There is a simple description of the Laplace eigenvalues of $G \,\square\, H$ in terms of the Laplace eigenvalues of G and H.

Theorem 3.3 *The Laplace eigenvalues of the Cartesian product $G \,\square\, H$ are precisely the numbers*

$$\lambda_i(G) + \lambda_j(H),$$

for $i = 1, 2, \ldots, |V(G)|$ and $j = 1, 2, \ldots, |V(H)|$.

The proof of this result relies on the fact that the Laplace matrix is

$$\mathbf{L}(G \,\square\, H) = \mathbf{L}(G) \otimes \mathbf{I}_{V(H)} + \mathbf{I}_{V(G)} \otimes \mathbf{L}(H),$$

where $\otimes$ denotes the Kronecker product of matrices (see Introduction).

Theorem 3.3 holds also for weighted graphs G and H, if the weight of an edge of $G \,\square\, H$ joining the vertices (v_1, v_2) and (w_1, w_2) is taken to be equal to the weight of the edge $v_1 w_1$ of G if $v_2 = w_2$, and equal to the weight of $v_2 w_2$ in H if $v_1 = w_1$. As a consequence, we obtain

$$\lambda_2(G \,\square\, H) = \min\{\lambda_2(G), \lambda_2(H)\} \text{ and } \lambda_{\max}(G \,\square\, H) = \lambda_{\max}(G) + \lambda_{\max}(H). \quad (7)$$

Theorem 3.3 can be used to determine the Laplace spectrum of several well-known families of graphs. For example, the *d-dimensional cube Q_d* is the Cartesian product of d copies of K_2. Since the Laplace eigenvalues of K_2 are 0 and 2, the Laplace spectrum of Q_d consists of the numbers $0, 2, 4, \ldots, 2d$. The multiplicity of $2k$ in the spectrum of Q_d is $\binom{d}{k}$.

Similarly, the Laplace eigenvalues of the path P_k are (see [2])

$$\ell_i^{(k)} = 4 \sin^2 \left(\frac{\pi i}{2k} \right) \qquad (i = 0, 1, \ldots, k-1).$$

Therefore, the *grid graph $P_m \,\square\, P_n$* has Laplace eigenvalues

$$\lambda_{i,j} = \ell_i^{(m)} + \ell_j^{(n)} = 4 \sin^2 \left(\frac{\pi i}{2m} \right) + 4 \sin^2 \left(\frac{\pi j}{2n} \right).$$

It is natural to ask to what extent the Laplace eigenvalues determine the graph. Several operations are known that change the graph but not its eigenvalues. This gives rise to examples of families of *cospectral graphs* (see, for example, [34]).

Bounding the Laplace eigenvalues

There are various useful min-max formulas for the eigenvalues of a symmetric matrix. If $\mathbf{M}$ is a real symmetric matrix indexed by V, then

$$\begin{aligned}\lambda_1(\mathbf{M}) &= \min\left\{\frac{\langle \mathbf{Mf},\mathbf{f}\rangle}{\langle \mathbf{f},\mathbf{f}\rangle} : \mathbf{0}\neq \mathbf{f}\in \mathbf{R}^V\right\} \\ &= \min\{\langle \mathbf{Mf},\mathbf{f}\rangle : \mathbf{f}\in\mathbf{R}^V,\ \|\mathbf{f}\|=1\},\end{aligned} \tag{8}$$

and similarly

$$\lambda_{\max}(\mathbf{M}) = \max\{\langle \mathbf{Mf},\mathbf{f}\rangle : \mathbf{f}\in\mathbf{R}^V,\ \|\mathbf{f}\|=1\}. \tag{9}$$

The *Rayleigh characterization* (8) has a generalization, the min-max characterization of the kth smallest eigenvalue $\lambda_k(\mathbf{M})$, known also as the *Courant-Fisher formula*:

$$\lambda_k(\mathbf{M}) = \min_U \max_{\mathbf{f}}\{\langle \mathbf{Mf},\mathbf{f}\rangle : \|\mathbf{f}\|=1,\ \mathbf{f}\in U\}, \tag{10}$$

where the minimum is taken over all k-dimensional subspaces U of $\mathbf{R}^V$. Another way of expressing (10) is

$$\lambda_k(\mathbf{M}) = \min\{\langle \mathbf{Mf},\mathbf{f}\rangle : \|\mathbf{f}\|=1,\ \mathbf{f}\perp\mathbf{f}^i,\ 1\le i<k\}, \tag{11}$$

where $\mathbf{f}^1,\mathbf{f}^2,\ldots,\mathbf{f}^{k-1}$ are pairwise orthogonal eigenvectors of $\lambda_1,\lambda_2,\ldots,\lambda_{k-1}$, respectively.

Among the Laplace eigenvalues of a connected graph G, the most important are the extreme non-zero eigenvalues: the second smallest eigenvalue $\lambda_2(G)$ and the largest eigenvalue $\lambda_{\max}(G)$. Theorem 3.2 shows that $\lambda_2(G) = n - \lambda_{\max}(\overline{G})$, and it is thus not surprising that the importance of each of these eigenvalues implies that of the other. For a (weighted) graph G with Laplace matrix $\mathbf{L}$, (11) implies that

$$\lambda_2(G) = \min\{\langle \mathbf{Lf},\mathbf{f}\rangle : \|\mathbf{f}\|=1,\ \mathbf{f}\perp\mathbf{j}\}, \tag{12}$$

since $\mathbf{j}$ is an eigenvector corresponding to $\lambda_1(G)$: notice that $\mathbf{f}$ is orthogonal to $\mathbf{j}$ if and only if the sum of its coordinates is 0.

Expression (12) can be used to get combinatorial upper bounds on $\lambda_2(G)$. For example:

Lemma 3.4 *For any non-adjacent vertices s and t in G,*

$$\lambda_2(G) \le \tfrac{1}{2}(\deg s + \deg t).$$

Proof Let $\mathbf{f} \in \mathbf{R}^V$ be defined by $\mathbf{f}_v = 1$ if $v = s$, $\mathbf{f}_v = -1$ if $v = t$, and $\mathbf{f}_v = 0$ otherwise. Since $\mathbf{f} \perp \mathbf{j}$, (12) yields

$$\lambda_2(G) \le \frac{\langle \mathbf{L}(G)\mathbf{f}, \mathbf{f} \rangle}{\langle \mathbf{f}, \mathbf{f} \rangle} = \frac{\sum_{vw \in E} a_{vw}(\mathbf{f}_v - \mathbf{f}_w)^2}{\sum_{v \in V} \mathbf{f}_v^2} = \frac{\deg s + \deg t}{2}. \qquad \blacksquare$$

The proof of Lemma 3.4 illustrates a basic technique for exploiting expressions such as (12). By inserting an appropriate function into the expression, we can give the right-hand side a meaningful combinatorial interpretation, and this enables us to relate combinatorial properties to Laplace eigenvalues. To overcome the orthogonality restriction on $\mathbf{f}$, Fiedler [18] transformed (12) to a more suitable expression that does not require $\mathbf{f}$ to be orthogonal to $\mathbf{j}$.

Theorem 3.5 *For any graph G of order n,*

$$\lambda_2(G) = 2n \cdot \min \left\{ \frac{\sum_{vw \in E} a_{vw}(\mathbf{f}_v - \mathbf{f}_w)^2}{\sum_{v \in V} \sum_{w \in V} (\mathbf{f}_v - \mathbf{f}_w)^2} : \mathbf{f} \ne c\mathbf{j}, \text{ for } c \in \mathbf{R} \right\} \tag{13}$$

and

$$\lambda_{\max}(G) = 2n \cdot \max \left\{ \frac{\sum_{vw \in E} a_{vw}(\mathbf{f}_v - \mathbf{f}_w)^2}{\sum_{v \in V} \sum_{w \in V} (\mathbf{f}_v - \mathbf{f}_w)^2} : \mathbf{f} \ne c\mathbf{j}, \text{ for } c \in \mathbf{R} \right\}. \tag{14}$$

The eigenvalues $\lambda_2(G)$ and $\lambda_{\max}(G)$ can be bounded in terms of the maximum and minimum degrees of G.

Theorem 3.6 *For any graph G of order n,*

$$\lambda_2(G) \le \frac{n}{n-1}\,\delta(G) \quad \text{and} \quad \frac{n}{n-1}\,\Delta(G) \le \lambda_{\max}(G) \le 2\Delta(G).$$

If G is an unweighted graph, the last inequality of Theorem 3.6 can be strengthened to $\lambda_{\max}(G) \le \max\{\deg v + \deg w : vw \in E\}$. If G is connected, then the equality holds if and only if G is bipartite semi-regular (see [2]). Theorem 3.2 implies that $\lambda_{\max}(G) \le n$, with equality if and only if the complement of G is disconnected.

Let G be a (weighted) graph, and let $G' = G + e$ be the graph obtained from G by adding an edge e to G (possibly increasing the weight of an existing edge). Then $\mathbf{L}(G')$ and $\mathbf{L}(G)$ differ by a positive semi-definite matrix of rank 1. It follows by the Courant-Weyl inequalities (see, for example, [11, Thm. 2.1]) that the following is true.

Theorem 3.7 *The eigenvalues of G and $G' = G + e$ interlace:*

$$0 = \lambda_1(G) = \lambda_1(G') \leq \lambda_2(G) \leq \lambda_2(G') \leq \lambda_3(G) \leq \cdots \leq \lambda_n(G) \leq \lambda_n(G').$$

Observe that $\sum_{i=1}^{n}(\lambda_i(G') - \lambda_i(G)) = 2$, by (4), so that at least one inequality $\lambda_i(G) \leq \lambda_i(G')$ is strict. By inserting more than one edge we may lose the interlacing of the eigenvalues.

Theorem 3.8 *Suppose that G_1 and G_2 are graphs with the same vertex-set. Let $G = G_1 \cup G_2$ be the graph whose edge-weights are the sums of edge-weights of G_1 and G_2. Then*

- $\lambda_2(G) \geq \lambda_2(G_1) + \lambda_2(G_2)$;
- $\max\{\lambda_{\max}(G_1), \lambda_{\max}(G_2)\} \leq \lambda_{\max}(G) \leq \lambda_{\max}(G_1) + \lambda_{\max}(G_2)$.

Eigenvalues of the transition Laplacian

The eigenvalues $\lambda_i' = \lambda_i(\mathbf{L}'(G))$ of the transition Laplace matrix behave similarly to the eigenvalues of $\mathbf{L}(G)$: they are all real and non-negative, $\lambda_1' = 0$, and $\lambda_2' > 0$ if and only if G is connected.

Theorem 3.9 *Let G be a graph of order n with p isolated vertices. Then*

- $\sum_{i=1}^{n} \lambda_i'(G) = n - p$;
- *if G is not the complete graph, then $\lambda_2'(G) \leq 1$;*
- $\lambda_n'(G) \leq 2$, *and equality holds if and only if G has a non-trivial bipartite component;*
- *if G is bipartite, then $\lambda_i'(G) = 2 - \lambda_{n-i+1}'(G)$ for $i = 1, 2, \ldots, n$.*

We refer to [8] for more details and further results.

4. Eigenvalues and vertex partitions of graphs

We begin this section with some notation. For each subset $S \subseteq V(G)$, let $\bar{S} = V(G) - S$ denote the complement of S in $V(G)$. Given sets of vertices $A, B \subseteq V(G)$, let $E(A, B)$ be the set of edges of G with one end in A and the other in B. We also let

$$e(A, B) = \sum_{v \in A} \sum_{w \in B} a_{vw}$$

denote the sum of the weights of the edges in $E(A, B)$, with all edges in $G[A \cap B]$ counted twice. Note that, for an unweighted graph, $e(S, \bar{S})$ counts the number of edges in the cut $E(S, \bar{S})$.

The partition problems discussed here are mainly concerned with finding an appropriate subset $S \subseteq V(G)$ for which the edge-cut $E(S, \bar{S})$ satisfies some specific extremal property. In particular, we focus our attention on the weight of a maximum cut, the bipartition width and the isoperimetric number. All of these problems are NP-hard, so that non-trivial bounds are desirable, and potentially very important.

The following lemma relates the weight of an edge-cut to the eigenvalues $\lambda_2(G)$ and $\lambda_{\max}(G)$.

Lemma 4.1 *Let G be a (weighted) graph of order n, and let $S \subseteq V(G)$ with $|S| = s$. Then*

$$\lambda_2(G)\,\frac{s(n-s)}{n} \le e(S, \bar{S}) \le \lambda_{\max}(G)\,\frac{s(n-s)}{n}.$$

Proof Let $\mathbf{f} \in \mathbf{R}^V$ be the characteristic function of S: $\mathbf{f}_v = 1$ if $v \in S$, and $\mathbf{f}_v = 0$ otherwise. Then

$$\sum_{v\in V}\sum_{w\in V}(\mathbf{f}_v - \mathbf{f}_w)^2 = 2s(n-s)$$

and

$$\sum_{vw\in E} a_{vw}(\mathbf{f}_v - \mathbf{f}_w)^2 = e(S, \bar{S}).$$

If $S \neq \emptyset$ and $S \neq V(G)$, then (13) implies that

$$\lambda_2(G) \le 2n \cdot \frac{e(S, \bar{S})}{2s(n-s)}.$$

This gives the lower bound, which obviously holds also for $S = \emptyset$ and $S = V(G)$. The upper bound is proved analogously, using (14). ∎

It is an immediate, and also an important, consequence of Lemma 4.1 that in a graph for which all non-trivial Laplace eigenvalues are close together (that is, $\lambda_{\max} - \lambda_2$ is small), the weights of all the edge-cuts $E(S, \bar{S})$ corresponding to vertex-sets S of the same cardinality are approximately the same. In particular, this property holds in random graphs. It is therefore not surprising that many algorithms dealing with edge-cuts perform well on randomly chosen graphs.

Lemma 4.1 also shows that graphs with large λ_2 have fast growth rate. This implies, in particular, that the diameter can be bounded as a function of $\lambda_2(G)$ (see, for example, [8] and [40] and the references therein).

The bipartition width

Loosely speaking, the bipartition width of a graph is the minimum number of edges in a cut that partitions the vertices into two sets of nearly equal size. More formally, for a graph G of order n, the *bipartition width* $\mathrm{bw}(G)$ is defined as

$$\mathrm{bw}(G) = \min\left\{e(S, \bar{S}) : S \subseteq V(G), |S| = \left\lfloor \tfrac{1}{2}n \right\rfloor\right\}.$$

It is known that, even for unweighted simple graphs, the problem of determining $\mathrm{bw}(G)$ is NP-hard (see, for example, [21, p. 210]).

Since all the sets over which the minimum is taken have size $\lfloor \frac{1}{2}n \rfloor$, Lemma 4.1 gives the following lower bound for $\mathrm{bw}(G)$.

Corollary 4.2 *Let G be a weighted graph of order n. If n is even, then*

$$\mathrm{bw}(G) \geq \tfrac{1}{4}n \cdot \lambda_2(G).$$

If n is odd, then

$$\mathrm{bw}(G) \geq \frac{n^2 - 1}{4n} \cdot \lambda_2(G).$$

We can further improve the bounds of Corollary 4.2 by introducing a correction function. A function $\mathbf{c} \in \mathbf{R}^V$ is called a *correction function* if $\mathbf{c} \perp \mathbf{j}$. The following bound was proved by Boppana [5].

Theorem 4.3 *Let G be a weighted graph of even order n. Then*

$$\mathrm{bw}(G) \geq \tfrac{1}{4}n \cdot \max_{\mathbf{c}} \min_{\mathbf{f}} \frac{\langle(\mathbf{L}(G) + \mathrm{diag}(\mathbf{c}))\mathbf{f}, \mathbf{f}\rangle}{\langle \mathbf{f}, \mathbf{f}\rangle},$$

where the maximum is taken over all correction functions $\mathbf{c} \in \mathbf{R}^V$ and the minimum is taken over all non-zero functions $\mathbf{f} \in \mathbf{R}^V$ with $\mathbf{f} \perp \mathbf{j}$.

Proof Let $S \subseteq V(G)$ be a set of cardinality $\frac{1}{2}n$ with $e(S, \bar{S}) = \mathrm{bw}(G)$, and let $\mathbf{g} \in \mathbf{R}^V$ be its signed characteristic function, defined by $\mathbf{g}_v = 1$ if $v \in S$ and $\mathbf{g}_v = -1$ if $v \in \bar{S}$. Since $|S| = |\bar{S}|$, we have $\mathbf{g} \perp \mathbf{j}$.

Take an arbitrary correction function $\mathbf{c} \in \mathbf{R}^V$. Since $\mathbf{c} \perp \mathbf{j}$, we have

$$\langle \mathrm{diag}(\mathbf{c})\mathbf{g}, \mathbf{g}\rangle = \sum_{v \in V} c_v \mathbf{g}_v^2 = \sum_{v \in V} c_v = 0. \tag{15}$$

Using (15), and applying (2), we deduce that

$$\begin{aligned}\frac{\langle(\mathbf{L}(G) + \mathrm{diag}(\mathbf{c}))\mathbf{g}, \mathbf{g}\rangle}{\langle \mathbf{g}, \mathbf{g}\rangle} &= \frac{\langle \mathbf{L}(G)\mathbf{g}, \mathbf{g}\rangle}{\langle \mathbf{g}, \mathbf{g}\rangle} = \frac{\sum_{vw \in E} a_{vw}(\mathbf{g}_v - \mathbf{g}_w)^2}{\sum_{v \in V} \mathbf{g}_v^2} \\ &= \frac{4e(S, \bar{S})}{n} = \frac{4}{n} \cdot \mathrm{bw}(G).\end{aligned}$$

Since $\mathbf{c}$ is arbitrary, the bound follows. ■

For computational purposes, it is convenient to express the bound of this theorem as a maximization of the smallest eigenvalue of an appropriate symmetric matrix. This can be done as follows. Let $\mathbf{Q} = (\mathbf{q}_1, \mathbf{q}_2, \ldots, \mathbf{q}_{n-1})$ be an $n \times (n-1)$ matrix whose columns $\mathbf{q}_i$ are pairwise orthogonal unit vectors for which $\mathbf{q}_i \perp \mathbf{j}\,(1 \le i < n)$. It is easy to see that, for each $\mathbf{x} \in \mathbf{R}^{n-1}$, we have $\langle \mathbf{Qx}, \mathbf{Qx}\rangle = \langle \mathbf{x}, \mathbf{x}\rangle$ and $\mathbf{Qx} \perp \mathbf{j}$. This implies the following result.

Theorem 4.4 *We have*

$$\mathrm{bw}(G) \ge \tfrac{1}{4}n \cdot \max_{\mathbf{c}} \lambda_{\min}(\mathbf{Q}^T(\mathbf{L}(G) + \mathrm{diag}(\mathbf{c}))\mathbf{Q}),$$

where the maximum is taken over all correction functions $\mathbf{c} \in \mathbf{R}^V$.

The bound of Theorem 4.4 can be formulated as a semi-definite program, and can therefore be computed to an arbitrary precision in polynomial time, using known polynomial-time methods for solving such programs. For more details, see the discussion at the end of the next section.

5. The max-cut problem and semi-definite programming

The *maximum cut* (or *max-cut*) *problem* is similar to the maximum version of the bipartition width problem, except that the restrictions on the sizes of the subsets over which the maximum is taken are omitted. More precisely, let $\mathrm{mc}(G)$ denote the maximum weight of an edge-cut in G:

$$\mathrm{mc}(G) = \max\{e(S, \bar{S}) : \emptyset \ne S \subset V(G)\}.$$

The problem of determining $\mathrm{mc}(G)$ is NP-hard (see [21, p. 210]). Moreover, it is known that there exists a constant $\varepsilon > 0$ for which there is no polynomial-time $(1-\varepsilon)$-approximation algorithm for the max-cut problem, unless $\mathbf{P} = \mathbf{NP}$ (see [3]). On the other hand, it is possible to find a 0.878-approximation to $\mathrm{mc}(G)$ in polynomial time [22].

Lemma 4.1 implies the following upper bound on $\mathrm{mc}(G)$, which was first observed by Mohar and Poljak (see [40]):

$$\mathrm{mc}(G) \le \tfrac{1}{4}n \cdot \lambda_{\max}(G). \tag{16}$$

Just as for the bipartition width problem, (16) can be further improved using correction functions. The corresponding optimized eigenvalue bound was introduced by Delorme and Poljak [13].

Theorem 5.1 *Let G be a weighted graph of order n. Then*

$$\mathrm{mc}(G) \le \tfrac{1}{4}n \cdot \min_{\mathbf{c}} \lambda_{\max}(\mathbf{L}(G) + \mathrm{diag}(\mathbf{c})), \tag{17}$$

where the minimum is taken over all correction functions $\mathbf{c} \in \mathbf{R}^V$.

The proof consists of steps similar to those in the proof of Theorem 4.3.

Minimizing the largest eigenvalue of an affine combination of symmetric matrices can be formulated as a *semi-definite program* (see, for example, [47]). Let $\mathbf{A}_0, \mathbf{A}_1, \ldots, \mathbf{A}_n \in \mathbf{R}^{m\times m}$ be symmetric matrices, and for $\mathbf{x} \in \mathbf{R}^n$, set

$$\mathbf{A}(\mathbf{x}) = \mathbf{A}_0 + \sum_{i=0}^{n} x_i \mathbf{A}_i.$$

The problem is to

$$\begin{cases} \text{minimize} & \lambda_{\max}(\mathbf{A}(\mathbf{x})) \\ \text{subject to} & \mathbf{x} \in U, \end{cases} \tag{LM1}$$

where U is a linear subspace of $\mathbf{R}^n$. We can translate this problem into a semi-definite program by introducing an auxiliary variable $t \in \mathbf{R}$:

$$\begin{cases} \text{minimize} & t \\ \text{subject to} & t\mathbf{I} - \mathbf{A}(\mathbf{x}) \succeq 0, \quad \mathbf{x} \in U, \end{cases} \tag{LM2}$$

where the notation $\mathbf{X} \succeq 0$ means that $\mathbf{X}$ is a positive semi-definite matrix. The first constraint in (LM2) is equivalent to $t \geq \lambda_{\max}(\mathbf{A}(\mathbf{x}))$.

In particular, the bound on the max-cut in Theorem 5.1 is of the form (LM1), where we take U to be the orthogonal complement of the vector $\mathbf{j}$. Also, the bound on the bipartition width in Theorem 4.4 can be viewed as a semi-definite program of the same form, since

$$\max_{\mathbf{x}\in U} \lambda_{\min}(\mathbf{A}(\mathbf{x})) = \max_{\mathbf{x}\in U}\{-\lambda_{\max}(-\mathbf{A}(\mathbf{x}))\} = -\min_{\mathbf{x}\in U} \lambda_{\max}(-\mathbf{A}(\mathbf{x})).$$

There exist efficient and practical algorithms for solving semi-definite programs. Given any $\varepsilon > 0$, the given semi-definite program can be solved within an additive error of ε in polynomial time, where ε is part of the input, its size being proportional to $\log(1/\varepsilon)$. For this purpose, one can use interior-point methods (see [41]).

Given a (weighted) graph G, the max-cut problem for G can be formulated as a quadratic integer program with variables y_v, $v \in V(G)$:

$$\begin{cases} \text{maximize} & \frac{1}{2} \sum_{vw\in E(G)} a_{vw}(1 - y_v y_w) \\ \text{subject to} & y_v = 1 \text{ or } -1, \quad \text{for every } v \in V(G). \end{cases} \tag{MC1}$$

If $(y_v)_{v\in V}$ is a feasible solution, then $1 - y_v y_w$ is equal to either 0 or 2. Given a solution of (MC1), the set S with the property that $\mathrm{mc}(G) = e(S, \bar{S})$ is determined by $S = \{v \in V(G) : y_v = 1\}$.

Goemans and Williamson [22] considered the following semi-definite relaxation of (MC1), where z_{vw} corresponds to the product $y_v y_w$ in (MC1):

$$\begin{cases} \text{maximize} & \frac{1}{2} \sum\limits_{vw \in E} a_{vw}(1 - z_{vw}) \\ \text{subject to} & z_{vv} = 1, \quad \text{for every } v \in V, \\ & \mathbf{Z} = [z_{vw}]_{v,w \in V} \succeq 0. \end{cases} \tag{MC2}$$

Poljak and Rendl [42] were the first to realize that the bound of the relaxation (MC2) is actually equivalent to the eigenvalue upper bound of Theorem 5.1. For a proof, which uses elegant duality theory of semi-definite programs, we refer to [42].

Theorem 5.2 *Let G be a weighted graph of order n. Then the value of an optimal solution to* (MC2) *is equal to the right-hand side of* (17).

Goemans and Williamson proved that (MC2) can be used to obtain a polynomial-time 0.878-approximation algorithm for the max-cut problem. Since the announcement of this algorithm, semi-definite programming has been successfully applied to the design of a number of other approximation algorithms, such as the max-k-cut (Frieze and Jerrum [20]), a 0.931-approximation algorithm for MAX 2-SAT (Feige and Goemans [16]), a 0.859-approximation algorithm for MAX DISCUT [16], and approximate colouring (Karger, Motwani and Sudan [30]). An application to the travelling salesman problem is described below in Section 7.

6. Isoperimetric inequalities

Isoperimetric problems are related to questions in which one considers the ratio between the surface area and the volume of some d-dimensional body. In graph theory, the natural analogue to the volume is the number of vertices or the sum of the degrees in a set S of vertices of the graph, while the counterpart of the surface area is the number $e(S, \bar{S})$ of edges with one end in S and the other end outside S. Problems in which one considers ratios of the form $e(S, \bar{S})/|S|$ are called *isoperimetric problems* for graphs; thus, isoperimetric properties concern the sizes of the neighborhood of a set of vertices. The related term 'expansion' usually means that the sizes of the neighborhood can be bounded from below by some function of the size of the subset. Such isoperimetric properties provide the foundation for many recent applications of graph theory to theoretical computer science, as mentioned in the introduction.

The *isoperimetric number* $i(G)$ of a graph G of order n (≥ 2) is defined as

$$i(G) = \min\left\{\frac{e(S, \bar{S})}{|S|} : S \subseteq V(G), 0 < |S| \leq \tfrac{1}{2}n\right\}.$$

Computationally, the problem of determining $i(G)$ is NP-hard (see [36]).

A straightforward application of Lemma 4.1 yields the following eigenvalue lower bound on $i(G)$.

Corollary 6.1 *Let G be a weighted graph of order n. Then $i(G) \geq \frac{1}{2}\lambda_2(G)$.*

It is difficult to obtain useful lower bounds on $i(G)$ by combinatorial means. Corollary 6.1, although easy to prove, gives a non-trivial lower bound on $i(G)$. For an example where this bound is tight, consider the d-dimensional cube Q_d as the product $K_2 \square Q_{d-1}$, and let S be the set of vertices in one copy of Q_{d-1}. Since $|S| = e(S, \bar{S}) = 2^{d-1}$, we have $i(Q_d) \leq 1$. On the other hand, as noted in Section 3, $\lambda_2(Q_d) = 2$. Using Corollary 6.1, we conclude that $i(Q_d) = 1$.

The following quantity is sometimes easier to deal with than $i(G)$:

$$i^*(G) = \min\left\{\frac{e(S, \bar{S})}{|S||\bar{S}|} \cdot \frac{|V|}{2} : \emptyset \neq S \subset V\right\}.$$

Clearly, $\frac{1}{2}i(G) \leq i^*(G) \leq i(G)$. It turns out that $i^*(G)$ satisfies a relation similar to (7):

$$i^*(G \square H) = \min\{i^*(G), i^*(H)\}.$$

The above inequality implies that

$$\tfrac{1}{2}\min\{i(G), i(H)\} \leq i(G \square H) \leq \min\{i(G), i(H)\}.$$

While the inequality on the right can be strict (see [36]), it is not known whether the factor $\frac{1}{2}$ on the left is best possible. We refer to [8, Sec. 2.6] and [26] for further results.

It is important to note that $i(G)$ also has upper bounds in terms of $\lambda_2(G)$. Such bounds are known as *Cheeger inequalities*, since they are discrete analogues of their continuous counterpart (see [6] and [7]) arising in the study of Laplace operators on Riemannian manifolds.

The classical form of Cheeger's bound adapted to graphs is

$$i(G) \leq \sqrt{2\Delta(G)\lambda_2(G)}. \tag{18}$$

An improvement of (18) was obtained by Mohar [36], and for weighted graphs in [39]:

Theorem 6.2 *Let G be a weighted graph of order n (≥ 3) with maximum degree Δ and second eigenvalue λ_2. Then*

$$i(G) \leq \sqrt{(2\Delta - \lambda_2)\lambda_2}.$$

Other discrete forms of Cheeger's inequality have been proved; see Theorem 8.4, and [8] and [9] and the references therein.

An important observation, following from the proof of Theorem 6.2, is that the partition based on the eigenfunction $\mathbf{f}$ of λ_2 is not too far from optimal. Constructing partitions based on the eigenfunctions of λ_2 has proved to be one of the most successful heuristics in parallel computation and for a divide-and-conquer approach for processor distribution.

7. The travelling salesman problem

The *travelling salesman problem* (TSP) is one of the best-known NP-hard combinatorial optimization problems, and there is an extensive literature on both its theoretical and practical aspects. In the *symmetric travelling salesman problem* (STSP), it is assumed that the cost of travelling between two points is the same in both directions.

We shall mention here only one approach, which uses semi-definite programming (see Section 5) to establish a lower bound on the length of an optimal tour. This bound is obtained by relaxing the STSP and invoking a 'branch-and-bound' algorithm. Semi-definite relaxations of the STSP were developed by Cvetković, Čangalović and Kovačević-Vujčić [10], and are based on the Laplace eigenvalues of graphs.

In what follows we assume that G is the complete graph of order n and that each edge vw has a cost c_{vw}, such that the matrix $\mathbf{C} = (c_{vw})$ is symmetric, with $c_{vv} = 0$. The STSP can now be formulated as follows: *find a Hamiltonian cycle in G of minimum cost.*

The next theorem gives a basis for a discrete semi-definite programming model of the STSP [10]. Recall from (5) that $\lambda_2(C_n) = 2 - 2\cos(2\pi/n)$.

Theorem 7.1 *Let H be a 2-regular spanning subgraph of G, and let $\mathbf{X} = \mathbf{L}(H) + \alpha\mathbf{J} - \beta\mathbf{I}$, where α and β are real parameters such that $\alpha > \lambda_2(C_n)/n$ and $0 < \beta \leq \lambda_2(C_n)$. Then H is a Hamiltonian cycle if and only if the matrix $\mathbf{X}$ is positive semi-definite.*

Proof Let $0 = \lambda_1 \leq \lambda_2 \leq \cdots \leq \lambda_n$ be the eigenvalues of $\mathbf{L}(H)$, and let $\mathbf{f}^1, \mathbf{f}^2, \ldots, \mathbf{f}^n$ be corresponding eigenvectors that are pairwise orthogonal, with

$\mathbf{f}^1 = \mathbf{j}$. Then $\mathbf{f}^1, \mathbf{f}^2, \ldots, \mathbf{f}^n$ are linearly independent eigenvectors of $\mathbf{X}$, with corresponding eigenvalues $n\alpha - \beta, \lambda_2 - \beta, \ldots, \lambda_n - \beta$. Since H is a union of disjoint cycles, either H is a Hamiltonian cycle with $\lambda_2 = \lambda_2(C_n)$, or H is disconnected with $\lambda_2 = 0$. In either case, $\lambda_2 - \beta$ is the smallest eigenvalue, since $n\alpha > \lambda_2(C_n) \geq \lambda_2$.

If H is a Hamiltionian cycle, then $\lambda_2 - \beta = \lambda_2(C_n) - \beta \geq 0$ and $\mathbf{X}$ is positive semi-definite. Conversely, if $\mathbf{X}$ is positive semi-definite, then $\lambda_2(H) = \lambda_2 \geq \beta > 0$, and so H is connected, and hence a Hamiltonian cycle. ■

A result that is similar to an extension of Theorem 7.1 for higher eigenvalues was used by Mohar [38] to derive a sufficient algebraic condition for non-Hamiltonicity. Based on that criterion, it can be shown, just by looking at its eigenvalues, that the Petersen graph is not Hamiltonian.

It follows from Theorem 7.1 that a spanning subgraph H of G is a Hamiltonian cycle if and only if its Laplacian $\mathbf{L}(H) = (l_{vw})$ satisfies the following conditions:

$$\begin{array}{ll} \text{each } l_{vv} = 2, \mathbf{X} = \mathbf{L}(H) + \alpha\mathbf{J} - \beta\mathbf{I} \succeq 0, & \text{for } \alpha > \lambda_2(C_n)/n \\ \text{and} \quad 0 < \beta \leq \lambda_2(C_n). & \end{array}$$

This result gives rise to a semi-definite relaxation of the STSP which has the following equivalent formulation in terms of Laplacians, where $\mathbf{L} = (l_{vw})$ is a symmetric matrix:

$$\begin{cases} \text{minimize} & \Phi(\mathbf{L}) = \sum\limits_{v\in V} \sum\limits_{w\in V} \left(-\frac{1}{2}c_{vw}\right) l_{vw}, \\ \text{subject to} & l_{vv} = 2, \sum_{w\in V} l_{vw} = 0, -1 \leq l_{vw} \leq 0 \text{ (for } v \neq w), \\ & \text{and } \lambda_2(\mathbf{L}) \geq \beta. \end{cases}$$

This semi-definite program can be solved in polynomial time and yields a good approximation to the optimum solution for the STSP.

8. Random walks on graphs

Isoperimetric properties and the eigenvalues treated in previous sections are closely related to the convergence rates of Markov chains. Several important randomized algorithms discovered in the last decade have increased the applicability of random walks and Markov chains to previously intractable problems. Additional reading on the results related to the presentation in this section can be found in [8] and [44].

Given a weighted graph G (possibly with loops), *a simple random walk* on G is a *Markov chain* with state space V and with *transition probabilities* p_{vw} of stepping from v to w, defined as follows:

$$p_{vw} = \frac{a_{vw}}{\deg v}.$$

The transition matrix $\mathbf{P}(G)$ is

$$\mathbf{P}(G) = \mathbf{D}^{-1}\mathbf{A}(G),$$

where, as before, $\mathbf{D}$ is the degree matrix. Although $\mathbf{P}$ is not necessarily symmetric, it has only real eigenvalues. This follows from the following result.

Theorem 8.1 $\mathbf{P}(G) = \mathbf{I} - \mathbf{D}^{-1/2}\mathbf{L}'(G)\mathbf{D}^{1/2} = \mathbf{I} - \mathbf{D}^{-1}\mathbf{L}(G)$.

Proof Since $\mathbf{I} - \mathbf{D}^{-1/2}\mathbf{L}'(G)\mathbf{D}^{1/2} = \mathbf{I} - \mathbf{D}^{-1}\mathbf{L}(G)$, the diagonal elements of the matrix are

$$1 - \deg v^{-1}\mathbf{L}_{vv} = 1 - (\deg v - a_{vv})/\deg v = a_{vv}/\deg v \quad (v \in V),$$

and the off-diagonal entries are

$$-(\deg v^{-1}(\mathbf{L}(G))_{vw} = a_{vw}/\deg v = p_{vw}\ (v, w \in V, v \neq w). \qquad \blacksquare$$

In particular, $\mathbf{I} - \mathbf{P}(G)$ and $\mathbf{L}'(G)$ have the same eigenvalues, so

$$\lambda_i(\mathbf{P}(G)) = 1 - \lambda'_{n-i+1}(G).$$

Theorem 8.1 also implies that $\mathbf{P}(G)$ (and thus also $\mathbf{P}(G)^T$) can be diagonalized, and that it has n pairwise orthogonal eigenvectors.

Rate of convergence of a random walk

For a random walk on G, let $\mathbf{x}^{(0)} : V \to \mathbf{R}^+$ be the initial probability distribution on $V(G)$. Then

$$\mathbf{x}^{(t)} = \mathbf{P}^T\mathbf{x}^{(t-1)} = (\mathbf{P}^T)^t\mathbf{x}^{(0)}$$

are the probabilities of where the Markov chain will be after t steps. If G is connected and non-bipartite, then $\lim_{t\to\infty}\mathbf{x}^{(t)}$ exists; it is denoted by $\mathbf{x}^{(\infty)}$, and is called the *stationary distribution*.

There is a close relationship between the eigenvalues of $\mathbf{P}(G)$ and the rate of convergence of a simple random walk on G to the stationary distribution.

Theorem 8.2 *Let G be a connected non-bipartite weighted graph, and let $\mathbf{x}^{(0)}$ be an initial probability distribution on $V(G)$. If*

$$\lambda = \max\{|\lambda_i(\mathbf{P}(G))| : \lambda_i(\mathbf{P}(G)) \neq 1\},$$

then, for a simple random walk on G,

$$\|\mathbf{x}^{(t)} - \mathbf{x}^{(\infty)}\| < \lambda^t.$$

Proof Let $\mathbf{Q} = \mathbf{P}(G)^T$, and for $i = 1, 2, \ldots, n$, let $\mathbf{y}^{(i)}$ be the eigenfunction of $\mathbf{Q}$ corresponding to the eigenvalue $\lambda_i = \lambda_i(\mathbf{Q}) = \lambda_i(\mathbf{P}(G))$. We may assume that $\|\mathbf{y}^{(i)}\| = 1$ and that the vectors $\mathbf{y}^{(i)}$ are pairwise othogonal. Recall that $\lambda_n(\mathbf{Q}) = 1$, and note that the assumptions in the theorem imply that $\lambda < 1$. Moreover, the Perron-Frobenius theorem shows that all the components of $\mathbf{y}^{(n)}$ are positive.

Let us write

$$\mathbf{x}^{(0)} = \sum_{i=1}^{n} \alpha_i \mathbf{y}^{(i)},$$

and observe that $\alpha_n = \langle \mathbf{x}^{(0)}, \mathbf{y}^{(n)} \rangle \neq 0$. Then

$$\mathbf{x}^{(t)} = \mathbf{Q}^t \mathbf{x}^{(0)} = \sum_{i=1}^{n} \alpha_i \lambda_i^t \mathbf{y}^{(i)} = \sum_{i=1}^{n-1} \alpha_i \lambda_i^t \mathbf{y}^{(i)} + \alpha_n \mathbf{y}^{(n)}.$$

Since $\lambda < 1$, $\mathbf{x}^{(t)}$ converges to $a_n \mathbf{y}^{(n)} = \mathbf{x}^{(\infty)}$, and so

$$\begin{aligned} \|\mathbf{x}^{(t)} - \mathbf{x}^{(\infty)}\|^2 &= \left\| \sum_{i=1}^{n-1} \alpha_i \lambda_i^t \mathbf{y}^{(i)} \right\|^2 = \sum_{i=1}^{n-1} \|\alpha_i \lambda_i^t \mathbf{y}^{(i)}\|^2 \\ &= \sum_{i=1}^{n-1} \alpha_i^2 \lambda_i^{2t} \|\mathbf{y}^{(i)}\|^2 \leq \lambda^{2t} \sum_{i=1}^{n-1} \alpha_i^2. \end{aligned}$$

Since also

$$\sum_{i=1}^{n-1} \alpha_i^2 < \sum_{i=1}^{n} \alpha_i^2 = \|\mathbf{x}^{(0)}\|^2 \leq 1,$$

we finally get $\|\mathbf{x}^{(t)} - \mathbf{x}^{(\infty)}\| < \lambda^t$. ■

As an example, let us consider a random walk on the d-dimensional hypercube Q_d. (Note that Q_d is a Cayley graph of the group $\mathbf{Z}_2^d$; Cayley graphs are discussed in Chapter 6.) Since Q_d is bipartite, we add a loop to each vertex, and denote the resulting graph by G. We would like to assign weights to the loops and edges of G in such a way that λ becomes as small as possible. Let α be the weight assigned to each loop, and β be the weight assigned to each edge of G. We also require that $\alpha + d\beta = 1$. Then

$$\mathbf{P}(G) = \alpha \mathbf{I} + \beta \mathbf{A}(Q_d).$$

From the Laplace eigenvalues of Q_d (see Section 3), we see that the eigenvalues of $\mathbf{P}(G)$ are $\lambda_i = \alpha + \beta(d - 2i)$, $i = 0, 1, \ldots, d$, with multiplicity $\binom{d}{i}$. Thus, $\lambda = \max\{\alpha + \beta(d-1), |\alpha - \beta d|\}$. By balancing both terms, we get equality when $\alpha = \beta = (d+1)^{-1}$. Hence,

$$\lambda = \frac{d-1}{d+1} = 1 - \frac{2}{d+1}.$$

By Theorem 8.2, the rate of convergence can be estimated by

$$\lambda^t = \left(\frac{d-1}{d+1}\right)^t = \left(1 - \frac{2}{d+1}\right)^t \leq \exp\left(-\frac{2t}{d+1}\right).$$

Similar estimates can be established for other Abelian groups (see, for example, [44]).

Perhaps the most useful distance between two probability distributions $\mathbf{x}$ and $\mathbf{y}$ of a Markov chain is the so-called 'chi-squared' distance, defined as follows:

$$\chi^2(\mathbf{x}, \mathbf{y}) = \sum_{w \in V(G)} \frac{(\mathbf{x}_w - \mathbf{y}_w)^2}{\mathbf{y}_w}.$$

Clearly, this definition makes sense only when $\mathbf{y}_v > 0$, for every $v \in V$. A proof similar to the proof of Theorem 8.2 gives the following result.

Theorem 8.3 *For a simple random walk on a non-bipartite weighted graph,*

$$\chi^2(\mathbf{x}^{(t)}, \mathbf{x}^{(\infty)}) \leq \lambda^t \cdot \chi^2(\mathbf{x}^{(0)}, \mathbf{x}^{(\infty)}).$$

The quantity $(1-\lambda)^{-1}$ is called the *mixing time* of the random walk, and is of interest since the above theorem implies that, in this number of steps, the chi-squared distance from the stationary distribution $\mathbf{x}^{(\infty)}$ is cut by a constant factor. The quantity $1-\lambda$ is related to the second smallest eigenvalue $\lambda_2(\mathbf{I} - \mathbf{P}) = \lambda_2(\mathbf{L}'(G))$.

The isoperimetric constant related to the transition Laplacian is also known as the *conductance* of the corresponding Markov chain. For a simple random walk on a graph G, the conductance is defined as

$$i_{\mathbf{P}}(G) = \min\left\{\frac{\sum_{v \in S}\sum_{w \notin S} p_{vw} \deg v}{\operatorname{vol}(S)} : S \subseteq V(G), 0 < \operatorname{vol}(S) \leq m\right\},$$

where

$$\operatorname{vol} S = \sum_{v \in S} \deg v \quad \text{and} \quad m = \tfrac{1}{2}\sum_{v \in V} \deg v = \tfrac{1}{2}\operatorname{vol} V.$$

If G is regular, then the conductance is just the isoperimetric number of G divided by the degree.

Theorem 8.4 *For the simple random walk on a graph G,*

$$\tfrac{1}{2} i_{\mathbf{P}}^2(G) \leq \lambda_2(\mathbf{I} - \mathbf{P}(G)) = \lambda_2(\mathbf{L}'(G)) \leq 2 i_{\mathbf{P}}(G).$$

For proofs we refer to [35]; see also [8].

Since $\lambda_2(\mathbf{I} - \mathbf{P}(G)) \geq 1 - \lambda$, Theorem 8.4 also gives a bound on the mixing time.

Rapidly mixing Markov chains proved to be an important tool in the design of polynomial-time randomized algorithms. For example, Dyer, Frieze and Kannan [15] found a polynomial-time randomized approximation algorithm for computing the volume of a convex set K, based on sampling random points in K. Lovász and Simonovits [32] improved this algorithm by providing random walk sampling with better mixing, and by improved isoperimetric estimates.

Another application, somewhat similar to the first, deals with approximate counting of perfect matchings in a bipartite graph, and therefore also with the approximation of the permanent of a (0, 1)-matrix (see Sinclair and Jerrum [46]).

References

1. N. Alon and V. D. Milman, λ_1, isoperimetric inequalities for graphs and superconcentrators, *J. Combin. Theory (B)* **38** (1985), 73–88.
2. W. N. Anderson and T. D. Morley, Eigenvalues of the Laplacian of a graph, *Linear and Multilinear Algebra* **18** (1985), 141–145.
3. S. Aurora, C. Lund, R. Motwani, M. Sudan and M. Szegedy, Proof verification and hardness of approximation problems, *Proc. 33rd FOCS* (1992), 14–23.
4. N. L. Biggs, *Algebraic Graph Theory* (2nd ed.), Cambridge Univ. Press, 1993.
5. R. B. Boppana, Eigenvalues and graph bisection: an average case analysis, *28th Annual Symp. Found. Comp. Sci.*, IEEE (1987), 280–285.
6. R. Brooks, Spectral geometry and the Cheeger constant, *Expanding Graphs* (ed. J. Friedman), Amer. Math. Soc. (1993), 5–19.
7. P. Buser, *Geometry and Spectra of Compact Riemann Surfaces,* Birkhäuser, 1992.
8. F. R. K. Chung, *Spectral Graph Theory,* Amer. Math. Soc., 1997.
9. Y. Colin de Verdière, *Spectres de Graphes,* Cours Spécialisés **4**, Soc. Math. France, 1998.
10. D. Cvetković, M. Čangalović and V. Kovačević-Vujčić, Semidefinite programming methods for the symmetric traveling salesman problem, *Integer Programming and Combinatorial Optimization* (eds. G. Cornuejols, R. E. Burkard and G. J. Woeginger), Springer (1999), 126–136.
11. D. M. Cvetković, M. Doob and H. Sachs, *Spectra of Graphs* (3rd ed.), Johann Ambrosius Barth Verlag, 1995.
12. D. M. Cvetković, M. Doob, I. Gutman and A. Torgašev, *Recent Results in the Theory of Graph Spectra, Ann. Discr. Math.* **36**, North-Holland, 1988.
13. C. Delorme and S. Poljak, Laplacian eigenvalues and the maximum cut problem, *Math. Programming* **62** (1993), 557–574.
14. W. E. Donath and A. J. Hoffman, Lower bounds for the partitioning of graphs, *IBM J. Res. Develop.* **17** (1973), 420–425.
15. M. E. Dyer, A. Frieze and R. Kannan, A random polynomial-time algorithm for approximating the volume of convex bodies, *J. Assoc. Comput. Mach.* **38** (1991), 1–17.

16. U. Feige and M. X. Goemans, Approximating the value of two prover proof systems, with applications to MAX 2SAT and MAX DICUT, *Proc. Third Israel Symp. on Theory of Computing and Systems* (1995), 182–189.
17. M. Fiedler, Algebraic connectivity of graphs. *Czech. Math. J.* **23 (98)** (1973), 298–305.
18. M. Fiedler, A property of eigenvectors of nonnegative symmetric matrices and its application to graph theory, *Czech. Math. J.* **25 (100)** (1975), 619–633.
19. P. W. Fowler, T. Pisanski and J. Shawe-Taylor, Molecular graph eigenvectors for molecular coordinates, *Graph drawing: GD'94* (ed. R. Tamassia), Springer-Verlag (1995), 282–285.
20. A. Frieze and M. Jerrum, Improved approximation algorithms for MAX k-CUT and MAX BISECTION, *Integer Programming and Combinatorial Optimization; Proc. 4th IPCO* (eds. E. Balas and J. Clausen), Springer-Verlag (1995), 1–13.
21. M. R. Garey and D. S. Johnson. *Computers and Intractability: A Guide to the Theory of NP-Completeness*, Freeman, 1979.
22. M. X. Goemans and D. P. Williamson, Improved approximation algorithms for maximum cut and satisfiability problems using semidefinite programming, *J. Assoc. Comp. Mach.* **42** (1995), 1115–1145.
23. C. D. Godsil and G. Royle, *Algebraic Graph Theory*, Springer-Verlag, 2001.
24. L. Hagen and A. B. Kahng, New spectral methods for ratio cut partitioning and clustering, *IEEE Trans. Computer-Aided Design* **11** (1992), 1074–1085.
25. C. Helmberg, B. Mohar, S. Poljak and F. Rendl, A spectral approach to bandwidth and separator problems in graphs, *Linear and Multilinear Algebra* **39** (1995), 73–90.
26. C. Houdré and P. Tetali, Concentration of measure for products with Markov kernels and graphs via functional inequalities, *Combin. Probab. Comput.* **10** (2001), 1–28.
27. B. Hendrickson and R. Leland, An improved spectral graph partitioning algorithm for mapping parallel computations, *SIAM J. Sci. Comput.* **16** (1995), 452–469.
28. A. J. Hoffman, On eigenvalues and colorings of graphs, *Graph Theory and its Applications* (ed. B. Harris), Academic Press (1970), 79–91.
29. M. Juvan and B. Mohar, Laplace eigenvalues and bandwidth-type invariants of graphs, *J. Graph Theory* **17** (1993), 393–407.
30. D. Karger, R. Motwani and M. Sudan, Approximate graph coloring by semidefinite programming, *Proc. 35th Ann. Symp. FOCS* (1994), 2–13.
31. L. Lovász, On the Shannon capacity of a graph, *IEEE Trans. Inform. Theory* **IT–25** (1979), 1–7.
32. L. Lovász and M. Simonovits, Random walks in a convex body and an improved volume algorithm, *Random Struct. Algorithms* **4** (1993), 359–412.
33. R. Merris, Laplacian matrices of graphs: a survey, *Linear Alg. Appl.* **197/198** (1994), 143–176.
34. R. Merris, Large families of Laplacian isospectral graphs, *Linear and Multilinear Algebra* **43** (1997), 201–205.
35. B. Mohar, Isoperimetric inequalities, growth, and the spectrum of graphs, *Linear Alg. Appl.* **103** (1988), 119–131.
36. B. Mohar, Isoperimetric numbers of graphs, *J. Combin. Theory (B)* **47** (1989), 274–291.
37. B. Mohar, The Laplacian spectrum of graphs, *Graph Theory, Combinatorics, and Applications* (eds. Y. Alavi *et al.*), Wiley (1991), 871–898.
38. B. Mohar, A domain monotonicity theorem for graphs and Hamiltonicity, *Discr. Appl. Math.* **36** (1992), 169–177.

39. B. Mohar, Some applications of Laplace eigenvalues of graphs, *Graph Symmetry: Algebraic Methods and Applications* (eds. G. Hahn and G. Sabidussi), NATO ASI Ser. C 497, Kluwer (1997), 225–275.
40. B. Mohar and S. Poljak, Eigenvalues in combinatorial optimization, *Combinatorial and Graph-Theoretical Problems in Linear Algebra* (eds. R. A. Brualdi, S. Friedland and V. Klee), IMA Volumes in Mathematics and its Applications **50**, Springer-Verlag (1993), 107–151.
41. Y. Nesterov and A. Nemirovskii, *Interior Point Polynomial Methods in Convex Programming,* SIAM, 1994.
42. S. Poljak and F. Rendl, Nonpolyhedral relaxations of graph-bisection problems, *SIAM J. Optim.* **5** (1995), 467–487.
43. F. Rendl and H. Wolkowicz, Applications of parametric programming and eigenvalue maximization to the quadratic assignment problem, *Math. Progr.* **53** (1992), 63–78.
44. J. S. Rosenthal, Convergence rates for Markov chains, *SIAM Review* **37** (1995), 387–405.
45. H. D. Simon, Partitioning of unstructured problems for parallel processing, *Computing Systems in Engineering* **2** (1991), 135–148.
46. A. J. Sinclair and M. R. Jerrum, Approximate counting, uniform generation and rapidly mixing Markov chains, *Inform. Comput.* **82** (1989), 93–133.
47. L. Vandenberghe and S. Boyd, Semidefinite programming, *SIAM Review* **38** (1996), 49–95.
48. H. S. Wilf, The eigenvalues of a graph and its chromatic number, *J. London Math. Soc.* **42** (1967), 330–332.

5

Automorphisms of graphs

PETER J. CAMERON

This chapter surveys automorphisms of finite graphs, concentrating on the asymmetry of typical graphs, prescribing automorphism groups (as either permutation groups or abstract groups), and special properties of vertex-transitive graphs and related classes. There are short digressions on infinite graphs and graph homomorphisms.

1. Graph automorphisms

An *automorphism* of a graph G is a permutation g of the vertex-set of G with the property that, for any vertices u and v, we have $vg \sim wg$ if and only if $v \sim w$. (As usual, vg denotes the image of the vertex v under the permutation g: see the Introduction for the terminology and main results of permutation group theory.)

This simple definition does not suffice for multigraphs, where we need to specify a permutation of the edges as well as a permutation of the vertices, to ensure that the multiplicity of edges between two vertices is preserved. Alternatively, a multigraph can be regarded as a *weighted graph,* where the weight $a_{v,w}$ is the number of edges from v to w; an automorphism must satisfy $a_{vg,wg} = a_{v,w}$: this gives a slightly different description of automorphisms, but the action on the set of vertices is the same. We consider only simple graphs here.

The set of all automorphisms of a graph G, with the operation of composition of permutations, is a *permutation group* on $V(G)$, a subgroup of the symmetric group on $V(G)$. This is the *automorphism group* of G, denoted by Aut(G). We describe any subgroup $\mathcal{H}$ of Aut(G) as a *group of automorphisms* of G, and refer to Aut(G) as the *full automorphism group.*

More generally, an *isomorphism* from a graph G to a graph H is a bijection f from the vertex-set of G to that of H such that $vf \sim wf$ (in H) if and only if $v \sim w$ (in G). We say that G and H are *isomorphic* (written $G \cong H$) if there is an isomorphism between them.

Among its other jobs, the automorphism group arises in the enumeration of graphs, specifically in the relation between counting labelled and unlabelled graphs. A *labelled graph* on n vertices is a graph whose vertex-set is $\{1, 2, \ldots, n\}$, while an *unlabelled graph* is simply an isomorphism class of n-element graphs.

Now the number of labellings of a given unlabelled graph G on n vertices is $n!/|\mathrm{Aut}(G)|$. For, a labelling is given by a bijective function F from $\{1, 2, \ldots, n\}$ to $V(G)$; there are $n!$ such functions, and two of them F_1 and F_2 define the same labelled graph if and only if there is an automorphism g such that $F_2(i) = F_1(i)g$, for all $i \in \{1, 2, \ldots, n\}$. Figure 1 shows the three labellings of the path P_3, a graph whose automorphism group has order 2.

The automorphism group is an algebraic invariant of a graph. Before giving some simple properties, we recall some terminology from the Introduction.

The *direct product* $\mathcal{G}_1 \times \mathcal{G}_2$ of two permutation groups $\mathcal{G}_1$ and $\mathcal{G}_2$ (acting on sets Ω_1 and Ω_2) is the permutation group on the disjoint union $\Omega_1 \cup \Omega_2$ whose elements are ordered pairs (g_1, g_2) for $g_i \in \mathcal{G}_i$; the action is given by

$$v(g_1, g_2) = \begin{cases} vg_1 & \text{if } v \in \Omega_1, \\ vg_2 & \text{if } v \in \Omega_2. \end{cases}$$

This notion extends to the direct product of any number of permutation groups.

If $\mathcal{G}_2$ is a permutation group on $\{1, 2, \ldots, n\}$, then the *wreath product* $\mathcal{G}_1 \wr \mathcal{G}_2$ is generated by the direct product of n copies of $\mathcal{G}_1$, together with the elements of $\mathcal{G}_2$ acting on these n copies of $\mathcal{G}_1$.

Fig. 1.

Finally, $\mathcal{S}_n$ denotes the symmetric group on $\{1, 2, \ldots, n\}$.

Theorem 1.1

(a) A graph and its complement have the same automorphism group.

(b) $\mathrm{Aut}(K_n) = \mathcal{S}_n$.

(c) Let the connected components of G consist of n_1 copies of G_1, n_2 copies of $G_2, \ldots, n_r$ copies of G_r, where $G_1, G_2, \ldots, G_r$ are pairwise non-isomorphic. Then

$$\mathrm{Aut}(G) = (\mathrm{Aut}(G_1) \wr \mathcal{S}_{n_1}) \times (\mathrm{Aut}(G_2) \wr \mathcal{S}_{n_2}) \times \cdots \times (\mathrm{Aut}(G_r) \wr \mathcal{S}_{n_r}).$$

In view of these results, we can reduce questions about automorphism groups to the case where the graphs are connected.

A recent survey with a somewhat different emphasis is that of Babai and Goodman [5]. In addition, no serious student should be without Hahn and Sabidussi's book [28], which contains surveys of aspects of graph symmetry.

2. Algorithmic aspects

Two algorithmic questions that arise from the above definitions are *graph isomorphism* and *finding the automorphism group*. The first is a decision problem.

Graph isomorphism
Instance: Graphs G and H
Question: Is $G \cong H$?

The second problem requires output. Note that a subgroup of $\mathcal{S}_n$ may be super-exponentially large in terms of n, but that any subgroup has a generating set of size $O(n)$, which specifies it in polynomial space.

Automorphism group
Instance: A graph G
Output: generating permutations for $\mathrm{Aut}(G)$

These two problems are closely related: indeed, the first has a polynomial reduction to the second. For, suppose that we are given two graphs G and H. By taking complements if necessary, we may assume that both G and H are connected. Now suppose that we can find generating permutations for $\mathrm{Aut}(K)$, where K is the disjoint union of G and H. Then G and H are isomorphic if and only if some generator interchanges the two connected components.

Conversely, if we can solve the graph isomorphism problem, we can at least check whether a graph has a non-trivial automorphism, by attaching distinctive 'gadgets' at each vertex and checking whether any pair of the resulting graphs are isomorphic. (Finding generators for the automorphism group may be more difficult.)

The exact status of these two problems is unresolved. They belong to a select group of problems that belong to NP but are not known either to belong to P or to be NP-complete. For some particular classes of graphs – notably, graphs of bounded valency [43] and graphs with bounded eigenvalue multiplicity [7] – the isomorphism problem is known to be polynomial; see Garey and Johnson [23] for the fundamentals of computational complexity.

In practice, these questions can be resolved for graphs with thousands of vertices. Chapter 10 gives an account of the algorithms used and their implementation.

It turns out that, for almost all graphs, the algorithmic questions can be answered very quickly. However, 'almost all' does not include some of the most interesting graphs, including strongly regular graphs (discussed in Chapter 8).

3. Automorphisms of typical graphs

The smallest graph whose automorphism group is trivial (apart from the 1-vertex graph) is shown in Fig. 2. However, small graphs are (as usual) not a reliable guide here. Erdős and Rényi [16] proved the following result.

Theorem 3.1 *Almost all graphs have no non-trivial automorphisms.*

Thus, the proportion of graphs on n vertices that have a non-trivial automorphism tends to 0 as $n \to \infty$: this is true whether we take labelled or unlabelled graphs. As noted in the introduction, this theorem implies that almost all graphs can be labelled in $n!$ different ways, so that the number of unlabelled graphs on n vertices is asymptotically $2^{n(n-1)/2}/n!$. (There are clearly $2^{n(n-1)/2}$ labelled graphs on the vertex-set $\{1, 2, \ldots, n\}$, since we can choose whether or not to join each pair of vertices by an edge.) There are now good estimates for the error term in the asymptotic expansion; it arises from graphs with non-trivial symmetry, and so these estimates quantify the theorem.

In fact, more is true. There are various methods for *canonically labelling* a graph – for example, choose the lexicographically least labelled graph in the isomorphism class. For almost all graphs, the canonical labelling is unique, and can

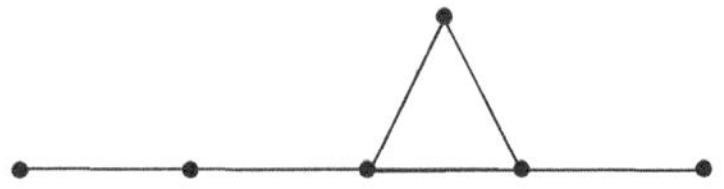

Fig. 2.

be found in polynomial time; for such graphs, we can verify efficiently that their automorphism groups are trivial. Typically, graphs with regularity properties, such as strongly regular graphs (Chapter 8), are hard for canonical labelling algorithms, even when their automorphism groups are trivial.

The theorem remains true for various special classes of graphs. These include regular graphs of fixed degree $k > 2$ – we can even allow the degree to grow, not too rapidly, with n(see [59]). They also include the prolific strongly regular graphs of Latin square or Steiner triple system type discussed in Chapter 8; this uses the fact that almost all of these structures have no non-trivial automorphisms.

Other methods of quantifying the theorem can be found. For example, any given graph can be altered so that some two vertices have the same neighbour sets by changing at most $n/2$ adjacencies. The resulting graph has an automorphism that interchanges the two vertices and fixes all the others. Erdős and Rényi [16] showed that, for almost all graphs, this is the 'shortest distance to symmetry'.

4. Permutation groups

The question, 'Which permutation groups are the full automorphism groups of graphs?', has no easy answer. Given a permutation group $\mathcal{G}$ on a set Ω, we can describe all of the graphs on which $\mathcal{G}$ acts, as follows. There is a coordinate-wise action of $\mathcal{G}$ on $\Omega \times \Omega$, given by $(v, w)g = (vg, wg)$. Let U be the set of all orbits of $\mathcal{G}$ on $\Omega \times \Omega$ that consist of pairs of *distinct* elements. There is a natural *pairing* of orbits in U, where an orbit O is paired with $O^* = \{(v, w) : (w, v) \in O\}$. Now let S be any subset of U that contains the orbit paired with each of its members, and define a graph $G(S)$ on the vertex-set Ω by the rule that $v \sim w$ in $G(S)$ if and only if $(v, w) \in O$ for some $O \in S$. Then $G(S)$ is a simple graph admitting $\mathcal{G}$ as a group of automorphisms; furthermore, every such graph arises by this construction.

The construction may easily be adapted to other classes of graphs. For directed graphs, we drop the requirement that S is closed under pairing; for graphs with loops, we include all orbits on $\Omega \times \Omega$, not just those consisting of pairs of distinct elements; and for multigraphs, we allow multisets of orbits. It is also very practical: instead of having to list all the edges in order to specify a graph, we need give only a set of orbit representatives.

A permutation group $\mathcal{G}$ on Ω is *2-closed* if every permutation of Ω that preserves all the $\mathcal{G}$-orbits in $\Omega \times \Omega$ belongs to $\mathcal{G}$. More generally, the *2-closure* of $\mathcal{G}$ is the group of permutations that preserve all the $\mathcal{G}$-orbits. These concepts were introduced by Wielandt [56].

Now, for any graph G, the group Aut(G) is 2-closed, for the edge-set of G is a union of orbits of Aut(G) and so is preserved by its 2-closure. The converse fails, but not too badly: in fact, a permutation group $\mathcal{G}$ is 2-closed if and only if it is the full automorphism group of an edge-coloured directed graph. (We associate a colour with each $\mathcal{G}$-orbit on pairs.) However, it is not easy to decide whether a given permutation group is 2-closed.

The construction of a graph from a permutation group has been reversed; for example, many of the sporadic simple groups discovered in the mid-twentieth century were constructed as groups of automorphisms of particular graphs. The simplest such construction is that of Higman and Sims [36]: the vertex-set consists of a special symbol ∞ together with the 22 points and 77 blocks of the *Witt design* $\mathcal{W}_{22}$. We join ∞ to all the points; we join a point and a block if they are incident, and we join two blocks if they are disjoint. It is clear that the automorphism group of $\mathcal{W}_{22}$ acts as a group of automorphisms of the graph, fixing ∞. It is not hard to show that the full automorphism group is transitive, and contains the Higman-Sims simple group as a subgroup of index 2.

In the case of the Fischer groups [18], the group and the graph are even more closely related. The vertices of the graph are the elements of a particular conjugacy class of involutions in the group, two vertices being joined whenever the involutions do not commute.

5. Abstract groups

If we consider groups as abstract algebraic objects, rather than as concrete permutation groups, a clear-cut result is possible. Frucht [20] proved the following basic result.

Theorem 5.1 *Every group is the automorphism group of some graph. Moreover, if the group is finite, then the graph can be taken to be finite.*

Subsequently, Frucht [21] showed that *every group is the automorphism group of a trivalent graph*, and this has inspired a large number of similar results. We call a class $\mathcal{C}$ of structures *universal* if every finite group is the automorphism group of a structure in $\mathcal{C}$. The combined results of Frucht, Sabidussi, Mendelsohn, Babai, Kantor and others show that the following types of graph or other structures are universal: regular graphs of degree k for any fixed $k > 2$; bipartite graphs; strongly regular graphs; Hamiltonian graphs; k-connected graphs for $k > 0$; k-chromatic graphs, for $k > 1$; switching classes of graphs; lattices; projective planes (possibly infinite); Steiner triple systems; symmetric designs (BIBDs).

For example, the universality of Steiner triple systems is shown by using a graph to construct a triple system with the same symmetry; then the universality of strongly regular graphs is shown by taking the line graphs of triple systems (see Chapter 8). The structures are finite in all cases except for projective planes; Even the question of whether every finite group is a subgroup of the automorphism group of a finite projective plane is still open.

Another class of results involves structures for which there is some obvious restriction on the automorphism group. For example, a tournament cannot admit an automorphism of order 2 (for such an automorphism would necessarily interchange the ends of some arc), and so the automorphism group has odd order. Moon [46] showed that every group of odd order is the automorphism group of a tournament, so this is the only restriction.

Pólya observed that not every group is the automorphism group of a tree. More precisely, the class of automorphism groups of trees is the smallest class that contains the trivial group and is closed under the direct product and the operation 'take the wreath product with the symmetric group S_n of degree n', for each $n > 1$.

Automorphism groups of trees are of further importance in group theory. Any finite tree has either a vertex or an edge which is fixed by all automorphisms, according as it is central or bicentral. Things are very different for infinite trees. We say that a group $\mathcal{G}$ acts *freely* on a tree T if only the identity element of $\mathcal{G}$ fixes any vertex or edge of T. Serre [49] has proved the following result.

Theorem 5.2 *A group is free if and only if it acts freely on a tree.*

This simple observation has led to the *Bass-Serre theory* in combinatorial group theory, describing certain group constructions (amalgamated free products and HNN-extensions) in terms of their actions on a tree; see Dicks and Dunwoody [15] for an account of this.

Usually such a precise description does not exist. For example, every composition factor of the automorphism group of a planar graph is either a cyclic group or an alternating group. More generally, Babai [2] showed the following result.

Theorem 5.3 *Given $k \geq 0$, there is a number $f(k)$ such that, if $\mathcal{G}$ is the automorphism group of any graph embeddable in the orientable surface of genus k, then every composition factor of $\mathcal{G}$ is cyclic, alternating, or of order at most $f(k)$.*

Still more generally, Babai showed that no class of finite graphs which is closed under subgraphs and contractions can be universal, except for the class of all graphs.

We can quantify Frucht's Theorem by asking the question:

> *given a group $\mathcal{G}$, what is the smallest number of vertices of a graph G with $\mathrm{Aut}(G) = \mathcal{G}$?*

This function behaves in a very erratic fashion. For example, the symmetric group $\mathcal{S}_n$ is the automorphism group of the null graph on n vertices, but the smallest graph whose automorphism group is the alternating group $\mathcal{A}_n$ has about 2^n vertices; the exact number was calculated by Liebeck [41]. Clearly this number is not smaller than the degree of the smallest faithful permutation representation of $\mathcal{G}$; such questions have been investigated in detail by Babai *et al.* [6].

Other measures of the 'size' of the graph can be used – for example, the number of edges, or the number of orbits of $\mathcal{G}$ on the set of vertices or edges: see [5] for a survey of results on these questions.

Given that almost all graphs admit only the identity automorphism, we might wonder whether Frucht's Theorem can be strengthened to state that almost all graphs that admit a given group $\mathcal{G}$ actually have $\mathcal{G}$ as their full automorphism group. This holds for some groups, but not all. Cameron [9] proved the following theorem:

Theorem 5.4 *Given a group $\mathcal{G}$, consider those n-vertex graphs whose automorphism group contains $\mathcal{G}$. The proportion of such graphs whose automorphism group is precisely $\mathcal{G}$ tends to a limit $a(\mathcal{G})$ as $n \to \infty$.*

The limit $a(\mathcal{G})$ is a rational number, but (unlike the case where $\mathcal{G}$ is the trivial group) is not necessarily 1. For example, if $\mathcal{G}$ is the dihedral group of order 10, then the limit is 1/3; this is because, of those graphs that admit $\mathcal{G}$, almost all have $\mathcal{G}$ acting on a set of five vertices and fixing the rest of the graph. (A random graph admitting $\mathcal{G}$ consists of a random graph on $n-5$ vertices, and five 'special' vertices on which $\mathcal{G}$ acts as the symmetry group of a pentagon, all joined to the same random subset of the other $n-5$ vertices.) The subgraph induced by the five special vertices may be complete, null, or a 5-cycle; only in the third case is $\mathcal{G}$ almost surely the full group.

In fact, $a(\mathcal{G}) = 1$ if and only if $\mathcal{G}$ is a direct product of symmetric groups. For abelian groups, $a(\mathcal{G}) = 1$ if $\mathcal{G}$ is an elementary abelian 2-group (possibly trivial), and $a(\mathcal{G}) = 0$ otherwise. For metabelian groups, the values of $a(\mathcal{G})$ are dense in the unit interval.

It would be interesting to know whether similar results hold under hypotheses that tend to work against such very local symmetries; for example, does a similar result hold for regular graphs?

6. Cayley graphs

A permutation group $\mathcal{G}$ is *regular* if it is transitive and only the identity stabilizes a point. Any regular action can be identified with the action of the group on itself by right multiplication, where the group element g induces the permutation $x \mapsto xg$ of $\mathcal{G}$. This is the action used by Cayley to show that every group is isomorphic to a permutation group.

The orbits on pairs of this group are parametrized by group elements: they have the form $O_g = \{(x, gx) : x \in \mathcal{G}\}$ for $g \in \mathcal{G}$. The orbit paired with O_g is $O_{g^{-1}}$.

So our description of a $\mathcal{G}$-invariant graph can be specialized in the following way.

> Let S be a subset of $\mathcal{G}$ that is closed under taking inverses and does not contain the identity. The *Cayley graph* Cay($\mathcal{G}$, S) has vertex-set $\mathcal{G}$, and edges $\{x, sx\}$, for all $s \in S$, $x \in \mathcal{G}$. Cayley graphs are discussed in Chapter 6.

Since $\mathcal{G}$ is a subgroup of Aut(Cay($\mathcal{G}$, S)), this Cayley graph is necessarily vertex-transitive (see Section 7). It is connected if and only if the set S generates $\mathcal{G}$. Note that many authors use a different convention, in which the action of the group is by left multiplication, and the edges have the form $\{x, xs\}$, for $s \in S$, $x \in \mathcal{G}$. The difference is immaterial.

A *graphical regular representation* of a group $\mathcal{G}$ is defined to be a graph for which the regular action of $\mathcal{G}$ is the full automorphism group – that is, a Cayley graph Cay($\mathcal{G}$, S) with Aut(Cay($\mathcal{G}$, S)) $= \mathcal{G}$. A considerable amount of effort went into the determination of groups which have such representations, and the problem was finally solved by Hetzel [35] for soluble groups, and Godsil [25] in general. First note that an abelian group of exponent greater than 2 never has a graphical regular representation, since any Cayley graph for an abelian group admits the automorphism $g \mapsto g^{-1}$. A *generalized dicyclic group* is a group that has a cyclic subgroup $\mathcal{H}$ of index 2 and an element g of order 4 for which $g^{-1}hg = h^{-1}$, for all $h \in \mathcal{H}$. The quaternion group of order 8 is an example.

The basic theorem is as follows:

Theorem 6.1 *A finite group has a graphical regular representation if and only if it is not an abelian group of exponent greater than* 2, *a generalized dicyclic group, or one of thirteen exceptional groups with order at most* 32.

What can be said about *random Cayley graphs* for $\mathcal{G}$, obtained by including inverse pairs of non-identity elements in S with probability $\frac{1}{2}$? Babai and Godsil [4] conjectured that, except for the two infinite classes in the theorem, almost all Cayley graphs for $\mathcal{G}$ are graphical regular representations – that is, the probability that a random Cayley graph is such a representation tends to 1 as $|\mathcal{G}| \to \infty$. They proved that this is true in some cases – for example, non-abelian nilpotent groups of odd order.

Random Cayley graphs have other useful properties; for example, they are often expanders (see Alon and Roichman [1]).

7. Vertex-transitive graphs

A graph G is *vertex-transitive* if the automorphism group of G acts transitively on the vertex-set of G; for example, the Petersen graph is vertex-transitive.

Every vertex-transitive graph has a description as a *Schreier coset graph*, generalizing the representation of a Cayley graph discussed above: we replace the group elements by cosets of a subgroup $\mathcal{H}$ of $\mathcal{G}$ as vertices of the graph, and for adjacency, we replace an inverse-closed set of elements by an inverse-closed set of double cosets of $\mathcal{H}$ in $\mathcal{G}$. Sabidussi [47] used this representation to show that any vertex-transitive graph has a 'multiple' that is a Cayley graph. (Here, a *multiple* of a graph is obtained by replacing each vertex by an independent set of size k, and each edge by all possible edges between the corresponding cocliques, for some k.)

Not every vertex-transitive graph is a Cayley graph. The smallest counterexample is the Petersen graph: it has no automorphism of order 2 that fixes no vertex. McKay and Praeger [44] have considered the class of vertex-transitive graphs that are not Cayley graphs.

Marušič and Jordan independently conjectured that any vertex-transitive graph has a group of automorphisms that acts *semi-regularly* on vertices – that is, the stabilizer of any vertex is the identity, but the subgroup is not required to be transitive. This conjecture was extended by Klin, who conjectured that any 2-closed permutation group contains such a subgroup. This conjecture is still open, although Giudici [24] has made substantial progress on it recently.

Obviously, all vertex-transitive graphs are regular. However, they also have some special properties that are not shared by all regular graphs. From the work of Mader, Watkins, Little, Grant, Holton, Babai and others, we take the following list. We need one further definition: a graph is *vertex-primitive* if no equivalence relation on the vertex-set is preserved by all automorphisms, apart from the trivial relations (equality and the 'universal' equivalence).

Theorem 7.1 *Let G be a connected k-regular vertex-transitive graph of order n. Then*

- *G is $\lfloor \frac{2}{3}(k+1) \rfloor$-connected (and k-connected if it is vertex-primitive);*
- *G is k-edge-connected;*
- *G has a 1-factor if n is even;*
- *G has a cycle of length at least $\sqrt{6n}$;*
- *the product of the clique number and the independence number of G is at most n.*

It has been conjectured that, with finitely many exceptions, every connected vertex-transitive graph is Hamiltonian. The Petersen graph is one of these exceptions: it has a Hamiltonian path but no Hamiltonian cycle. Only four exceptional graphs are currently known, and all have Hamiltonian paths.

The *Hadwiger number* of a graph is the smallest number k for which some component of the graph can be contracted to the complete graph K_k. A graph is *toroidal* if it is embeddable in the torus, and is *ring-like* if the vertices can be partitioned into sets $S_0, S_1, \ldots, S_{n-1}$, such that all edges join vertices in the same set or cyclically in consecutive sets, and the automorphism group induces a cyclic or dihedral group on this family of sets. Using these concepts, Babai [3] and Thomassen [50] have recently obtained structure theorems for connected vertex-transitive graphs with prescribed Hadwiger number.

Theorem 7.2 *Every sufficiently large connected vertex-transitive graph with Hadwiger number k is either toroidal or ring-like, with the cardinalities of the sets S_i bounded by a function of the Hadwiger number in the ring-like case.*

Clearly, arbitrarily large toroidal vertex-transitive graphs can be obtained as quotients of plane lattices – for example, rectangular grids with opposite sides identified. The proof of this substantial result involves many geometrical ideas, including isoperimetric inequalities for the hyperbolic plane. A related result of Thomassen [51] shows that there are only finitely many vertex-transitive graphs of given genus $g \geq 3$.

Two properties that are weaker than vertex-transitivity, but stronger than regularity, are walk-regularity and neighbourhood-regularity. The first of these is touched on in Chapter 8; here we consider the second.

Let a graph H be given. A graph G is *locally* H if, for each vertex $v \in V(G)$, the induced subgraph on the set of neighbours of v is isomorphic to H. A graph is *neighbourhood-regular* if it is locally H, for some H.

The problem of deciding, for a given graph H, whether there is a graph that is locally H, is recursively unsolvable (see [8] and [57]). Nevertheless, there are a number of positive results. For example:

- For some graphs H, all graphs that are locally H have been determined; see Hall [32] for locally Petersen graphs, for example.
- If H is regular and connected with girth at least 6, then there are infinite graphs that are locally H; see Weetman [54].
- If H is regular with diameter 2, and if it satisfies some extra conditions, then every locally H graph is finite; see Weetman [55].

A property of graphs that does not obviously relate to symmetry, but turns out to imply vertex-transitivity, is compactness. The basic results on this concept are due to Tinhofer [52]. To define it, we note that any permutation g of $\{1, 2, \ldots, n\}$ can be represented by a *permutation matrix* $\mathbf{P}(g)$, and that $g \in \mathrm{Aut}(G)$ if and only if $\mathbf{P}(g)$ commutes with $\mathbf{A}$, where $\mathbf{A}$ is the adjacency matrix of G.

A matrix $\mathbf{M}$ is *doubly stochastic* if its entries are non-negative and each of its row and column sums is 1. By a theorem of Birkhoff (see Chapter 2, Theorem 5.2), any doubly stochastic matrix is a convex combination of permutation matrices.

A graph G with adjacency matrix $\mathbf{A}$ is *compact* if every doubly stochastic matrix $\mathbf{M}$ that commutes with $\mathbf{A}$ is a convex combination of permutation matrices that commute with $\mathbf{A}$ – that is, automorphisms of $\mathbf{A}$.

The set of doubly stochastic matrices that commute with $\mathbf{A}$ is a polytope. If G is compact, its automorphisms are precisely the extreme points of this polytope, and can be found efficiently by linear programming. Birkhoff's theorem shows that the complete graph is compact.

For arbitrary graphs the meaning of compactness is somewhat mysterious, but for regular graphs, we have the following result. Note that if a graph is compact, then so is its complement.

Theorem 7.3 *If G is a compact connected regular graph, then any two vertices of G can be interchanged by an automorphism of G. In particular, G is vertex-transitive.*

The converse of this theorem is false. If G is compact and regular of degree k, then $(1/k)\mathbf{A}$ is a doubly stochastic matrix that commutes with $\mathbf{A}$, and is thus a convex combination of automorphisms. Each such automorphism g has the property that $vg \sim v$ for all vertices v. However, many vertex-transitive graphs, such as the Petersen graph, have no non-identity automorphisms with this property. For more on compact graphs, see Godsil [14]. A more general concept called *weak compactness* has been considered by Evdokimov *et al.* [17].

8. Higher symmetry

Symmetry conditions related to (and mostly stronger than) vertex-transitivity have received a lot of attention, often using group-theoretic techniques. One of the simplest is *edge-transitivity*, which usually implies vertex-transitivity. Indeed, an edge-transitive graph that is not vertex-transitive must be bipartite, with the orbits of the automorphism group as the bipartite blocks; the complete bipartite graph $K_{r,s}$ with $r \neq s$ is a simple example. There are also graphs that are vertex-transitive and edge-transitive but not *arc-transitive* (where an arc is a directed edge). For arc-transitive graphs, the connection between graphs and groups becomes particularly strong; we refer to Chapter 7 for a survey.

A strengthening of arc-transitivity is *distance-transitivity,* where we require the automorphism group to act transitively on pairs of vertices at distance i, for $i = 0, 1, \ldots, d$, where d is the diameter of the graph. A major research effort directed at the determination of all such graphs is discussed in Chapter 9.

An even stronger symmetry condition is homogeneity: a graph G is *homogeneous* if each isomorphism between (finite) induced subgraphs of G can be extended to an automorphism of G. All finite homogeneous graphs have been determined by Sheehan and Gardiner (see [22]).

Theorem 8.1 *A finite graph is homogeneous if and only if it is one of the following:*

- *a disjoint union of complete graphs of the same order;*
- *a regular complete multipartite graph;*
- *the* 5*-cycle* C_5*;*
- *the line graph of* $K_{3,3}$.

More generally, we say that a graph G is *t-homogeneous* if any isomorphism between induced subgraphs of order at most t extends to an automorphism of G. Now t-homogeneity obviously implies the combinatorial property $C(t)$ defined in Chapter 8. The list of graphs that satisfy $C(5)$ is the same as the list of homogeneous graphs in the preceding theorem, so the hypothesis can be weakened to 5-homogeneity.

9. Infinite graphs

We turn now to infinite graphs. Here, there are two very different areas of research – the first for locally finite graphs (in which each vertex has finite degree), and the second for general graphs (but usually requiring homogeneity or some model-theoretic notions).

For a locally finite graph, the notion of an *end* (introduced by Halin [30] for graphs, although used earlier for groups) is crucial. A *ray* is a one-way infinite path in a graph. König's Infinity lemma (see [39]) shows that any locally finite connected infinite graph contains a ray. Let $\mathcal{R}(G)$ be the set of rays in G. We define an equivalence relation $\equiv$ on $\mathcal{R}(G)$ by saying that $R_1 \equiv R_2$ if there is a ray R_3 that intersects both R_1 and R_2 in infinitely many vertices. The equivalence classes of $\equiv$ are the *ends* of G; we denote the set of ends by $\mathcal{E}(G)$.

It can be shown that the number of ends of a locally finite graph G is the supremum of the number of infinite components of $G - S$, taken over all finite subsets S of $V(G)$. (This result does not distinguish among infinite cardinals: we just say that the supremum of an unbounded set of natural numbers is ∞.) For example, the integer lattice graph $\mathbf{Z}^k$ has just one end for $k > 1$, but two ends for $k = 1$; an infinite tree with maximum degree greater than 2 has infinitely many (indeed, uncountably many) ends.

The main results connecting ends and automorphisms are the following theorems of Halin [31] and Jung [38], respectively.

Theorem 9.1 *Every automorphism of a locally finite connected infinite graph fixes either an end or a finite subgraph.*

Theorem 9.2 *Let G be a locally finite connected infinite graph, and suppose that* $\mathrm{Aut}(G)$ *has only finitely many orbits in $V(G)$. Then the number of ends of G is* 1, 2 *or* $2^{\aleph_0}$.

For a locally finite graph, we can also consider the rate of growth of the number a_n of vertices that are at distance at most n from a fixed vertex v. This number depends on the chosen vertex v, but if the distance from u to v is d, then

$$a_{n-d}(v) \leq a_n(u) \leq a_{n+d}(v).$$

Thus, the asymptotics of the rate of growth (for example, polynomial of degree k or exponential with constant c) do not depend on v, and we can talk of the *growth* of G. If $G = \mathrm{Cay}(\mathcal{G}, S)$, then the choice of a finite generating set S for $\mathcal{G}$ does not affect the asymptotics of growth, and we can again talk of the growth of $\mathcal{G}$. (Note, however, that a group can act vertex-transitively on each of two graphs with different growth.) The growth is polynomial of degree k for the integer lattice $\mathbf{Z}^k$, and exponential with constant $k-1$ for the k-regular tree.

This different behaviour is related to the number of ends.

Theorem 9.3 *Let G be a locally finite connected infinite graph whose automorphism group has only finitely many orbits.*

(a) If the growth of G is bounded by a polynomial, then

$$c_1 n^k \leq a_n \leq c_2 n^k,$$

where k is a positive integer and $c_1, c_2 > 0$.

(b) G has linear growth if and only if it has two ends.

(c) If G is a Cayley graph of $\mathcal{G}$, then the growth is polynomial if and only if $\mathcal{G}$ is nilpotent-by-finite.

(d) There exist groups whose growth is faster than polynomial but slower than exponential.

(e) If G has infinitely many ends, then it has exponential growth.

Here part (*c*) is a celebrated theorem of Wolf [58] and Gromov [27] on groups of polynomial growth; see Trofimov [53] for an extension to vertex-transitive graphs. Part (*b*) follows from Gromov and from Seifter and Trofimov [48], and an example for part (*d*) is the *Grigorchuk group* [26]. For a survey of vertex-transitive graphs with polynomial growth, see [37]. There are also connections between the growth of a graph and harmonic analysis; Lubotzky [42] gives an account of this material.

Macpherson [45] has determined the locally finite infinite graphs that are distance-transitive. For any integers $s, t > 1$, there is an infinite tree that is semi-regular, with degrees s and t in the two bipartite blocks. Let $M(s, t)$ be the graph with the bipartite block of degree s as its vertex-set; and where two vertices are adjacent if they lie at distance 2 in the tree.

Theorem 9.4 *Every locally finite distance-transitive infinite graph is isomorphic to* $M(s, t)$, *for some* $s, t > 1$.

No such result holds without local finiteness. Some examples to illustrate this are given in Cameron [12].

Turning to arbitrary infinite graphs, we first describe the paradoxical result of Erdős and Rényi [16]: up to isomorphism there is *a unique countable random graph* – that is, a graph R such that any countable random graph is isomorphic to R with probability 1. Moreover, R is highly symmetric – indeed, it is homogeneous, as defined in the preceding section – so the typical asymmetry of finite graphs does not hold in the countably infinite! A survey of this remarkable graph R appears in [11].

Other aspects of the theory are also very different in the countably infinite case, largely as a result of the random graph R. For example, in any group $\mathcal{G}$, the *square-root* set is the set $\sqrt{g} = \{x \in \mathcal{G} : x^2 = g\}$; it is *non-principal* if $g \neq 1$. The hypotheses of the following theorem of Cameron and Johnson are very mild.

Theorem 9.5 *Let* $\mathcal{G}$ *be a countable group that is not the union of a finite number of translates of non-principal square root sets. Then almost all random Cayley graphs for* $\mathcal{G}$ *are isomorphic to* R.

The classification of homogeneous graphs was extended to the infinite case by Lachlan and Woodrow [40]. A graph is *universal* K_n*-free* if it contains no complete graph of size n but contains every K_n-free graph as an induced subgraph. For each $n \geq 2$, there is just one countable homogeneous universal K_n-free graph; these graphs were first constructed by Henson [34], but their existence and uniqueness follow from a general construction method of Fraïssé [19]. We denote this unique graph by H_n.

Theorem 9.6 *Every countable homogeneous graph is one of the following:*

- *a disjoint union of complete graphs of the same size, or its complement (a regular complete multipartite graph);*
- *Henson's graph* H_n *(for* $n \geq 3$*), or its complement;*
- *the random graph* R.

10. Graph homomorphisms

In this final section we turn to graph homomorphisms. In general, homomorphisms are more revealing of graph structure than automorphisms, and the theory has developed in surprising directions. The material here is based on Hahn and Tardif [29], to which we refer for more details and references. In this section, we consider only finite simple graphs.

A *homomorphism* from a graph G to a graph H is a function f from $V(G)$ to $V(H)$ such that, if $v \sim w$ in G, then $vf \sim wf$ in H. Thus, an isomorphism is a bijective homomorphism whose inverse is also a homomorphism. We write $G \to H$ if there is a homomorphism from G to H.

If $G \to H$ and $H \to G$, then we say that G and H are *homomorphically equivalent,* and write $G \leftrightarrow H$. We write the homomorphic equivalence class of G as $[G]$. The set of such equivalence classes is partially ordered by the rule that $[G] \preceq [H]$ if $G \to H$. Much is known about this partial order: it is a lattice order, and it is dense – that is, if $[G] \prec [H]$, then there exists a graph for which $[G] \prec [K] \prec [H]$.

A proper *vertex-colouring* of G with r colours, being a map from $V(G)$ to $\{1, 2, \ldots, r\}$ such that adjacent vertices have distinct images, is a homomorphism from G to the complete graph K_r; for example, every bipartite graph has a homomorphism onto K_2. We think of $G \to H$ as saying that 'G has a H-colouring'. Thus, the existence and enumeration questions for homomorphisms from G to H generalize the chromatic number and chromatic polynomial of G, and so we expect these questions to be hard!

The *independence ratio* $i(G)$ of a graph G is the ratio of the cardinality of the largest independent set of vertices in G to the total number of vertices of G. The *odd girth* $g_o(G)$ of G is the length of a shortest odd cycle in G, and $\omega(G)$ and $\chi(G)$ denote the clique number and the chromatic number of G. The following result gives a necessary condition for the existence of a homomorphism from G to H:

Theorem 10.1 *Suppose that $G \to H$. Then*
- $\omega(G) \geq \omega(H)$, $\chi(G) \leq \chi(H)$ *and* $g_o(G) \geq g_o(H)$;
- *if H is vertex-transitive, then* $i(G) \geq i(H)$.

A *retraction* of a graph G is a homomorphism f from G onto an induced subgraph H of G for which the restriction of f to $V(H)$ is the identity map; the subgraph H is then called a *retract* of G. It is easy to see that some power of every homomorphism from G to G is a retraction. Any retraction of a connected graph can be expressed as the composition of a sequence of foldings, where a *folding* is a homomorphism that identifies one pair of vertices. The result of Sabidussi [47] mentioned in Section 7 shows that every vertex-transitive graph is a retract of a Cayley graph.

Retracts play an important role in topology, based in part on the fact that every retract H of a graph G is an *isometric subgraph* of G – that is, the distance between two vertices of H is the same in both graphs.

A graph G is a *core* if it has no non-trivial retraction – that is, if every homomorphism is an automorphism. If H is a retract of G and is itself a core, then we say that H is a *core of* G. The following result holds.

Theorem 10.2

(a) If $G \leftrightarrow H$, then any core of G is isomorphic to any core of H.

(b) In particular, all cores of G are isomorphic, and we can speak of the core of G.

(c) Up to isomorphism, the core of G is the smallest graph in $[G]$.

(d) The core of a vertex-transitive graph is vertex-transitive.

Many cores are known – examples include all circulants of prime order, and the *Kneser graph* $K(r, s)$ whose vertices correspond to the s-subsets of $\{1, 2, \ldots, r\}$ with $r > 2s$ and where two vertices are adjacent if the corresponding sets are disjoint (the Petersen graph is $K(5, 2)$). For a survey of cores and their properties, see Hell and Nešetřil [33].

References

1. N. Alon and Y. Roichman, Random Cayley graphs and expanders, *Random Structures Appl.* **5** (1994), 271–284.
2. L. Babai, Automorphism groups of graphs and edge-contraction, *Discrete Math.* **8** (1974), 13–20.
3. L. Babai, Vertex-transitive graphs and vertex-transitive maps, *J. Graph Theory* **15** (1991), 587–627.
4. L. Babai and C. D. Godsil, On the automorphism groups of almost all Cayley graphs, *Europ. J. Combin.* **3** (1982), 9–15.
5. L. Babai and A. J. Goodman, On the abstract group of automorphisms, *Coding Theory, Design Theory, Group Theory* (eds. D. Jungnickel and S. A. Vanstone), Wiley (1993), 121–143.
6. L. Babai, A. J. Goodman and L. Pyber, On faithful permutation representations of small degree, *Comm. Algebra* **21** (1993), 1587–1602.
7. L. Babai, D. Yu. Grigoryev and D. M. Mount, Isomorphism of graphs with bounded eigenvalue multiplicity, *Proc. 14th ACM STOC* (1982), 310–324.
8. V. K. Bulitko, O grafach s zadanymi okruzeniami veršin (in Russian), *Trudy Mat. Inst. im. Steklova* **133** (1973), 78–94.
9. P. J. Cameron, On graphs with given automorphism group, *Europ. J. Combin.* **1** (1980), 91–96.

10. P. J. Cameron, 6-transitive graphs, *J. Combin. Theory (B)* **28** (1980), 168–179.
11. P. J. Cameron, The random graph, *The Mathematics of Paul Erdős* (eds. J. Nešetřil and R. L. Graham), Springer (1996), 331–351.
12. P. J. Cameron, A census of infinite distance-transitive graphs, *Discrete Math.* **192** (1998), 11–26.
13. P. J. Cameron, *Permutation Groups,* Cambridge Univ. Press, 1999.
14. A. Chan and C. D. Godsil, Symmetry and eigenvectors, *Graph Symmetry: Algebraic Methods and Applications* (eds. G. Hahn and G. Sabidussi), Kluwer (1997), 75–106.
15. W. Dicks and M. J. Dunwoody, *Groups Acting on Graphs,* Cambridge Univ. Press, 1989.
16. P. Erdős and A. Rényi, Asymmetric graphs, *Acta Math. Acad. Sci. Hungar.* **14** (1963), 295–315.
17. S. Evdokimov, M. Karpinski and I. Ponomarenko, Compact cellular algebras and permutation groups, *Discrete Math.* **197/198** (1999), 247–267.
18. B. Fischer, Finite groups generated by 3-transpositions, I, *Invent. Math.* **13** (1971), 232–246.
19. R. Fraïssé, Sur certains relations qui généralisent l'ordre des nombres rationnels, *C. R. Acad. Sci. Paris* **237** (1953), 540–542.
20. R. Frucht, Herstellung von Graphen mit vorgegebener abstrakter Gruppe, *Compositio Math.* **6** (1938), 239–250.
21. R. Frucht, Graphs of degree 3 with given abstract group, *Canad. J. Math.* **1** (1949), 365–378.
22. A. D. Gardiner, Homogeneous graphs, *J. Combin. Theory (B)* **20** (1976), 94–102.
23. M. R. Garey and D. S. Johnson, *Computers and Intractibility: A Guide to the Theory of NP-Completeness,* W. H. Freeman, 1979.
24. M. Guidici, Quasiprimitive groups with no fixed-point-free elements of prime order, *J. London Math. Soc. (2)* **67** (2003), 73–84.
25. C. D. Godsil, GRR's for non-solvable groups, *Colloq. Math. Soc. János Bolyai* **25** (1981), 221–239.
26. R. I. Grigorchuk, On the Burnside problem for periodic groups, *Funkcional Anal. i Prilozen* **14** (1980), 53–54; transl. *Functional Anal. Appl.* **14** (1980), 41–43.
27. M. Gromov, Groups of polynomial growth and expanding maps, *Publ. Math. IHES* **53** (1981), 53–78.
28. G. Hahn and G. Sabidussi (eds.), *Graph Symmetry: Algebraic Methods and Applications,* Kluwer, 1997.
29. G. Hahn and C. Tardif, Graph homomorphisms: Structure and symmetry, *Graph Symmetry: Algebraic Methods and Applications* (eds. G. Hahn and G. Sabidussi), Kluwer (1997), 107–166.
30. R. Halin, Über unendliche Wege in Graphen, *Math. Ann.* **157** (1964), 125–137.
31. R. Halin, Automorphisms and endomorphisms of infinite locally finite graphs, *Abh. Math. Sem. Hamburg* **39** (1973), 251–283.
32. J. I. Hall, Locally Petersen graphs, *J. Graph Theory* **4** (1980), 173–187.
33. P. Hell and J. Nešetřil, The core of a graph, *Discrete Math.* **109** (1992), 117–126.
34. C. W. Henson, A family of countable homogeneous graphs, *Pacific J. Math.* **38** (1971), 69–83.
35. D. Hetzel, *Über reguläre graphische Darstellungen von auflösbare Grupen*, Diplomarbeit, Techn. Univ. Berlin, 1976.

36. D. G. Higman and C. C. Sims, A simple group of order 44,352,000, *Math. Z.* **105** (1968), 110–113.
37. W. Imrich and N. Seifter, A survey on graphs with polynomial growth, *Discrete Math.* **95** (1991), 101–117.
38. H. A. Jung, A note on fragments of infinite graphs, *Combinatorica* **1** (1981), 285–288.
39. D. König, *Theorie der endlichen und unendlichen Graphen,* Akad. Verlagsgesellschaft, 1936; English translation: *Theory of Finite and Infinite Graphs* (transl. R. McCoart), Birkhäuser, 1990.
40. A. H. Lachlan and R. E. Woodrow, Countable ultrahomogeneous undirected graphs, *Trans. Amer. Math. Soc.* **262** (1980), 51–94.
41. On graphs whose full automorphism group is an alternating group or a finite classical group, *Proc. London Math. Soc. (3)* **47** (1983), 337–362.
42. A. Lubotzky, *Discrete Groups, Expanding Graphs, and Invariant Measures,* Birkhäuser, 1994.
43. E. M. Luks, Isomorphism of graphs of bounded valence can be tested in polynomial time, *J. Comp. Sys. Sci.* **25** (1982), 42–65.
44. B. D. McKay and C. E. Praeger, Vertex-transitive graphs which are not Cayley graphs, I, *J. Austral. Math. Soc. (A)* **56** (1994), 53–63.
45. H. D. Macpherson, Infinite distance-transitive graphs of finite valency, *Combinatorica* **2** (1982), 63–69.
46. J. W. Moon, Tournaments with a given automorphism group, *Canad. J. Math.* **16** (1964), 485–489.
47. G. Sabidussi, Vertex-transitive graphs, *Monatsh. Math.* **68** (1964), 426–438.
48. N. Seifter and V. I. Trofimov, Automorphism groups of graphs with quadratic growth, *J. Combin. Theory (B)* **71** (1997), 205–210.
49. J.-P. Serre, *Trees,* Springer, 1980.
50. C. Thomassen, The Hadwiger number of infinite vertex-transitive graphs, *Combinatorica* **12** (1992), 481–491.
51. C. Thomassen, Embeddings of graphs, *Discrete Math.* **124** (1994), 217–228.
52. G. Tinhofer, Graph isomorphism and theorems of Birkhoff type, *Computing* **36** (1986), 285–300.
53. V. I. Trofimov, Graphs with polynomial growth, *Mat. Sbornik* **123** (1984), 407–421; transl. *Math. USSR Sbornik* **51** (1985), 405–417.
54. G. M. Weetman, A construction of locally homogeneous graphs, *J. London Math. Soc. (2)* **50** (1994), 68–86.
55. G. M. Weetman, Diameter bounds for graph extensions, *J. London Math. Soc. (2)* **50** (1994), 209–221.
56. H. Wielandt, *Permutation Groups through Invariant Relations and Invariant Functions,* Lecture Notes, Ohio State University, 1969.
57. P. M. Winkler, Existence of graphs with a given set of r-neighbourhoods, *J. Combin. Theory (B)* **34** (1983), 165–176.
58. J. Wolf, Growth of finitely generated solvable groups and curvature of Riemannian manifolds, *J. Diff. Geom.* **2** (1968), 421–446.
59. N. C. Wormald, Models of random regular graphs, *Surveys in Combinatorics, 1999* (eds. J. D. Lamb and D. A. Preece), London Math. Soc. Lecture Notes Series **267**, Cambridge Univ. Press (1999), 239–298.

6

Cayley graphs

BRIAN ALSPACH

Cayley graphs are constructed using groups. This chapter discusses algebraic and graph-theoretic aspects of finite Cayley graphs. The algebraic aspects include recognition, isomorphism, prevalence, enumeration and automorphisms. The graph-theoretic aspects include notable members of the family, subgraphs, factorization, hamiltonicity and embeddings.

1. Introduction

There are several constructions that produce families of vertex-transitive graphs. This chapter deals with the family of *Cayley graphs,* probably the most widely known and extensively studied family of vertex-transitive graphs. The construction for Cayley graphs is based on groups. We restrict ourselves to finite groups, but the basic construction is the same for infinite groups. While Cayley graphs on finite groups and those on infinite groups share a variety of features, there are aspects of Cayley graphs on finite groups that do not carry over to those on infinite groups, and *vice versa*.

Let $\mathcal{G}$ be a finite group with identity 1. Let $\mathcal{S}$ be a subset of $\mathcal{G}$ satisfying $1 \notin \mathcal{S}$ and $\mathcal{S} = \mathcal{S}^{-1}$ – that is, $s \in \mathcal{S}$ if and only if $s^{-1} \in \mathcal{S}$. The *Cayley graph* Cay($\mathcal{G}$; $\mathcal{S}$) on $\mathcal{G}$ with *connection set* $\mathcal{S}$ is defined as follows:

- the vertices are the elements of $\mathcal{G}$;
- there is an edge joining g and h if and only if $h = sg$ for some $s \in \mathcal{S}$.

The set of all Cayley graphs on $\mathcal{G}$ is denoted by Cay($\mathcal{G}$).

When $\mathcal{G}$ is an Abelian group, we use additive notation for the group operation. Hence, we write $\mathcal{S} = -\mathcal{S}$ for the connection set, and $h = s + g$ (for some $s \in \mathcal{S}$) for adjacency.

Cayley digraphs are defined in a similar way, but with the condition $\mathcal{S} = \mathcal{S}^{-1}$ removed. Many results about Cayley graphs carry over immediately to Cayley digraphs, but there are aspects of Cayley digraphs that distinguish them from Cayley graphs. This chapter does not specifically address Cayley digraphs, except briefly in Section 9.

2. Recognition

When $\mathcal{G}$ is a finite group and $g \in \mathcal{G}$, define g_R acting on $\mathcal{G}$ by $hg_R = hg$, for all $h \in \mathcal{G}$; clearly, g_R is a permutation of the elements of $\mathcal{G}$. Define the group $\mathcal{G}_R$ by $\mathcal{G}_R = \{g_R : g \in \mathcal{G}\}$. It is easy to see that $\mathcal{G}_R$ acts transitively on the elements of $\mathcal{G}$, and that g_R is an automorphism of any Cayley graph on $\mathcal{G}$. These observations prove the following result, thereby providing a large class of vertex-transitive graphs.

Theorem 2.1 *Every Cayley graph is vertex-transitive.*

The Cayley graphs on cyclic groups have played a special role in the study of Cayley graphs. They are widely known as *circulant graphs*, because their adjacency matrices are circulant matrices. We use the notation Circ(n; $\mathcal{S}$) to denote the circulant graph of order n with connection set $\mathcal{S}$. An example is given in Fig. 1.

How difficult is it to recognize whether an arbitrary graph is a Cayley graph, or a vertex-transitive graph for that matter? A celebrated theorem of Sabidussi addresses this recognition question. Before proceeding to the theorem, a definition is required.

Let $\mathcal{G}$ be a transitive permutation group acting on a finite set Ω. It is easy to show that the following three conditions are equivalent:

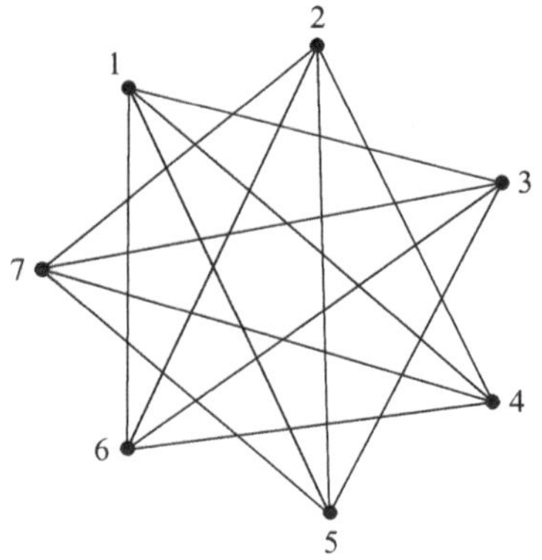

Fig. 1. The circulant graph Circ(7; {2, 3, 4, 5})

- the only element of $\mathcal{G}$ that fixes an element of Ω is the identity permutation;
- $|\mathcal{G}| = |\Omega|$;
- for any $\omega_1, \omega_2 \in \Omega$, there is a unique element $g \in \mathcal{G}$ satisfying $\omega_1 g = \omega_2$.

A transitive permutation group $\mathcal{G}$ that satisfies any one of the these conditions is said to be *regular*. We now state the fundamental result of Sabidussi [48].

Theorem 2.2 *A graph G is a Cayley graph if and only if* Aut(G) *contains a regular subgroup.*

Here is an outline of the proof. To prove the necessity, note that $\mathcal{G}_R$ is regular and that $\mathcal{G}_R \leq \text{Aut}(G)$, for any $G \in \text{Cay}(\mathcal{G})$. To prove the sufficiency, assume that Aut(G) contains a regular subgroup $\mathcal{G}$. Label an arbitrary vertex of G with the identity element 1 of $\mathcal{G}$. For each of the remaining vertices $v \in V(G)$, there is a unique permutation $g_v \in \mathcal{G}$ that maps the vertex labelled 1 to v. Label v with the group element g_v. It is then easy to show that G is a Cayley graph on $\mathcal{G}$.

Essentially all of the work that has been done on establishing whether particular graphs are Cayley graphs employs Theorem 2.2, and the way in which the theorem is used depends on what we are given. Sometimes we are provided a graph construction whereby the automorphism group of the graph may be computed. We may then determine whether the group contains a regular subgroup. At other times we may not be able to produce the automorphism group easily, but still may be able to show that a regular subgroup cannot occur. In other words, we are dealing with cases in which our description of the graph is special and contains considerable information regarding recognition.

In the case of arbitrary graphs, we assume that all we have is a description of the graph as an adjacency matrix, or as a list of the edges in the graph. Recognition now means that we are asking about the existence of an algorithm for determining whether an arbitrary graph is a Cayley graph. There are good algorithms for determining whether a graph is vertex-transitive, and they work on graphs with thousands of vertices. They produce generators for the automorphism group from which verifying vertex-transitivity is easy. However, in trying to recognize whether or not a vertex-transitive graph is a Cayley graph, we are left with the problem of determining whether the automorphism group contains a regular subgroup.

There has been some work on finding efficient algorithms for Cayley graph recognition, but the classes studied have been limited to circulant tournaments and special circulant graphs. Efficiency is gained by working with association schemes arising from the graphs (see [43]). We then must show that the association scheme itself is cyclic, which may be done efficiently, so that Theorem 2.2 is still present, again demonstrating the fundamental nature of Sabidussi's Theorem.

We shall ignore the computational problem of recognizing whether an arbitrary graph is a Cayley graph. Instead, we always assume that Cayley graphs have been described in terms of the groups on which they are built, together with the connection sets. For most problems this is not a drawback.

3. Special examples

There are situations involving graphs for which Cayley graphs appear in a natural way. Often they arise from a small select collection of Cayley graphs. In this section we present some notable Cayley graphs that appear frequently in the literature. We define them in as straightforward a manner as possible, but point out that usually they may be defined as Cayley graphs in many ways. This aspect will be discussed in Section 4.

The complete graphs and their complements are Cayley graphs. In particular, K_n is a Cayley graph on any group $\mathcal{G}$ of order n, where the connection set is the set of non-identity elements of the group.

The complete multipartite graph $K_{r(s)}$, with r parts each of cardinality s, is also a Cayley graph. It can be achieved by using the circulant graph of order rs with the connection set consisting of all the elements that are not congruent to 0 (modulo s).

Of course, complete graphs and complete multipartite graphs appear all the time, but for the most part their appearances are not related to the fact that they are Cayley graphs. The rest of the special Cayley graphs we discuss owe much of their interest to the fact they are Cayley graphs.

The k-dimensional cube Q_k is the Cayley graph defined on the elementary Abelian 2-group $\mathbf{Z}_2^k$, where the connection set is the standard generating set for

$\mathbf{Z}_2^k$. The graph Q_k has generated a long history of interesting questions, and has become even more important in recent years because of the role it plays in computer architecture.

The graph formed on the finite field $\mathbf{F}_q$, where $q \equiv 1 \pmod 4$ and the connection set is the set of quadratic residues in $\mathbf{F}_q$, is called a *Paley graph*; an example appears in Fig. 2. Here we define edges additively, since we are using an additive group as the underlying vertex-set. Paley graphs have many interesting properties.

Let n be even and let $S = \{\pm 1, n/2\}$. The circulant graph of order n with connection set S is known as the *Möbius ladder* of order n. These graphs play a role in topological graph theory.

Problems arising in computer science have brought about a resurgence of interest in 'grid-like' graphs. The corresponding Cayley graphs are the Cartesian products of cycles of some fixed length $\ell \geq 3$. These graphs are realized as Cayley graphs by using the group $\mathbf{Z}_\ell^n$, where the connection set is the set of standard generators of the group.

4. Prevalence

The family of Cayley graphs provides us with a straightforward construction for vertex-transitive graphs. A natural first question is to ask whether the family of Cayley graphs encompasses all finite vertex-transitive graphs. We turn to the Petersen graph P for a negative answer.

It is easy to show that P is vertex-transitive. To show that it is not a Cayley graph, we show that Aut(P) does not contain a regular subgroup and apply Theorem 2.2.

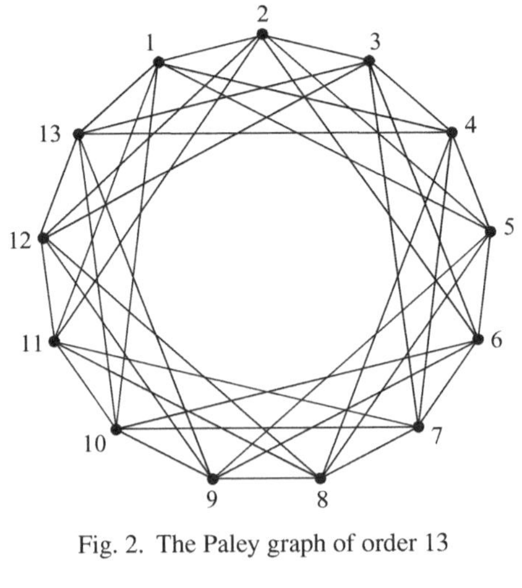

Fig. 2. The Paley graph of order 13

We can prove this naïvely using some graph-theoretic facts about P. If $\mathcal{G}$ is a regular subgroup of Aut(P) and

$$\rho = (v_0\, v_1\, v_2\, v_3\, v_4)(w_0\, w_1\, w_2\, w_3\, w_4)$$

is an element of order 5 in $\mathcal{G}$, then by using the facts that P is connected, is regular of degree 3, is not bipartite, has girth 5, and by relabelling (if necessary), we may assume that $v_i v_{i+1}$, $w_i w_{i+2}$ and $v_i w_i$ are edges, for $i \equiv 0, 1, 2, 3, 4$ (modulo 5). Let τ be the element of $\mathcal{G}$ for which $v_0\tau = w_0$. Then τ must have order 2 and satisfy $\tau^{-1}\langle\rho\rangle\tau = \langle\rho\rangle$. This requires that τ cannot preserve edges of P, so that no such τ can exist. Hence, Aut(P) contains no regular subgroup, and we conclude that P is not a Cayley graph.

We pursue the preceding example in two directions. The essential reason that the Petersen graph is not a Cayley graph is that no involution interchanges the two orbits of $\langle\rho\rangle$. We examine this more carefully.

With the same labelling as before, the mapping defined by $f : v_i \to w_{2i}$ and $f : w_i \to v_{2i}$ is an automorphism of P. Note that f has order 4, and that $2^2 = -1$ in $\mathbf{F}_5$. This is the key to what follows.

For any prime $p \equiv 1 \pmod 4$, -1 is a quadratic residue modulo p. Let $\alpha^2 \equiv -1 \pmod p$. Form the circulant graph H_0 on $\mathbf{Z}_p$ with connection set $\{1, -1\}$ (recall that we use additive notation for Abelian groups) and vertex-set $\{v_0, v_1, \ldots, v_{p-1}\}$, and the circulant graph H_1 on $\mathbf{Z}_p$ with connection set $\{\alpha, -\alpha\}$ and vertex-set $\{w_0, w_1, \ldots, w_{p-1}\}$. Define the cubic graph GP(p,α) as the vertex-disjoint union of H_0 and H_1, together with the perfect matching $v_i w_i$, for $i = 0, 1, \ldots, p-1$; GP(p, α) is sometimes called a *generalized Petersen graph*; GP(13, 5) appears in Fig. 3.

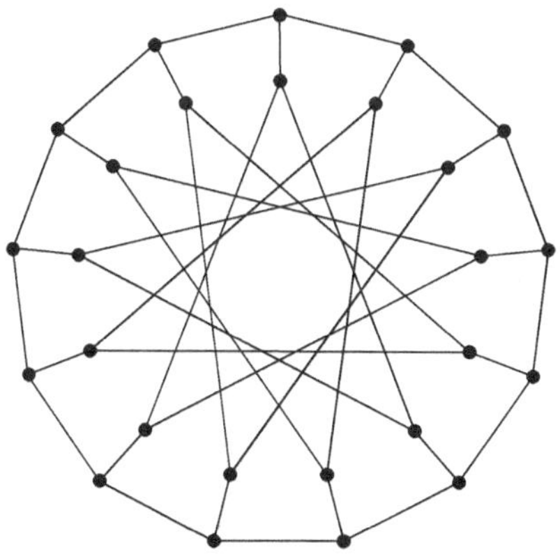

Fig. 3. The graph GP(13, 5)

The graph $\mathrm{GP}(p, \alpha)$ can be shown to be a non-Cayley vertex-transitive graph, in a manner similar to that used for P. In contrast, it is known that every vertex-transitive graph of order $2p$, when p is a prime satisfying $p \equiv 3 \pmod 4$, is a Cayley graph. In the same vein, Marušič [35] proved that all vertex-transitive graphs of orders p^2 and p^3, where p is a prime, are Cayley graphs. This prompted him to pose the problem of determining the set

$$\mathbf{NC} = \{n\text{: there exists a non-Cayley vertex-transitive graph of order } n\}.$$

An important elementary feature of **NC** is that $k\mathbf{NC} = \mathbf{NC}$, for all positive integers k. This follows from the observation that if G is a non-Cayley vertex-transitive graph of order n, then the graph obtained by taking k vertex-disjoint copies of G is a non-Cayley vertex-transitive graph of order kn. The complement of the latter graph provides a connected example.

Since $k\mathbf{NC} = \mathbf{NC}$, for all positive integers k, Marušič's problem is solved if we can find all the 'minimal' elements of **NC**: 'minimal' in this context means that $n \in \mathbf{NC}$, but no proper divisor of n belongs to **NC**.

It is convenient to consider the number of distinct prime factors when looking for minimal elements. When there is only one prime factor – that is, n is a prime power – the answer is completely settled. Every vertex-transitive graph of prime order is a circulant graph, and, as stated above, Marušič [35] showed that p^2 and p^3 never belong to **NC**.

It was known that $16 \in \mathbf{NC}$, and for any odd prime p, McKay and Praeger [37] showed by construction that $p^4 \in \mathbf{NC}$. Their construction is easy to describe. Let $G(p)$ be the graph with vertex-set $\{v_{i,j} : i \in \mathbf{Z}_p, j \in \mathbf{Z}_{p^3}\}$ and edge-set

$$\{v_{i,j}v_{i,j+pk},\ v_{i,j}v_{i+1,j},\ v_{i,j}v_{i+1,j+pa^i},\ v_{i,j}v_{i+1,j+a^{rp+i}}\},$$

where $i \in \mathbf{Z}_p$, $j \in \mathbf{Z}_{p^3}$, $k \in \mathbf{Z}_{p^2}$, $r \in \mathbf{Z}_p$, and $a = p + 1$. Then $G(p)$ does the job.

Thus, for any prime p, $p^e \in \mathbf{NC}$ if and only if $e \geq 4$. This implies that $n \in \mathbf{NC}$ whenever n is divisible by p^4, for some prime p. However, the latter result was superseded by a much stronger result obtained by McKay and Praeger [37]. They proved that any positive integer divisible by a square, other than 12, is in **NC**.

The preceding results reduce the problem of trying to characterize membership in **NC** to the consideration of square-free integers. We already have seen the answer when $n = 2p$, where p is an odd prime. The next theorem [38] characterizes membership in **NC** for a product of two distinct primes.

Theorem 4.1 *Let p and q be distinct primes, with $p < q$. Then $pq \in \mathbf{NC}$ if and only if one of the following holds:*
- *p^2 divides $q - 1$;*
- *$q = 2p - 1 > 3$ or $q = (p^2 + 1)/2$;*
- *$q = 2^t + 1$, and either p divides $2^t - 1$ or $p = 2^{t-1} - 1$;*

- $q = 2^t - 1$ *and* $p = 2^{t-1} + 1$;
- $p = 7$ *and* $q = 11$.

The situation for a product of three distinct primes becomes more complicated, but the answer is complete nonetheless (see [26]). We separate even and odd values.

Theorem 4.2 *Let p and q be odd primes, with $p < q$. Then $2pq \in$ **NC** if and only if one of the following holds:*
- p^2 *divides* $q - 1$;
- $p \equiv 1 \pmod 4$ *or* $q \equiv 1 \pmod 4$;
- $p = 7$ *and* $q = 11$ *or* 19;
- $p \equiv q \equiv 3 \pmod 4$, *$p$ divides $q - 1$, and p^2 does not divide $q - 1$;*
- $p \equiv q \equiv 3 \pmod 4$ *and* $p = (q + 1)/4$.

Theorem 4.3 *Let p, q, r be distinct odd primes, with $p < q < r$. Then $pqr \in$ **NC** if and only if at least one of pq, pr or qr is a member of **NC**, or none of pq, pr and qr is a member of **NC** but one of the following holds:*
- $pqr = (2^{2^t} + 1)(2^{2^{t+1}} + 1)$, *for some t;*
- $pqr = (2^{d\pm1} + 1)(2^d - 1)$, *for some prime d;*
- $pq = 2r + 1, 2r - 1$ *or* $(r + 1)/2$;
- $pq = (r^2 + 1)/2$ *or* $pr = (q^2 + 1)/2$;
- $pq = (r^2 - 1)/x$ *or* $pr = (q^2 - 1)/x$, *where* $x = 24, 48$ or 120;
- $ab = 2^t + 1$ *and c divides* $2^t - 1$, *where* $\{a, b, c\} = \{p, q, r\}$;
- *the largest power of p dividing $q - 1$ is p^p, and the largest power of q dividing $r - 1$ is q^q;*
- $q = (3p + 1)/2$ *and* $r = 3p + 2$, *or* $q = 6p - 1$ *and* $r = 6p + 1$;
- $q = (r - 1)/2$ *and p divides $r + 1$, where $p > q$ when* $p = (r + 1)/2$;
- $p = (k^{d/2} + 1)/(k + 1), q = (k^{d/2} - 1)/(k - 1), r = (k^{d-1} - 1)/(k - 1)$, *where $k, d - 1$ and $d/2$ are primes, and $p > q$ may hold;*
- $p = (k^{(d-1)/2} + 1)/(k + 1), q = (k^{(k-1)/2} - 1)/(k - 1), r = (k^d - 1)/(k - 1)$, *where k, d and $(d - 1)/2$ are primes, and $p > q$ may hold;*
- $p = k^2 - k + 1, q = (k^5 - 1)/(k - 1), r = (k^7 - 1)/(k - 1)$, *where k is prime;*
- $p = 3, q = (2^d + 1)/3, r = 2^d - 1$, *where d is prime;*
- $p = (2^d + 1)/3, q = 2^d - 1, r = 2^{2d\pm2} + 1$, *where $d = 2^t \pm 1$ is prime;*
- $p = 5, q = 11$ *and* $r = 19$;
- $p = 7, q = 73$ *and* $r = 257$.

So far there is no characterization of the members of **NC** that are products of four distinct primes. This leaves us with the following intriguing question. Is there a number $k > 0$ for which every product of k distinct primes is in **NC**? With the little evidence currently available, a guess in either direction appears hazardous.

The other direction that we wish to pursue, regarding the Petersen graph P as a non-Cayley vertex-transitive graph, depends on viewing P in a different way. Let the ten 2-element subsets of a 5-set correspond to the vertices of a graph, and let two vertices be adjacent whenever their corresponding subsets are disjoint; then the resulting graph is isomorphic to P.

The preceding is an interpretation of P as a *Kneser graph* (see Chapter 5). The general definition of Kneser graphs is as follows. Let r and s be positive integers satisfying $r \geq 2$ and $s \geq 2r + 1$. Let the vertices of the graph $K(s, r)$ correspond to the r-subsets of an s-set, and let two vertices be adjacent if and only if their corresponding r-subsets are disjoint; then the Petersen graph is $K(5, 2)$. It is easy to see that $K(s, r)$ is vertex-transitive; in fact, the automorphism group of $K(s, r)$ is the action on the r-subsets induced by the action of $\mathcal{S}_s$ on $\{1, 2, \ldots, s\}$. Using this fact, Godsil [20] proved that the only exceptions to $K(s, r)$ being a non-Cayley vertex-transitive graph occur when $r = 2$ and s is a prime power congruent to $3 \pmod 4$, or $r = 3$ and $s = 8$ or 32. The fact that $\binom{s}{r} \in$ **NC**, with the exceptions just mentioned, is not particularly useful for determining **NC**.

5. Isomorphism

Some of the most interesting and deepest work on Cayley graphs has revolved around the question of trying to determine when two Cayley graphs are isomorphic. For the most part, the question has been confined to Cayley graphs on the same group, but there has been some work examining the isomorphism of Cayley graphs on different groups. For a recent excellent survey that goes into far more detail than this section, see Li [30]. Some preliminary definitions and results are required.

For historical reasons, as well as the levels of difficulty involved, much of the work on isomorphism problems has been concentrated on Cayley graphs on Abelian groups. In fact, circulant graphs by themselves have garnered considerable attention.

Let p be a prime and let $\mathbf{Z}_p^*$ denote the multiplicative group of units of $\mathbf{Z}_p$. Define the permutation $T_{a,b}$ on $\mathbf{Z}_p$ for $a \in \mathbf{Z}_p^*$ and $b \in \mathbf{Z}_p$, by $xT_{a,b} = ax + b$. Using the notation $\mathcal{G} \equiv \mathcal{H}$ for equivalent permutation groups, we have the following classical theorem of Burnside.

Theorem 5.1 *A transitive permutation group $\mathcal{G}$ of prime degree p is either doubly transitive or $\mathcal{G} \equiv \{T_{a,b} : a \in \mathcal{H} < \mathbf{Z}_p^*, b \in \mathbf{Z}_p\}$.*

Burnside's theorem has some interesting applications for circulant graphs of prime order. If $\mathrm{Aut}(G)$ is doubly transitive for a graph G, then G is either complete or has no edges. Thus, Burnside's theorem reveals considerable information about other circulant graphs of prime order.

The first significant isomorphism result about Cayley graphs was the following theorem of Turner [52].

Theorem 5.2 *Let p be a prime. Two circulant graphs* $\mathrm{Circ}(p; \mathcal{S})$ *and* $\mathrm{Circ}(p; \mathcal{S}')$ *of order p are isomorphic if and only if* $\mathcal{S}' = a\mathcal{S}$, *for some* $a \in \mathbf{Z}_p^*$.

Here is an outline of the proof. If $\mathcal{S}' = a\mathcal{S}$ for some $a \in \mathbf{Z}_p^*$, then it is easy to verify that $T_{a,0}$ is an isomorphism from $\mathrm{Circ}(p; \mathcal{S})$ to $\mathrm{Circ}(p; \mathcal{S}')$. On the other hand, let $\mathrm{Circ}(p; \mathcal{S})$ and $\mathrm{Circ}(p; \mathcal{S}')$ be isomorphic circulant graphs on $\mathbf{Z}_p$ via an isomorphism φ. If $\mathrm{Circ}(p; \mathcal{S})$ is complete or has no edges, then the result follows trivially. In all other cases, Theorem 5.1 implies that their automorphism groups contain a unique subgroup of order p. Without loss of generality, we may assume that the subgroup is generated by the p-cycle $(0\ 1\ 2 \cdots p-1)$. Using the fact that the p-cycle $(\varphi(0)\,\varphi(1)\,\varphi(2)\ \cdots\ \varphi(p-1))$ generates the same subgroup, we easily find an $a \in \mathbf{Z}_p^*$ satisfying $\mathcal{S}' = a\mathcal{S}$.

Ádám posed the question of whether two circulant graphs $\mathrm{Circ}(n; \mathcal{S})$ and $\mathrm{Circ}(n; \mathcal{S}')$ are isomorphic if and only if there exists an $a \in \mathbf{Z}_n^*$ satisfying $\mathcal{S}' = a\mathcal{S}$, where $\mathbf{Z}_n^*$ denotes the multiplicative group of units in $\mathbf{Z}_n$.

It is easy to describe a counter-example to Ádám's conjecture. For $n = 25$, let

$$\mathcal{S} = \{1, 4, 5, 6, 9, 11, 14, 16, 19, 20, 21, 24\}$$

and

$$\mathcal{S}' = \{1, 4, 6, 9, 10, 11, 14, 15, 16, 19, 21, 24\}.$$

The two circulant graphs $\mathrm{Circ}(25, \mathcal{S})$ and $\mathrm{Circ}(25, \mathcal{S}')$ are isomorphic, since each is a wreath product of a 5-cycle with a 5-cycle. On the other hand, it is easy to see that there is no $a \in \mathbf{Z}_{25}^*$ for which $\mathcal{S}' = a\mathcal{S}$.

Theorem 5.2 set the stage for research on isomorphisms of Cayley graphs. The observation that multiplication by $a \in \mathbf{Z}_n^*$ is an automorphism of the underlying additive group indicates the direction to follow.

Let $\mathcal{G}$ be a finite group. Suppose that two Cayley graphs $\mathrm{Cay}(\mathcal{G}; \mathcal{S})$ and $\mathrm{Cay}(\mathcal{G}; \mathcal{S}')$ on $\mathcal{G}$ are isomorphic if and only if there exists a group automorphism $\alpha \in \mathrm{Aut}(\mathcal{G})$ such that $\mathcal{S}' = \mathcal{S}\alpha$; then we say that $\mathcal{G}$ is a *CI-group*. Rephrasing Theorem 5.2 in terms of this definition, we see that $\mathbf{Z}_p$ is a CI-group, where p is a prime. The problem of characterizing finite CI-groups then became, and still remains, one of the major problems on Cayley graphs. Some notable results have been obtained, and we examine some of them now.

Elementary Abelian p-groups have been studied in some detail. Godsil [21] proved that $\mathbf{Z}_p^2$ is a CI-group, while Dobson [15] proved that $\mathbf{Z}_p^3$ is a CI-group, for each prime p. Hirasaka and Muzychuk [25] proved that $\mathbf{Z}_p^4$ is a CI-group, for each prime p. These results lend credence to an earlier conjecture of Babai and Frankl [7] that $\mathbf{Z}_p^n$ is a CI-group for all primes p and all $n \geq 1$. However, Nowitz [44] proved that $\mathbf{Z}_2^6$ is not a CI-group by explicitly constructing two isomorphic Cayley graphs on $\mathbf{Z}_2^6$ between which no group automorphism acts as an isomorphism.

Nowitz's example is rather intriguing. Its degree is 32, so the degree of the complements, which are themselves counter-examples to the Babai-Frankl conjecture, is 31. Does there exist an example on $\mathbf{Z}_2^6$ with smaller degree?

It is known that $\mathbf{Z}_2^5$ is a CI-group. In other words, 6 is the smallest rank for which an elementary 2-group is not a CI-group. Nowitz's example suggests that there is a smallest rank r_p for which the elementary p-group of rank r_p is not a CI-group, but all of those with rank smaller than r_p are CI-groups. A recent result of Muzychuk [42] shows that such a rank does exist.

Theorem 5.3 *If p is a prime and $n \geq 2p - 1 + \binom{2p-1}{p}$, then the elementary p-group $\mathbf{Z}_p^n$ is not a CI-group.*

Note that the bound given in this theorem is tight for $p = 2$. This makes the problem of determining the behaviour of the function r_p tantalizing indeed. Also, must a counter-example to the Babai-Frankl conjecture of minimal rank have the property that the graphs and their complements are dense, as is the case for Nowitz's example?

Another remarkable result about CI-groups is the following theorem of Li [29].

Theorem 5.4 *If $\mathcal{G}$ is a finite CI-group, then $\mathcal{G}$ is solvable.*

It is feasible to determine the CI-groups completely because the list of potential CI-groups is known. However, given the long struggle to specify which cyclic groups are CI-groups, this is likely to be difficult. Muzychuk completed the determination of the cyclic CI-groups in two excellent papers [40], [41]. His results may be stated as follows.

Theorem 5.5 *The cyclic group $\mathbf{Z}_n$ is a CI-group if and only $n = 8, 9, 18$ or $2^e m$, where m is odd and square-free and $e = 0, 1$ or 2.*

It may be the case that a group $\mathcal{G}$ is not a CI-group, but many of the Cayley graphs on $\mathcal{G}$ behave nicely with regard to isomorphism. Accordingly, we say that a particular Cayley graph $\mathrm{Cay}(\mathcal{G}; \mathcal{S})$ is a *CI-graph* if $\mathrm{Cay}(\mathcal{G}; \mathcal{S}) \cong \mathrm{Cay}(\mathcal{G}; \mathcal{S}')$ implies that there exists an $\alpha \in \mathrm{Aut}(\mathcal{G})$ satisfying $\mathcal{S}' = \mathcal{S}\alpha$. Thus, a group $\mathcal{G}$ is a CI-group if and only if every Cayley graph on $\mathcal{G}$ is a CI-graph.

A useful criterion for deciding whether or not a Cayley graph is a CI-graph is provided by the following theorem of Babai [6].

Theorem 5.6 *Let G be a Cayley graph on the finite group G. Then G is a CI-graph if and only if all regular subgroups of* Aut(G) *isomorphic to G are conjugate in* Aut(G).

There are many other questions that we can ask about CI-graphs. For example, we can ask about degree conditions that guarantee that Cayley graphs on some group are CI-graphs. We can ask about conditions on the automorphism groups that guarantee that Cayley graphs are CI-graphs. We can ask about various restrictions on the connection sets that guarantee that the corresponding Cayley graphs are CI-graphs. One such restriction is a conjecture of Toida that any circulant graph of order n whose connection set is contained in the group of units of $\mathbf{Z}_n$ is a CI-graph. This conjecture has been proved recently (see [16]).

The last isomorphism topic we discuss is when a given graph can be represented as a Cayley graph in more than one way. As mentioned earlier, the complete graph K_n can be represented as a Cayley graph on any group of order n. Similarly, the complete multipartite graph $K_{r(s)}$ can be represented as a Cayley graph on any group of order rs that possesses a subgroup of order s.

There are more interesting results than the two simple examples above. We say that two Cayley graph representations Cay($\mathcal{G}_1$, $\mathcal{S}_1$) and Cay($\mathcal{G}_2$, $\mathcal{S}_2$) are *equivalent* when there exists an isomorphism $\varphi\colon \mathcal{G}_1 \to \mathcal{G}_2$ for which $\mathcal{S}_1\varphi = \mathcal{S}_2$. There are then four non-equivalent representations of the 3-dimensional cube Q_3 as Cayley graphs: one representation is on $\mathbf{Z}_2^3$, one is on $\mathbf{Z}_2 \times \mathbf{Z}_4$, and two are on the dihedral group of order 8. Moreover, for $k = 4$, 5 and 6, Dixon [14] has proved that there are 14, 45 and 238 non-equivalent representations (respectively) of Q_k.

Morris [39] has determined when circulants of odd prime power order have representations as Cayley graphs on other Abelian p-groups.

6. Enumeration

Turner's motivation for proving Theorem 5.2 was his interest in trying to enumerate the vertex-transitive graphs of order n. His theorem sets the stage for an elementary enumeration of the vertex-transitive graphs of prime order. Since no primes belong to **NC**, every vertex-transitive graph of prime order is a Cayley graph. On the other hand, for each prime p, the group $\mathbf{Z}_p$ is the unique group of order p. Thus, all vertex-transitive graphs of prime order are circulant graphs.

Using Theorem 5.2, we see that the isomorphism of circulant graphs of prime order is completely described by the action of a group on the set of all possible connection sets. However, the latter action is induced by the action of a group on the set $\{1, 2, \ldots, p-1\}$, and this is the canonical setting for the employment of Pólya's famous enumeration theorem. Since i belongs to a connection set if and only if $p-i$ also belongs to the connection set, and since multiplication by a and $-a$ are the same, the action we have is the cyclic group of order $(p-1)/2$ on a set with $(p-1)/2$ elements. Since $\{i, p-i\}$ either belongs, or does not belong, to a connection set, we obtain the following result of Turner [52].

Theorem 6.1 *If p is an odd prime, then the number of isomorphism classes of vertex-transitive graphs of order p is*

$$\frac{2}{p-1}\sum_{d}\Phi(d)2^{(p-1)/2d},$$

where the summation runs over all divisors d of $(p-1)/2$.

One of the motivations for classifying CI-groups is to extend Theorem 6.1 to other families of Cayley graphs. Whenever we have a CI-group $\mathcal{G}$, we may employ Pólya's theorem to enumerate the Cayley graphs on $\mathcal{G}$. In order to do this, we have to determine the so-called *cycle index* of $\mathrm{Aut}(\mathcal{G})$ acting on the set of unordered pairs $\{g, g^{-1}\}$ of non-identity elements of $\mathcal{G}$.

Some work in this direction has been carried out by Alspach and Mishna [5]. Theorem 5.5 classifies the cyclic CI-groups and the circulant graphs of the appropriate orders are enumerated in [5]. The Cayley graphs on the elementary Abelian p-groups $\mathbf{Z}_p^2$ have also been enumerated.

Even though the cyclic group of order p^2 is not a CI-group when p is a prime bigger than 3, the circulant graphs of order p^2 have been enumerated (see [28]).

7. Automorphisms

The problem of determining the full automorphism group of a Cayley graph is difficult in general. We first examine the special case of prime order circulants, where the answer is completely known. The tool that plays an essential role is Theorem 5.1.

Suppose that p is a prime, and that we are given the circulant graph $G = \mathrm{Circ}(p; \mathcal{S})$. It is now more convenient to think of circulant graphs as Cayley graphs on the additive group $\mathbf{Z}_p$. The graph is either the complete graph or its complement if and only if $\mathcal{S}$ is all of $\mathbf{Z}_p^*$ or $\emptyset$, respectively. The resulting automorphism group is $\mathcal{S}_p$, and this situation is easy to recognize.

When $\emptyset \subset \mathcal{S} \subset \mathbf{Z}_p^*$, Theorem 5.1 tells us that $\mathrm{Aut}(G)$ has the form $\{T_{a,b} : a \in \mathcal{H} < \mathbf{Z}_p^*, b \in \mathbf{Z}_p\}$. This implies that the stabilizer of the vertex labelled 0 is $T_{a,0}$, with $a \in \mathcal{H} < \mathbf{Z}_p$. Thus, if there is an edge joining 0 and k in G, then there is an edge joining 0 and all of $k\mathcal{H}$, and so the connection set $\mathcal{S}$ is a union of cosets of the multiplicative subgroup $\mathcal{H}$ of $\mathbf{Z}_p^*$. On the other hand, if $\mathcal{S}$ is a union of cosets of the subgroup $\mathcal{H}$ of $\mathbf{Z}_p^*$, but not a union of cosets of any supergroup of $\mathcal{H}$, then the stabilizer of 0 is $\{T_{a,0} : a \in \mathcal{H} < \mathbf{Z}_p^*\}$ and we know precisely what $\mathrm{Aut}(G)$ is.

If $G = \mathrm{Circ}(p; \mathcal{S})$ and $\emptyset \subset \mathcal{S} \subset \mathbf{Z}_p^*$, then let $e(\mathcal{S})$ denote the maximum even order subgroup $\mathcal{H}$ of $\mathbf{Z}_p^*$ for which $\mathcal{S}$ is a union of cosets of $\mathcal{H}$. Note that this is defined because $i \in \mathcal{S}$ if and only if $p - i \in \mathcal{S}$.

The above comments provide an outline of a proof of the following result in [3].

Theorem 7.1 *Let* $G = \mathrm{Circ}(p; \mathcal{S})$ *be a circulant graph of prime order. If* $\mathcal{S} = \emptyset$ *or* $\mathbf{Z}_p^*$, *then* $\mathrm{Aut}(G) = \mathcal{S}_p$. *Otherwise,* $\mathrm{Aut}(G) \equiv \{T_{a,b} \colon a \in e(\mathcal{S}), b \in \mathbf{Z}_p\}$.

We know that the automorphism group of any Cayley graph on a finite group $\mathcal{G}$ contains the left regular representation of $\mathcal{G}$. From the well-known orbit-stabilizer theorem for permutation groups, we know that the order of the automorphism group of a Cayley graph on $\mathcal{G}$ is the product of $|\mathcal{G}|$ and the order of the stabilizer of any vertex. What makes Theorem 7.1 work is the fact that Theorem 5.1 provides an exact description of possible stabilizers for circulant graphs of prime order.

A natural situation to consider is when the stabilizer of any vertex is the identity, implying that the full automorphism group of the Cayley graph on $\mathcal{G}$ is the left regular representation of $\mathcal{G}$. In this case, we say that $\mathrm{Cay}(\mathcal{G}, \mathcal{S})$ is a *graphical regular representation* of $\mathcal{G}$. The first problem arising in this context is the determination of the groups that admit a graphical regular representation.

It was known that Abelian groups of exponent greater than 2 and generalized dicyclic groups do not admit graphical regular representations. Hetzel [24] and Godsil [19] then independently completed the classification by showing that there are only finitely many additional groups that do not admit graphical regular representations. These additional groups are all of order 32 or less.

A Cayley graph G on a group $\mathcal{G}$ is *normal* when $\mathcal{G}_R$ is a normal subgroup of $\mathrm{Aut}(G)$. If $\mathcal{S}$ is the connection set for G and $\mathrm{Aut}(\mathcal{G})_{\{s\}} = \{\sigma \in \mathrm{Aut}(\mathcal{G}) \colon \mathcal{S}\sigma = \mathcal{S}\}$, then $\mathrm{Aut}(G) = \mathcal{G}_R \mathrm{Aut}(\mathcal{G})_{\{s\}}$ when G is normal. This means that the automorphism group of G is as small as it can be, because the latter product is always a subgroup of $\mathrm{Aut}(G)$. This is a new area for investigation.

8. Subgraphs

We now consider the murky area of the interaction between subgraph structures and symmetry. There are some problems for which vertex-transitivity has a significant impact, and there are others for which it is unclear what the impact is. We first look at an example of the former.

Recall that a *1-factor* of a graph G is a spanning subgraph in which each vertex has degree 1. Since all vertices of a vertex-transitive graph G are the same under the automorphism group, they must all lie in the same class of the Gallai-Edmonds decomposition of G. The next theorem then follows easily.

Theorem 8.1 *Let G be a connected vertex-transitive graph. If G has even order, then it has a 1-factor. If G has odd order, then $G - v$ has a 1-factor, for each vertex $v \in V(G)$.*

If G is a connected vertex-transitive graph of degree d, then G is d-edge-connected. Thus, if G is a connected vertex-transitive graph of degree d, then for any two distinct vertices v and w of G, there are d edge-disjoint paths whose terminal vertices are v and w.

The situation for vertex-connectivity is not as straightforward as for edge-connectivity. Watkins [53] proved the following theorem.

Theorem 8.2 *If G is a connected vertex-transitive regular graph of degree d, and if $\kappa(G)$ denotes the connectivity of G, then $\kappa(G) > 2d/3$. Also, for each $\epsilon > 0$, there exists a connected d-regular vertex-transitive graph H for which*

$$\kappa(H) < \left(\epsilon + \tfrac{2}{3}\right) d.$$

Even more interesting than the above lower bound on the connectivity as a proportion of the degree are the methods introduced in [53] (and independently in [34]) to prove the result. For each cutset of cardinality $\kappa(G)$, we make a list of the components that result when we delete the cutset. Any resulting component of minimum cardinality, taken over all cutsets of cardinality $\kappa(G)$, is called an *atom*. Mader and Watkins have developed some interesting results for the atoms in any connected vertex-transitive graph whose degree exceeds its connectivity.

Probably the most important feature possessed by the atoms is that they form a system of blocks of imprimitivity for the automorphism group. This lies at the heart of the work on atoms.

We see from Theorem 8.2 that there are connected vertex-transitive graphs whose connectivity is close to two-thirds of its degree. Is this still the case for Cayley graphs? Consider the circulant graph G of order $5d$ whose connection set S comprises all integers from $\{1, 2, \ldots, 5d - 1\}$ that are congruent to 0, 1 or

2 (modulo 5). The circulant graph G has connectivity $2d$ and degree $3d - 1$. Thus, the ratio is as close to $\frac{2}{3}$ as we wish, and there is no improvement in Theorem 8.2 by restricting ourselves to Cayley graphs.

There has been some work on those Cayley graphs whose connectivity is maximum. A connection set $\mathcal{S}$ is *quasi-minimal* if the elements of $\mathcal{S}$ can be ordered as $s_1, s_2, \ldots, s_t$, in such a way that

- if $|s_i| > 2$, then s_i^{-1} is either s_{i-1} or s_{i+1};
- if $\mathcal{S}_i$ is the set $\{s_1, s_2, \ldots, s_i\}$, then, for each i such that $|s_i| = 2$, $\langle \mathcal{S}_i \rangle$ is a proper supergroup of $\langle \mathcal{S}_{i-1} \rangle$, and for each i such that $|s_i| > 2$ and $s_i^{-1} = s_{i-1}$, $\langle \mathcal{S}_i \rangle$ is a proper supergroup of $\langle \mathcal{S}_{i-2} \rangle$.

The next theorem is proved using atoms in [4].

Theorem 8.3 *If $\mathcal{S}$ is a quasi-minimal generating set of the group $\mathcal{G}$, then the Cayley graph* Cay$(\mathcal{G}; \mathcal{S})$ *has connectivity* $|\mathcal{S}|$.

In [10] it is shown that a connected circulant regular graph of degree at least 3 contains cycles of all possible even lengths $4, 6, \ldots$. In addition, if a circulant graph has girth 3, then there are cycles of all possible lengths.

There have been studies regarding universality (see [11] and [22]). If H is a graph of order n, then for any group $\mathcal{G}$ of order at least cn^2, there is a Cayley graph G on $\mathcal{G}$ for which H is an induced subgraph of G. In a different direction, for each positive integer r, there exists a Paley graph that contains all graphs of order r as induced subgraphs.

9. Hamiltonicity

Hamiltonicity refers to graph properties that are related to Hamilton cycles and Hamilton paths, such as the existence and enumeration of Hamilton paths and cycles, whether the graph is Hamilton-connected or Hamilton-laceable, the existence of Hamilton decompositions, and the structure of the Hamilton space.

The interest in Hamilton cycles in Cayley graphs grew out of a question posed by Lovász. He asked whether every connected vertex-transitive graph has a Hamilton path. In fact, there are only four known non-trivial connected vertex-transitive graphs that do not possess Hamilton cycles. None of these is a Cayley graph, which naturally leads one to ask whether every connected Cayley graph has a Hamilton cycle. This is the first topic we address.

The following result of Chen and Quimpo [12] is one of the best results on the topic. We need two definitions before stating it. A graph G is *Hamilton-connected* if, for any two distinct vertices v, w in G, there is a Hamilton path whose terminal vertices are v and w. Since a vertex-transitive bipartite graph cannot be Hamilton-connected, we change the definition for bipartite graphs accordingly. A bipartite graph, with parts A and B satisfying $|A| = |B|$, is *Hamilton-laceable* if, for any vertices $v \in A$ and $w \in B$, there is a Hamilton path whose terminal vertices are v and w.

Theorem 9.1 *Let G be a connected Cayley graph on a finite Abelian group. If G is bipartite and has degree at least 3, then G is Hamilton-laceable. If G is not bipartite and has degree at least 3, then G is Hamilton-connected.*

One immediate corollary of the *Chen-Quimpo theorem* is the fact that each edge of a connected Cayley graph on an Abelian group of order 3 or more lies in a Hamilton cycle. Theorem 9.1 has been extended to Hamiltonian groups.

This theorem also provides a useful tool for establishing the existence of Hamilton cycles in Cayley graphs with special structural properties. For example, there are situations when the vertex-set can be partitioned in such a way that the subgraphs induced on the parts are Cayley graphs on Abelian groups, even though the entire graph is not. Theorem 9.1 gives us a lot of freedom in finding Hamilton paths in the subgraphs. The problem then becomes one of linking the paths to form a Hamilton cycle in the original graph.

There also has been some success in looking for Hamilton cycles in Cayley graphs by considering the order. One general question that arose was whether every connected Cayley graph of prime power order has a Hamilton cycle. Witte [55] provided a dramatic answer by proving the corresponding much stronger result for Cayley digraphs.

Theorem 9.2 *Every connected Cayley digraph on a group of order p^e, where p is a prime and $e \geq 1$, has a Hamilton directed cycle.*

There have been many particular Cayley graphs that have been shown to have Hamilton cycles. There is insufficient space to include most of them, and it is not clear that these special results give much of an indication towards resolving the general problem anyway. In fact, there is so little current evidence that conjecturing one way or the other about the general problem seems no better than guessing. We include one special case, however, because the proof of the existence of a Hamilton cycle in this case is difficult (see [13]).

Theorem 9.3 *The Cayley graph on the symmetric group $S_n (n \geq 3)$ with connection set $\{(1\,2), (1\,2 \cdots n), (n\; n-1\; \cdots\; 1)\}$ has a Hamilton cycle.*

10. Factorization

We saw earlier that every connected Cayley graph of even order has a 1-factor. In fact, something much stronger may be the case. A *1-factorization* of a graph is a partition of the edge-set into 1-factors. No example is known of a connected Cayley graph of even order that is not 1-factorizable. We now present some results about 1-factorizable Cayley graphs, starting with one of Stong [50].

Theorem 10.1 *A connected Cayley graph on the group $\mathcal{G}$ has a 1-factorization if one of the following holds:*
- $|\mathcal{G}| = 2^k$, *for some integer k;*
- $\mathcal{G}$ *is an Abelian group of even order;*
- $\mathcal{G}$ *is dihedral or dicyclic.*

We say that the connection set $\mathcal{S}$ is a *minimal generating Cayley set* for $\mathcal{G}$, if $\mathcal{S}$ generates $\mathcal{G}$, but $\mathcal{S} - \{s, s^{-1}\}$ generates a proper subgroup of $\mathcal{G}$ for each $s \in \mathcal{S}$. In addition to proving Theorem 10.1, Stong [50] proved that Cayley graphs whose connection sets are minimal generating Cayley sets have 1-factorizations whenever the group is a nilpotent group of even order, or the group contains a proper Abelian normal subgroup of index 2^k, or the group has order $2^m p^k$ for some prime p satisfying $p > 2^m$.

Cubic graphs are of interest in their own right. If we seek Cayley graphs without Hamilton cycles, then it seems natural to believe that sparser graphs are less likely to have a Hamilton cycle. Any cubic graph with a Hamilton cycle clearly has a 1-factorization, but the converse is not true in general. So checking cubic Cayley graphs for 1-factorizations should be easier than checking them for Hamilton cycles. Recently, Potočnik has shown that any cubic Cayley graph G whose automorphism group has a soluble subgroup that acts transitively on the vertex-set of G has a 1-factorization. In particular, this means that any connected cubic Cayley graph on a soluble group of even order has a 1-factorization.

A *Hamilton decomposition* of a graph G is a partition of the edge-set into Hamilton cycles when the degree is even, or a partition into Hamilton cycles and a 1-factor when the degree is odd. Many examples of graphs with Hamilton decompositions are representable as Cayley graphs on Abelian groups. This leads naturally to ask whether all connected Cayley graphs on Abelian groups have Hamilton decompositions.

Theorem 9.1 gives an affirmative answer for graphs of degree 2 or 3. The first progress came in [8], where it was proved that the answer is 'yes' for degree 4. It is then easy to show that the answer is 'yes' for degree 5, and most of the cases for degree 6 are settled. The best result on Hamilton decompositions of Cayley graphs on Abelian groups is the next theorem, obtained by Liu [31], [32]. He has also established Hamilton decompositions for certain special cases.

Theorem 10.2 *If* $G = \mathrm{Cay}(\mathcal{G}, \mathcal{S})$ *is a connected Cayley graph on an Abelian group* $\mathcal{G}$*, and if* $\mathcal{S}$ *is a minimal generating Cayley set, then* G *has a Hamilton decomposition.*

An *isomorphic factorization* of a graph G is a partition of the edge-set E of G for which the subgraphs induced by the edges in each part are isomorphic. In general, given a graph G and a divisor d of $|E|$, we are interested in whether there is an isomorphic factorization of G into d subgraphs. Some early work on isomorphic factorizations dealt with specific graphs that happen to be Cayley graphs. This leads naturally to the following question: if d is a divisor of $|E|$ and G is a circulant graph, a Cayley graph, or a vertex-transitive graph, does there exist an isomorphic factorization of G into d subgraphs?

One contribution towards these questions is the following result of Fink [17].

Theorem 10.3 *If* T *is any tree with* k *edges, then the* k*-cube* Q_k *has an isomorphic factorization by* T*. Furthermore, there is an isomorphic factorization for which each copy of* T *is an induced subgraph.*

11. Embeddings

There are a long history and an extensive literature on embedding graphs in surfaces. The book of White [54] and a recent excellent survey [47] provide a good starting point for this topic.

Cayley graphs arise naturally in the context of graph embeddings. Cayley diagrams arose more than a hundred years ago out of an interest in trying to find a way to 'picture' groups. These are Cayley digraphs in which the arcs are coloured according to the group element that generates the arc. When we try to draw the diagrams with no arcs crossing, the colours of the arcs are immaterial, as is whether there are arcs in both directions between two vertices. Thus, we define the *genus* of a group $\mathcal{G}$ as the smallest genus g with the property that there exists a connected Cayley graph G on $\mathcal{G}$ which has an embedding into an orientable surface of genus g.

As early as 1896, planar groups were characterized by Maschke [36]. The toroidal groups were classified into seventeen infinite families, together with some sporadic cases, by Proulx [46]. Tucker [51] proved that, for any genus $g \geq 2$, there are only finitely many groups of genus g. The genus of most Abelian groups was found by Jungerman and White [27].

A *Cayley map* is an embedding of a Cayley graph into an orientable surface with the same cyclic rotation scheme at each vertex. This term is relatively recent – apparently first appearing in [9] – but the idea has been around for more than one hundred years. Cayley maps play a significant role in the proof of the Heawood map colouring problem. They also play a role in the determination of the genus of a group. The long paper [47] unifies some earlier scattered work on Cayley maps, and provides a general framework for studying them.

12. Applications

There are many meaningful applications of Cayley graphs, and we conclude with a quick outline of a few of them. Any problem for which graphs are used as a model and the use of edges is being optimized provides a natural setting for Cayley graphs. For example, suppose that we are to construct a network for which the number of direct links we may use is restricted, but we want to maximize the probability that the network remains connected after some links or vertices of the network are deleted. Then there is a strong tendency for the graph to be either vertex-transitive or 'close to it'. Consequently, Cayley graphs – in particular, the k-dimensional cube Q_k – have been extensively studied by researchers working with networks. A recent book on this topic is by Xu [56], and a fundamental paper is [1].

Let C be a circle with circumference r. An *r-circular colouring* of a graph G is a mapping $\mathbf{c}$ that assigns to each vertex v of G an open arc $v\mathbf{c}$ of unit length on C, with the property that $v\mathbf{c} \cap w\mathbf{c} = \emptyset$ whenever v and w are adjacent vertices of G. The *circular chromatic number* of G is $\inf\{r : G \text{ is } r\text{-circular colourable}\}$. Circulant graphs have played a role in the development of circular chromatic numbers of graphs. For a general survey of circular chromatic numbers, see [57].

If G is a graph, and if A is a subset of $V(G)$, then let $N(A) = \{v \in V(G) - A : v$ is adjacent to some vertex of $A\}$. The graph G is an *(n, d, c)-expander* if G has order n, maximum degree d, and there exists a constant $c > 0$ such that $|N(A)| \geq c|A|$, for all $A \subset V(G)$ satisfying $|A| \leq n/2$. The construction of expanders and families of expanders with c fixed is of considerable interest, and Cayley graphs play a significant role; two excellent references are [2] and [33].

Acknowledgements: I wish to thank Luis Goddyn, Cai Heng Li, Brendan McKay and Mikhail Muzychuk for valuable conversations.

References

1. S. Akers and B. Krishnamurthy, A group-theoretic model for symmetric interconnection networks, *IEEE Trans. Comput.* **38** (1989), 555–566.
2. N. Alon, Tools from higher algebra, *Handbook of Combinatorics Vol II* (eds. R. L. Graham, M. Grötschel and L. Lovász), MIT Press and North-Holland (1995), 1749–1783.

3. B. Alspach, Point-symmetric graphs and digraphs of prime order and transitive permutation groups of prime degree, *J. Combin. Theory* (*B*) **15** (1973), 12–17.
4. B. Alspach, Cayley graphs with optimal fault tolerance, *IEEE Trans. Computers* **41** (1992), 1337–1339.
5. B. Alspach and M. Mishna, Enumeration of Cayley graphs and digraphs, *Discrete Math.* **256** (2002), 527–539.
6. L. Babai, Isomorphism problem for a class of point-symmetric structures, *Acta Math. Acad. Sci. Hungar.* **29** (1977), 329–336.
7. L. Babai and P. Frankl, Isomorphisms of Cayley graphs I, *Combinatorics* (Proc. Fifth Hungarian Colloq., Keszthely, 1976), *Colloq. Math. Soc. János Bolyai* **18**, North-Holland (1978), 35–52.
8. J.-C. Bermond, O. Favaron and M. Maheo, Hamiltonian decompositions of Cayley graphs of degree 4, *J. Combin. Theory* (*B*) **46** (1989), 142–153.
9. N. Biggs, Cayley maps and symmetrical maps, *Proc. Cambridge Philos. Soc.* **72** (1972), 381–386.
10. Z. Bogdanowicz, Pancyclicity of connected circulant graphs, *J. Graph Theory* **22** (1996), 167–174.
11. B. Bollobás and A. Thomason, Graphs which contain all small graphs, *Europ. J. Combin.* **2** (1981), 13–15.
12. C. C. Chen and N. Quimpo, On strongly hamiltonian abelian group graphs, *Combinatorial Mathematics VIII*, Lecture Notes in Mathematics **884**, Springer-Verlag (1981), 23–34.
13. R. Compton and S. G. Williamson, Doubly adjacent Gray codes for the symmetric group, *Linear and Multilinear Algebra* **35** (1993), 237–293.
14. J. D. Dixon, Groups with a Cayley graph isomorphic to a hypercube, *Bull. Austral. Math. Soc.* **55** (1997), 385–393.
15. E. Dobson, Isomorphism problem for Cayley graphs of Z_p^3, *Discrete Math.* **147** (1995), 87–94.
16. E. Dobson and J. Morris, Toida's Conjecture is true, *Electron. J. Combin.* **9** (2002), R35.
17. J. Fink, On the decomposition of n-cubes into isomorphic trees, *J. Graph Theory* **14** (1990), 405–411.
18. G. Gamble and C. E. Praeger, Vertex-primitive groups and graphs of order twice the product of two distinct odd primes, *J. Group Theory* **3** (2000), 247–269.
19. C. D. Godsil, *Graphs with Regular Groups*, Ph.D. thesis, University of Melbourne, 1979.
20. C. D. Godsil, More odd graph theory, *Discrete Math.* **32** (1980), 205–207.
21. C. D. Godsil, On Cayley graph isomorphisms, *Ars Combin.* **15** (1983), 231–246.
22. C. D. Godsil and W. Imrich, Embedding graphs in Cayley graphs, *Graphs Combin.* **3** (1987), 39–43.
23. A. Hassani, M. Iranmanesh and C. E. Praeger, On vertex-imprimitive graphs of order a product of three distinct primes, *J. Combin. Math. Combin. Comput.* **28** (1998), 187–213.
24. D. Hetzel, *Über reguläre graphische Darstellungen von auflösbaren Gruppen*, Ph.D. thesis, Technische Universität Berlin, 1976.
25. M. Hirasaka and M. Muzychuk, An elementary abelian group of rank 4 is a CI-group, *J. Combin. Theory* (*A*) **94** (2001), 339–362.

26. M. Iranmanesh and C. E. Praeger, On non-Cayley vertex-transitive graphs of order a product of three primes, *J. Combin. Theory (B)* **81** (2001), 1–19.
27. M. Jungerman and A. White, On the genus of finite Abelian groups, *Europ. J. Combin.* **1** (1980), 243–251.
28. M. Klin, V. Liskovets and R. Pöschel, Analytical enumeration of circulant graphs with prime-squared number of vertices, *Sém. Lothar. Combin.* **36** (1996), B36d, 36 pp. (electronic).
29. C. H. Li, Finite CI-groups are soluble, *Bull. London Math. Soc.* **31** (1999), 419–423.
30. C. H. Li, On isomorphisms of finite Cayley graphs – a survey, *Discrete Math.* **256** (2002), 301–334.
31. J. Liu, Hamiltonian decompositions of Cayley graphs on abelian groups of odd order, *J. Combin. Theory (B)* **66** (1996), 75–86.
32. J. Liu, Hamiltonian decompositions of Cayley graphs on abelian groups of even order, *J. Combin. Theory (B)* **88** (2003), 305–321.
33. A. Lubotzky, *Cayley Graphs: Eigenvalues, Expanders and Random Walks*, London Math. Soc. Lecture Notes **218** (1995), 155–189.
34. W. Mader, Eine Eigenschaft der Atome endlicher Graphen, *Arch. Math. (Basel)* **22** (1971), 333–336.
35. D. Marušič, Vertex-transitive graphs and digraphs of order p^k, *Ann. Discrete Math.* **27** (1985), 115–128.
36. W. Maschke, The representation of finite groups, *Amer. J. Math.* **18** (1896), 156–194.
37. B. D. McKay and C. E. Praeger, Vertex-transitive graphs which are not Cayley graphs, I, *J. Graph. Theory* **22** (1996), 321–334.
38. A. Miller and C. E. Praeger, Non-Cayley vertex-transitive graphs of order twice the product of two odd primes, *J. Algebraic Combin.* **3** (1994), 77–111.
39. J. Morris, Isomorphic Cayley graphs on nonisomorphic groups, *J. Graph Theory* **31** (1999), 345–362.
40. M. Muzychuk, Ádám's conjecture is true in the square-free case, *J. Combin. Theory (A)* **72** (1995), 118–134.
41. M. Muzychuk, On Ádám's conjecture for circulant graphs, *Discrete Math.* **167** (1997), 497–510.
42. M. E. Muzychuk, An elementary abelian group of large rank is not a CI-group, *Discrete Math.* **264** (1–3) (2003), 167–186.
43. M. Muzychuk and G. Tinhofer, Recognizing circulant graphs in polynomial time: An application of association schemes, *Electron. J. Combin.* **8** (2001), R26.
44. L. Nowitz, A non-Cayley-invariant Cayley graph of the elementary Abelian group of order 64, *Discrete Math.* **110** (1992), 223–228.
45. P. Potočnik, Edge-colourings of cubic graphs admitting a solvable vertex-transitive group of automorphisms, preprint.
46. V. Proulx, Classification of the toroidal groups, *J. Graph Theory* **2** (1978), 269–273.
47. R. B. Richter, J. Širáň, R. Jajcay, T. Tucker and M. Watkins, Cayley maps, preprint.
48. G. Sabidussi, On a class of fixed-point-free graphs, *Proc. Amer. Math. Soc.* **9** (1958), 800–804.
49. Á. Seress, On vertex-transitive, non-Cayley graphs of order *pqr*, *Discrete Math.* **182** (1998), 279–292.
50. R. Stong, On 1-factorizability of Cayley graphs, *J. Combin. Theory (B)* **39** (1985), 298–307.

51. T. Tucker, The number of groups of a given genus, *Trans. Amer. Math. Soc.* **258** (1980), 167–179.
52. J. Turner, Point-symmetric graphs with a prime number of points, *J. Combin. Theory* **3** (1967), 136–145.
53. M. Watkins, Connectivity of transitive graphs, *J. Combin. Theory* **8** (1970), 23–29.
54. A. White, *Graphs of Groups on Surfaces*, Mathematics Studies **188**, North-Holland, 2001.
55. D. Witte, Cayley digraphs of prime-power order are hamiltonian, *J. Combin. Theory (B)* **40** (1984), 107–112.
56. J. Xu, *Topological Structure and Analysis of Interconnection Networks*, Kluwer, 2001.
57. X. Zhu, Circular chromatic number: a survey, *Discrete Math.* **229** (2001), 371–410.

7

Finite symmetric graphs

CHERYL E. PRAEGER

The study of finite symmetric graphs goes back to work of Tutte dating from 1947. Because of their transitivity properties, symmetric graphs are regular, their connected components are isomorphic to one another, and they have other regularity in their structure. They are thus useful in many applications. This chapter traces two broad approaches to the study of symmetric graphs, a local analysis focusing on the structure of a vertex stabilizer, and a global approach that gives information about the full group of automorphisms. These different approaches are explained with reference to distance-transitive graphs, s-arc transitive graphs, and other families of symmetric graphs.

1. Introduction

A graph G is *symmetric* if its automorphism group Aut(G) is transitive on vertices and on ordered pairs of adjacent vertices – that is, on the *arcs* of G; symmetric graphs are also called *arc-transitive graphs.* They have been studied for over fifty years, beginning with two important papers by Tutte on finite cubic symmetric graphs in 1947 and 1959, and a paper by Sabidussi in 1964 which suggested a method for constructing all symmetric graphs from a small amount of information about their automorphism groups (see [32, Secs. 9 and 11]). As a consequence of their transitivity properties, symmetric graphs are regular, their connected components are isomorphic to one another, and they have other regularity in their structure. Perhaps for these reasons, they are useful in many applications, such as modelling interconnection networks (see [20]).

The theory of finite symmetric graphs has two basic threads, one focusing on a broad global analysis of classes of groups or graphs, and the other concentrating on tight local analysis of specific combinatorial or group-theoretic situations. The theoretical frameworks adopted, as well as the statements of results and the methodology, all involve a combination of combinatorics and abstract group theory, often requiring an application of the finite simple group classification. The purpose of this chapter is to trace these developmental threads for finite symmetric graphs, in order to understand the power of each, and to appreciate the ways in which they intertwine in elucidating structure, and in characterizing classes of these graphs. In Section 3 we discuss the fundamental links between graph theory and group theory which underpin the theory of symmetric graphs.

A 'local analysis' of a symmetric graph is a study of the combinatorial structure or symmetry properties 'close' to a vertex. Sometimes local properties, either combinatorial or group-theoretic, influence the structure of the whole graph, and a major problem is to determine the extent to which this occurs. Tutte's theorem, discussed in Section 2, provided the first successful local analysis of a family of symmetric graphs and was enormously influential on subsequent research. Further examples of local analyses of symmetric graphs are given in Section 6.

In contrast to this, 'global analysis' of symmetric graphs focuses on aspects of graph structure that affect the whole graph. The global approach leads naturally to a consideration of quotients of symmetric graphs, and thence to a study of symmetric graphs admitting a vertex-primitive or vertex-biprimitive automorphism group (see Section 4). The most complete manifestation of this approach has been the classification programme for finite distance-transitive graphs, described briefly in Section 5 and at greater depth in Chapter 9. However, the method of passing to primitive quotients is unsatisfactory for studying many families of symmetric graphs that are given by a local defining property, such as s-arc transitivity. For several of these families, the appropriate quotients to study are normal quotients, and it is essential for an understanding of these families to look closely at the quasiprimitive and bi-quasiprimitive graphs in the family. This approach is illustrated for 2-arc transitive graphs in Section 7.

For yet other families of symmetric graphs, information about quotients (whether primitive or normal) is insufficient to describe the overall structure of typical graphs in the family. A geometric approach that injects additional combinatorial information, and offers a framework for studying imprimitive symmetric graphs, is discussed in Section 9.

Most investigations of symmetric graphs begin with a given graph G and a given group $\mathcal{G}$ acting symmetrically (that is, vertex-transitively and arc-transitively) on G, and we call $(G, \mathcal{G})$ a symmetric pair. The difficult problem of determining the full automorphism group of G, or indeed any subgroup of $\mathrm{Aut}(G)$ containing $\mathcal{G}$, often needs to be addressed. Some effective strategies for this are discussed for primitive and quasiprimitive symmetric graphs in Section 8. Finally, in Section 10, we mention some classes of vertex-transitive graphs that are close to being symmetric.

2. s-arc transitive graphs

In this section, by way of introduction to the subject, we describe Tutte's seminal discoveries about the structure of cubic s-arc transitive graphs. For a positive integer s, an *s-arc* in a graph G is an $(s+1)$-tuple $(v_0, v_1, \ldots, v_s)$ of vertices with the properties that $v_{i-1}v_i$ is an edge of G, for $1 \le i \le s$, and $v_{i-1} \ne v_{i+1}$ for $1 \le i \le s-1$. The graph G is *s-arc transitive* if $\mathrm{Aut}(G)$ is transitive on s-arcs. If $\mathcal{G}$ is a subgroup of $\mathrm{Aut}(G)$, then $(G, \mathcal{G})$ is an *s-arc transitive pair* if $\mathcal{G}$ is transitive on both vertices and s-arcs of G. Thus, the 1-arc transitive pairs are precisely the symmetric pairs, and indeed all s-arc transitive pairs are symmetric pairs.

Lemma 2.1 *Let G be a graph, let $\mathcal{G}$ be a subgroup of* $\mathrm{Aut}(G)$, *and let s be a positive integer. If $(G, \mathcal{G})$ is an s-arc transitive pair, then $(G, \mathcal{G})$ is a symmetric pair.*

Proof Let $(G, \mathcal{G})$ be an s-arc transitive pair. Then $\mathcal{G}$ is vertex-transitive, by definition, so G is regular of degree k, say. If $k = 0$, then G has no arcs, and if $k = 1$, then each component of G is a complete graph K_2: in these cases, $(G, \mathcal{G})$ is trivially a symmetric pair. So let $k \ge 2$. If $s > 1$, then each 1-arc (v_0, v_1) of G can be extended to an s-arc $(v_0, v_1, \ldots, v_s)$. Since $\mathcal{G}$ is transitive on s-arcs, it follows that $\mathcal{G}$ is transitive on 1-arcs. Thus, for every s, $(G, \mathcal{G})$ is a 1-arc transitive pair, and as observed above, the 1-arc transitive pairs are precisely the symmetric pairs. ■

The property of being s-arc transitive is of special interest for $s \ge 2$. Many families of examples are known, and we list a few of them below. The vertices of the *odd graph* O_n are the $(n-1)$-element subsets of $\{1, 2, \ldots, 2n-1\}$, with two vertices being adjacent if and only if they are disjoint; O_n is regular of degree n, and $\mathrm{Aut}(O_n) = \mathcal{S}_{2n-1}$ (see Chapter 5 and [7, Sec. 9.1]).

Theorem 2.2

(a) For $n \geq 3$ and $s \geq 1$, the cycle C_n is s-arc transitive;

(b) For $n \geq 4$, the complete graph K_n is 2-arc transitive, but not 3-arc transitive;

(c) For $n \geq 3$, the complete bipartite graph $K_{n,n}$ is 3-arc transitive, but not 4-arc transitive;

(d) For an odd integer $n \geq 3$, O_n is 3-arc transitive, but not 4-arc transitive.

There are also examples of cubic graphs that are 4-arc transitive but not 5-arc transitive (for example, the Coxeter graph on 28 vertices), and 5-arc transitive but not 6-arc transitive (for example, the (3, 8)-cage on 30 vertices) – see [7, Secs. 12.3 and 6.9].

Tutte proved that there are no 6-arc transitive cubic graphs (see [32, Thm. 9.2]). His arguments were an elegant combination of group theory and combinatorial reasoning.

Theorem 2.3 (Tutte's theorem) *If G is an s-arc transitive cubic graph, then $s \leq 5$.*

Given that a graph G is symmetric, the fact that it is s-arc transitive can be decided by examining the stabilizer $(\mathrm{Aut}(G))_v$ in $\mathrm{Aut}(G)$ of a vertex v; precisely, we require that $(\mathrm{Aut}(G))_v$ be transitive on the s-arcs with first vertex v. Thus, s-arc transitivity may be regarded as a local property. Tutte's result demonstrates that in a symmetric graph a local property can sometimes control the global structure.

3. Group-theoretic constructions

In 1964, Sabidussi proved that each symmetric graph G is isomorphic to one of the symmetric coset graphs in the Introduction (Section 3) – namely, if $\mathcal{G}$ is a subgroup of $\mathrm{Aut}(G)$ and G is $\mathcal{G}$-symmetric, then $G \cong G(\mathcal{G}, \mathcal{H}, \mathcal{H}g\mathcal{H})$, where $\mathcal{H} = \mathcal{G}_v$ for some vertex v, and $g \in \mathcal{G}$ with $vg \sim v$ and $g^{-1} \in \mathcal{H}g\mathcal{H}$ (see the Introduction, Theorem 3.5).

Studying symmetric graphs as coset graphs has proved useful in classification problems and focuses attention on a given subgroup of automorphisms. It thus leads naturally to the study of symmetric pairs $(G, \mathcal{G})$, consisting of a symmetric graph G and a subgroup $\mathcal{G}$ of $\mathrm{Aut}(G)$ that acts symmetrically on G. Usually information about symmetric pairs is required up to equivalence, where equivalence of symmetric pairs is defined in the Introduction (Section 3), and we note that, for a $\mathcal{G}$-symmetric graph G, $(G, \mathcal{G})$ is equivalent to $(G(\mathcal{G}, \mathcal{H}, \mathcal{H}g\mathcal{H}), \mathcal{G})$.

Shortly after Sabidussi's work, a seminal paper of Sims [35] gave a second group-theoretic characterization of symmetric graphs in terms of orbital graphs.

For a transitive permutation group $\mathcal{G}$ and a non-diagonal self-paired orbital O, the orbital graph $G(O)$, defined in the Introduction (Section 3), is undirected and is $\mathcal{G}$-symmetric, so $(G(O), \mathcal{G})$ is a symmetric pair. Moreover, up to equivalence, each symmetric pair arises in this way. Also, a finite transitive permutation group $\mathcal{G}$ has at least one non-diagonal self-paired orbital if and only if $|\mathcal{G}|$ is even (see [41, Thm. 16.5]).

Sims demonstrated a crucial connection between the primitivity of a transitive group $\mathcal{G}$ and the connectivity of its orbital graphs, showing that $\mathcal{G}$ is primitive if and only if *all* its non-diagonal orbital graphs are connected – see the Introduction, Section 3. However, as the following example demonstrates, if $\mathcal{G}$ is not primitive, it is possible that some of its non-diagonal orbital graphs may still be connected.

Example Let $V = \{1, 2, 3, 4, 5, 6\}$ and $\mathcal{G} = \langle(123456)\rangle \cong \mathbf{Z}_6$. There are five non-diagonal orbitals: if O is the orbital containing (1, 2) or (1, 6), then $G(O)$ is the directed 6-cycle; if O is the orbital containing (1, 3) or (1, 5), then $G(O)$ has two components, each of which is the directed 3-cycle; if O is the orbital containing $(1, 4)$, then $G(O)$ has three components, each of which is the undirected graph K_2.

Imprimitive permutation groups

Let $\mathcal{G}$ be a permutation group on V. A partition $\mathcal{P}$ of V is *$\mathcal{G}$-invariant* if the elements of $\mathcal{G}$ permute the parts of $\mathcal{P}$ set-wise – that is, $pg \in \mathcal{P}$ for all $p \in \mathcal{P}$ and $g \in \mathcal{G}$, where $pg = \{vg : v \in p\}$. The $\mathcal{G}$-invariant partitions of V are in one-one correspondence with the $\mathcal{G}$-invariant equivalence relations on V – namely, a $\mathcal{G}$-invariant partition $\mathcal{P}$ corresponds to the equivalence relation $\sim$, where $v \sim w$ if and only if v and w lie in the same part of $\mathcal{P}$. In particular, the relation of equality corresponds to the partition with all parts of size 1, and the 'universal relation' corresponds to the partition with a single part. These two partitions are called the *trivial partitions* of V, and both are $\mathcal{G}$-invariant for each permutation group $\mathcal{G}$. A transitive permutation group $\mathcal{G}$ is *primitive* if the only $\mathcal{G}$-invariant partitions are the trivial ones; otherwise $\mathcal{G}$ is *imprimitive*. This definition of a primitive permutation group is equivalent to the one given in the Introduction (Section 3).

If $\mathcal{P}$ is invariant under a transitive permutation group $\mathcal{G}$, then $\mathcal{G}$ induces a transitive permutation group $\mathcal{G}^{\mathcal{P}}$ on $\mathcal{P}$ (by defining $g : p \to pg$), and the set-wise stabilizer $\mathcal{G}_p$ of $p \in \mathcal{P}$ induces a transitive permutation group $\mathcal{G}_p^p$ on p. Moreover, if $p, q \in \mathcal{P}$, then the transitive groups $\mathcal{G}_p^p$ and $\mathcal{G}_q^q$ are equivalent. Thus, if the set V is finite, then each non-trivial $\mathcal{G}$-invariant partition $\mathcal{P}$ gives rise to two smaller transitive permutation groups (up to equivalence) – namely, $\mathcal{G}^{\mathcal{P}}$ and $\mathcal{G}_p^p$. We shall be especially interested in the cases where one or other of these transitive groups is as small as possible.

A $\mathcal{G}$-invariant partition $\mathcal{P}$ is *maximal* if $\{V\}$ is the only $\mathcal{G}$-invariant partition of V refined by $\mathcal{P}$; in this case, the group $\mathcal{G}^{\mathcal{P}}$ is primitive. Similarly, $\mathcal{P}$ is *minimal* if its only $\mathcal{G}$-invariant refinement is $\{\{v\} : v \in V\}$; in this case, the group $\mathcal{G}_p^p$ is primitive. We shall show in the next section that, if $(G, \mathcal{G})$ is a symmetric pair and $\mathcal{P}$ is a non-trivial $\mathcal{G}$-invariant partition of the vertex-set, then the two smaller transitive groups associated to $\mathcal{P}$ correspond to two smaller symmetric pairs. Thus, we shall be especially interested in symmetric pairs $(G, \mathcal{G})$ for which $\mathcal{G}$ is vertex-primitive.

The most useful tool for studying finite primitive permutation groups is the O'Nan-Scott Theorem (see [9, Sec. 3]):

Theorem 3.1 (O'Nan-Scott theorem) *Let $\mathcal{G}$ be a subgroup of* $\mathrm{Sym}(V)$. *If $\mathcal{G}$ is a finite primitive group, then $\mathcal{G}$ belongs to one of several disjoint types of primitive groups.*

The number of types of finite primitive groups varies for different statements of this theorem. A version that is often used for studying finite symmetric graphs involves eight types of primitive groups, and each of these types corresponds to an analogous type of quasiprimitive permutation group, where, as defined in the Introduction (Section 3), $\mathcal{G}$ is quasiprimitive if all its non-trivial normal subgroups are transitive. A brief description of the O'Nan-Scott types of primitive groups (and their generalizations for quasiprimitive groups), suitable for applications to symmetric graphs, is given in Theorem 7.2 (see also [31, Sec. 5]). Many of the types are described in terms of the structure and action of a minimal normal subgroup of such a group $\mathcal{G}$, or of its *socle* (the product of its minimal normal subgroups), denoted by $\mathrm{soc}(\mathcal{G})$.

Two of the 'O'Nan-Scott types' occur frequently as types of primitive automorphism groups of graphs – namely, types AS and HA. A primitive or quasiprimitive group $\mathcal{G}$ is said to have type AS, or to be *almost simple*, if $\mathcal{G}$ is a subgroup of $\mathrm{Aut}(\mathcal{H})$, for some non-abelian simple subgroup $\mathcal{H}$ of $\mathcal{G}$: here, $\mathrm{soc}(\mathcal{G}) = \mathcal{H}$ is the unique minimal normal subgroup and is transitive. A primitive group $\mathcal{G}$ is said to have type HA, or to be of *affine type*, if $\mathcal{G}$ has a non-trivial abelian normal subgroup: here, $\mathcal{G}$ is a group of affine transformations of a finite vector space and $\mathrm{soc}(\mathcal{G})$ is its unique minimal normal subgroup and is elementary abelian and *regular* (transitive with trivial stabilizers); in fact, $\mathrm{soc}(\mathcal{G})$ is the group of translations.

Amalgams and their completions

A group amalgam for a graph aims to capture essential local information about the symmetry properties of the graph. Several definitions appear in the literature, and we follow the treatment in [21, Sec. 1.3]. A finite *amalgam $\mathcal{A}$ of rank n* is a set consisting of n groups $\mathcal{H}_i$ $(1 \le i \le n)$ with the property that, for each $i \ne j$, the intersection $\mathcal{H}_i \cap \mathcal{H}_j$ is a subgroup of both $\mathcal{H}_i$ and $\mathcal{H}_j$.

There is no assumption that the $\mathcal{H}_i$ are subgroups of some common group. Instead, a group $\mathcal{G}$ is called a *completion* of $\mathcal{A}$ if there exists a map $\varphi : \cup \mathcal{H}_i \to \mathcal{G}$ for which the restriction of φ to each $\mathcal{H}_i$ is a group homomorphism, and the image of φ generates $\mathcal{G}$. If φ is one-one, then the completion is *faithful*; for example, if $\mathcal{A}$ is defined as a set of subgroups of a group, then the subgroup generated by the subgroups in $\mathcal{A}$ is a faithful completion of $\mathcal{A}$ relative to the inclusion map. For every amalgam $\mathcal{A}$, there exists a universal completion $U(\mathcal{A})$ (relative to a map ψ) with the property that, for any completion $\mathcal{G}$ (relative to a map φ), there is a unique homomorphism $\chi : U(\mathcal{A}) \to \mathcal{G}$ for which φ is the composition of ψ and χ. The *universal completion* $U(\mathcal{A})$ is defined as the group with generating set $\cup \mathcal{H}_i$ and with relations all the equalities of the form $z = xy$, satisfied by elements x, y and z of some $\mathcal{H}_i$. Sometimes (see [21], for example) it is possible to find not only $U(\mathcal{A})$, but also all finite completions of $\mathcal{A}$.

For a symmetric pair $(G, \mathcal{G})$ and an edge $e = vw$, the amalgam $\mathcal{A} = \{\mathcal{G}_v, \mathcal{G}_e\}$ contains the information that $\mathcal{G}_v$ is transitive on the set of vertices adjacent to v, and that some element of $\mathcal{G}$ interchanges v and w. Thus, $\mathcal{A}$ contains the information that $\mathcal{G}$ is symmetric on G, and the set-wise stabilizer in $\mathcal{G}$ of the component of G containing v is a faithful completion of $\mathcal{A}$. The universal completion $U(\mathcal{A})$ acts symmetrically on an infinite regular tree $T(\mathcal{A})$ with vertex-stabilizer and edge-stabilizer giving the same amalgam $\mathcal{A}$. Further, any symmetric pair $(H, \mathcal{H})$ with amalgam $\mathcal{A}$ is a quotient of $(T(\mathcal{A}), U(\mathcal{A}))$, relative to some graph homomorphism $T(\mathcal{A}) \to H$ and group homomorphism $U(\mathcal{A}) \to \mathcal{H}$. There are infinitely many such pairs $(H, \mathcal{H})$, even with H finite, and one is usually interested in symmetric pairs with specific additional properties.

Amalgams of rank greater than 2 carry information about additional combinatorial structure of the graph, and are key tools for analysing certain classes of symmetric graphs (see, for example, [21, Ch. 9]). Such amalgams have applications beyond graph theory, especially to the study of diagram geometries and of groups that preserve them (see [21] and [29]).

For a symmetric pair $(G, \mathcal{G})$ with G a cubic graph, and an edge e incident with a vertex v, the structure of possible amalgams $\{\mathcal{G}_v, \mathcal{G}_e\}$ was deduced from early work of Tutte and Wong (see Biggs [3, Ch. 18]). This led to the determination of presentations for universal completions of such pairs. According to Biggs [5], John Conway discovered such presentations but did not publish them, and 'also inaugurated a programme for constructing such [symmetric pairs] from the presentations'; this programme was developed further by Biggs in [4]. Systematic investigations of these presentations were undertaken in [14] and [40], and these enabled symmetric pairs for cubic graphs of small girth to be classified (see [28]), and also led to constructions of arbitrarily large 4-arc transitive and 5-arc transitive finite cubic graphs (see [12] and [13]).

4. Quotient graphs and primitivity

For every graph G and every partition $\mathcal{P}$ of its vertex-set V, the *quotient graph* $G_{\mathcal{P}}$ is the graph with vertex-set $\mathcal{P}$ for which $p, p' \in \mathcal{P}$ are joined by an edge if and only if there exist $v \in p$ and $v' \in p'$ that are adjacent in G. Also, for each $p \in \mathcal{P}$, let $G[p]$ the induced subgraph of G on p. Here is a simple example.

Example Let $G = C_6$ on $V = \{1, 2, 3, 4, 5, 6\}$ with edges $12, 23, \ldots, 61$, and let $\mathcal{P} = \{p_1, p_2, p_3\}$ where $p_1 = \{1, 4\}$, $p_2 = \{2, 5\}$, $p_3 = \{3, 6\}$. Then $G_{\mathcal{P}} \cong C_3$ and, for each i, $G[p_i]$ is a null graph with two vertices.

In this example, $\mathcal{P}$ is invariant under $\mathcal{G} = \mathrm{Aut}(G) = \langle(123456), (26)(35)\rangle \cong D_{12}$, and we have $\mathcal{G}^{\mathcal{P}} \cong \mathcal{S}_3$, and $\mathcal{G}_p^p \cong \mathcal{S}_2$, for each $p \in \mathcal{P}$. All the pairs $(G, \mathcal{G})$, $(G_{\mathcal{P}}, \mathcal{G}^{\mathcal{P}})$, $(G[p], \mathcal{G}_p^p)$ are symmetric pairs (the last somewhat vacuously, since $G[p]$ has no edges). This is a rather general fact, as we see in the next result.

Theorem 4.1 *Let $(G, \mathcal{G})$ be a symmetric pair, let $\mathcal{P}$ be a $\mathcal{G}$-invariant partition of V, and let $p, p' \in \mathcal{P}$. Then $(G_{\mathcal{P}}, \mathcal{G}^{\mathcal{P}})$, $(G[p], \mathcal{G}_p^p)$ and $(G[p'], \mathcal{G}_{p'}^{p'})$ are all symmetric pairs, and $(G[p], \mathcal{G}_p^p)$ and $(G[p'], \mathcal{G}_{p'}^{p'})$ are equivalent.*

Proof If (p, p') and (q, q') are arcs of $G_{\mathcal{P}}$, then there exist $v \in p, v' \in p'$ and $w \in q, w' \in q'$ for which (v, v') and (w, w') are arcs of G. The group $\mathcal{G}$ contains an element that maps (v, v') to (w, w'), and hence maps (p, p') to (q, q'). The rest of the proof that $(G_{\mathcal{P}}, \mathcal{G}^{\mathcal{P}})$ and $(G[p], \mathcal{G}_p^p)$ are symmetric pairs is straight-forward.

For the equivalence, if $g \in \mathcal{G}$ maps p to p', then the restriction of g to p induces an isomorphism from $G[p]$ to $G[p']$, while the conjugation action of g induces a group isomorphism from $\mathcal{G}_p^p$ to $\mathcal{G}_{p'}^{p'}$, and these together define the required equivalence. ■

If $(G, \mathcal{G})$ is a symmetric pair and $\mathcal{G}$ is primitive on vertices, then we say that $(G, \mathcal{G})$ is a *primitive symmetric pair*. By choosing $\mathcal{P}$ to be a maximal $\mathcal{G}$-invariant partition, we can ensure that $(G_{\mathcal{P}}, \mathcal{G}^{\mathcal{P}})$ is primitive. Similarly by choosing $\mathcal{P}$ to be a minimal $\mathcal{G}$-invariant partition, we can ensure that $(G[p], \mathcal{G}_p^p)$ is primitive. However, when $\mathcal{G}$ is symmetric on G, either all edges have their ends in the same block of $\mathcal{P}$, or all edges have their ends in different blocks of $\mathcal{P}$. Thus, one of $G_{\mathcal{P}}$ and $G[p]$ is a null graph; in particular, if G is connected, then $G[p]$ is null.

Suppose now that G is connected, so that $G[p]$ is a null graph. Several strategies have proved effective for investigating this situation. First, it is sometimes possible to introduce some combinatorial structure on p, other than $G[p]$, to aid in the understanding of $(G, \mathcal{G})$. This was the case for distance-transitive graphs, which will be discussed in Section 5, and is also an integral part of the geometric approach to be described in Section 9.

Secondly, for some families of symmetric graphs, we can extract crucial information about $(G, \mathcal{G})$ from its primitive quotients $(G_{\mathcal{P}}, \mathcal{G}^{\mathcal{P}})$, and in these cases it is of fundamental importance to study primitive symmetric pairs in the family. In such a study, the O'Nan-Scott theorem is essential: see the case of distance-transitive graphs in Section 5, below.

In one rather trivial case, a primitive quotient gives little useful information – namely, if G is bipartite and $\mathcal{P}$ is the bipartition, then $(G_{\mathcal{P}}, \mathcal{G}^{\mathcal{P}})$ is equivalent to $(K_2, \mathcal{S}_2)$. In this case, it would be helpful to examine quotients $(G_{\mathcal{P}}, \mathcal{G}^{\mathcal{P}})$ where the only $\mathcal{G}$-invariant partitions refined by $\mathcal{P}$ are $\{V\}$ and the bipartition of V. We call a permutation group with this property *biprimitive*. Possible structures for biprimitive groups can be inferred from the O'Nan-Scott theorem. However, satisfactory analyses of families of biprimitive symmetric pairs seem difficult to achieve.

Often we wish to study a family of symmetric graphs with an additional defining property, sometimes combinatorial and sometimes group-theoretic. Unfortunately, the extra defining property is rarely inherited by a quotient graph $G_{\mathcal{P}}$, so studying the vertex-primitive graphs in the family does not help us to understand the structure of typical graphs in the family; for example, if $(G, \mathcal{G})$ is 2-arc transitive and $\mathcal{P}$ is $\mathcal{G}$-invariant, then $(G_{\mathcal{P}}, \mathcal{G}^{\mathcal{P}})$ is not in general 2-arc transitive (see Section 7). However, some families of symmetric pairs (such as the family of 2-arc transitive pairs) which are not closed under forming primitive quotients are closed under forming normal quotients. This will be discussed further in Section 7.

5. Distance-transitive graphs

Let G be a connected graph of diameter $d \geq 1$, and let $\mathcal{G}$ be a subgroup of Aut(G). Then the pair $(G, \mathcal{G})$ is *distance-transitive* if $\mathcal{G}$ is transitive on $G_i = \{(v, w)\colon d(v, w) = i\}$, for each $i = 0, 1, \ldots, d$. A pair $(G, \mathcal{G})$ is symmetric if and only if $\mathcal{G}$ is transitive on G_0 and G_1, and so distance-transitive pairs are symmetric. An equivalent defining condition for distance-transitivity is that $(G, \mathcal{G})$ is distance-transitive if and only if $\mathcal{G}$ is transitive on V and, for a given $v \in V$ and for each $i = 1, 2, \ldots, d$, $\mathcal{G}_v$ is transitive on $N_i(v) = \{w : d(v, w) = i\}$. A distance-transitive pair $(G, \mathcal{G})$ of degree at least 3 has at most two non-trivial $\mathcal{G}$-invariant vertex partitions (see [7, Thm. 4.2.1 ff.]). If G is bipartite, then the bipartition is

$\mathcal{G}$-invariant. If G is *antipodal* – that is, if $v \neq w$ and $d(u, v) = d(u, w) = d$ implies that $d(v, w) = d$ – then $\mathcal{G}$ leaves invariant the *antipodal partition* consisting of the antipodal blocks, where the antipodal block containing v is $\{v\} \cup N_d(v)$. It is possible for a distance-transitive pair to be both bipartite and antipodal.

Example 1 For $k \geq 2$, the complete bipartite graph $K_{k,k}$ is regular of degree k, has diameter 2, and is distance-transitive, bipartite and antipodal with $\mathrm{Aut}(G) = \mathcal{S}_k \wr \mathcal{S}_2$. Here the antipodal partition is the same as the bipartition.

Example 2 For $d \geq 2$, the *d-dimensional cube* Q_d is the graph with vertex-set $\mathbf{Z}_2^d$, d-tuples v and w being joined by an edge if and only if they differ in exactly one entry. It has degree and diameter d, and is distance-transitive, bipartite and antipodal, with $\mathrm{Aut}(Q_d) = \mathcal{S}_2^d : \mathcal{S}_d = \mathcal{S}_2 \wr \mathcal{S}_d$.

Suppose that $(G, \mathcal{G})$ is distance-transitive of diameter d, and let $d' = 2\lfloor d/2 \rfloor$. If G is bipartite and $v \in V$, let G^+ be the graph with vertex-set $V^+ := \{v\} \cup N_2(v) \cup \cdots \cup N_{d'}(v)$, the part of the bipartition containing v, whose edges are the pairs from V^+ at distance 2 in G. Then G^+ is called a *bipartite half* of G, $(G^+, \mathcal{G}^+)$ is distance-transitive, where $\mathcal{G}^+$ is the index-2 subgroup of $\mathcal{G}$ that fixes V^+ set-wise, and G is called a *bipartite double* of G^+. If G is antipodal of diameter at least 3, with antipodal partition $\mathcal{P}$, then the *antipodal quotient* $(G_{\mathcal{P}}, \mathcal{G}^{\mathcal{P}})$ is distance-transitive, and G is a *cover* of $G_{\mathcal{P}}$ – that is, the map $v \rightarrow p(v)$ that sends each vertex v to the part $p(v) \in \mathcal{P}$ containing it has the property that, for each v, the restriction to the neighbourhood $N(v)$ is a bijection whose image is the neighbourhood of $p(v)$ in $G_{\mathcal{P}}$. Thus, if $(G, \mathcal{G})$ is imprimitive, then it is bipartite or antipodal and we obtain a smaller distance-transitive pair. After at most two such reductions, we reach a primitive distance-transitive pair.

Thus the classification problem for finite distance-transitive pairs splits into two separate problems: the major one of classifying the finite primitive distance-transitive pairs, and the other of finding all distance-transitive pairs obtainable from the primitive ones by forming antipodal covers and bipartite doubles. There are many infinite families of primitive distance-transitive pairs (see [7]), and among these the Hamming graphs have a special role.

Example 3 For $d \geq 2$ and $q \geq 2$, the *Hamming graph* $H(d, q)$ is the graph with vertex-set $\mathbf{Z}_q^d$, where d-tuples are joined by an edge if and only if they differ in exactly one entry. The automorphism group $\mathrm{Aut}(H(d, q))$ is $\mathcal{S}_q \wr \mathcal{S}_d$ and is distance-transitive. For $q = 2$, $H(d, 2)$ is the d-cube Q_d and is bipartite and antipodal, while for $q \geq 3$, $(H(d, q), \mathcal{S}_q \wr \mathcal{S}_d)$ is primitive. Furthermore, for $q \geq 3$, the complement of $H(2, q)$ is primitive and distance-transitive.

The following theorem is due to Praeger, Saxl and Yokoyama (see [7, Thm. 7.7.1]).

Theorem 5.1 *Let $(G, \mathcal{G})$ be a primitive distance-transitive pair. Then either G is $H(d, q)$ or the complement of $H(2, q)$, or $\mathcal{G}$ is primitive of affine type* HA *or almost simple type* AS.

This reduction theorem led to a programme to classify the primitive distance-transitive pairs $(G, \mathcal{G})$ of almost simple and affine types – see Chapter 9, Sections 5 and 6.

6. Local characterizations

Here we discuss the celebrated theorem of Weiss, that the only finite 8-arc transitive graphs are cycles. We also mention several other characterizations of families of symmetric graphs by their local combinatorial or group-theoretic properties, including locally projective graphs and graphs associated with P-geometries.

s-arc transitive graphs and *Weiss's theorem*

As in Section 2, if $\mathcal{G}$ is a subgroup of Aut(G) and $s \geq 1$, then $(G, \mathcal{G})$ is an s-arc transitive pair if $\mathcal{G}$ is transitive on both vertices and s-arcs of G. We show below that 2-arc transitivity is characterized by a 2-transitive action of a vertex-stabilizer $\mathcal{G}_v$ on the neighbourhood $N(v)$. If $U \subseteq V$ and a subgroup $\mathcal{H}$ of Sym(V) leaves U invariant, then we denote by $\mathcal{H}^U$ the permutation group on U induced by $\mathcal{H}$.

Lemma 6.1 *Let G be a connected graph, let $\mathcal{G}$ be a subgroup of* Aut(G) *that is vertex-transitive on G, and let v be a vertex. Then $(G, \mathcal{G})$ is a 2-arc transitive pair if and only if $\mathcal{G}_v^{N(v)}$ is 2-transitive.*

Proof If $(G, \mathcal{G})$ is 2-arc transitive, then $\mathcal{G}_v$ is 2-transitive on $N(v)$, since each ordered pair (u, w) of distinct vertices of $N(v)$ corresponds to a 2-arc (u, v, w) with second vertex v. Conversely, let $\mathcal{G}_v$ be 2-transitive on $N(v)$. Then we can map an arbitrary 2-arc to one with second vertex v, since $\mathcal{G}$ is vertex-transitive. Since $\mathcal{G}_v$ is 2-transitive on $N(v)$, $\mathcal{G}_v$ is transitive on the 2-arcs with second vertex v. Thus $\mathcal{G}$ is transitive on 2-arcs. ∎

A consequence of the classification of the finite simple groups is that all of the finite 2-transitive permutation groups are known explicitly (see [9, Thm. 5.1]). Each is primitive and is of affine type HA or almost simple type AS. Building on work already published in numerous papers, the final stage of the proof of Weiss's theorem invoked the classification of 2-transitive groups to rule out certain group-theoretic configurations (see [9, Thm. 5.12]).

Theorem 6.2 *If $(G, \mathcal{G})$ is a finite s-arc transitive pair of degree at least 3, then* $s < 8$.

This result is best possible, since there exists an infinite family of 7-arc transitive graphs associated with the simple groups $G_2(3^n)$.

In addition to proving Theorem 6.2, Weiss showed that, if $s \geq 4$, then G has degree $q + 1$, for some prime power q, and $\mathcal{G}_v^{N(v)}$ is an almost simple 2-transitive group with socle PSL$(2, q)$. Further, if q is even, then $s \leq 5$, while if q is odd, then q is a power of 3 and $s = 7$. Weiss had previously classified all possible amalgams for 4-arc transitive graphs, and using this classification, Li [22] was able to classify all primitive and biprimitive 4-arc transitive pairs. His very interesting examples include, among others, pairs $(G, \mathcal{G})$ constructed in [36] and [39], where $\mathcal{G}$ is one of the sporadic simple groups J_3, Ru or Th, together with a new example where $\mathcal{G}$ is the Monster simple group.

Theorem 6.3 *The finite primitive and biprimitive 4-arc transitive pairs $(G, \mathcal{G})$ are known explicitly.*

2-arc transitive graphs and symmetric 2-designs

Investigations of finite 2-arc transitive graphs were inspired also by Cameron's re-interpretation and generalization of an old result of Manning on primitive permutation groups. Let $\mathcal{G}$ be a finite primitive, but not 2-transitive, permutation group on V. Manning proved that, if a stabilizer $\mathcal{G}_v$ is 2-transitive on an orbit of cardinality $k > 2$, then $\mathcal{G}_v$ has a second orbit of length $m > k$, where m is a divisor of $k(k-1)$ (see [41, 17.7]). If w is a point of the first $\mathcal{G}_v$-orbit, and if the $\mathcal{G}$-orbital O containing (v, w) is self-paired, then by Lemma 6.1, $(G(O), \mathcal{G})$ is a 2-arc transitive pair, where $G(O)$ is the orbital graph associated with O. Further, all 2-arc transitive pairs for which the group is primitive on vertices arise in this way. Cameron found a graph-theoretic meaning for the second $\mathcal{G}_v$-orbit, as the set $N_2(v)$ of vertices at distance 2 from v in $G(O)$ (see [8]).

Theorem 6.4 *Let $(G, \mathcal{G})$ be a finite 2-arc transitive pair of degree $k > 2$, for which $\mathcal{G}$ is primitive on vertices and G is not a complete graph. Then the number of vertices at distance 2 from a vertex v is $k(k-1)/\lambda$, where λ is the number of paths in G of length 2 between a given vertex in $N_2(v)$ and v.*

The largest value of λ for which an example is known is $\lambda = 6$; it is attained by a graph of order 100 and degree 22 admitting the Higman-Sims sporadic simple group. The following conjecture is due to Cameron.

Conjecture *If $(G, \mathcal{G})$ is a 2-arc transitive pair of degree $k > 2$ for which $\mathcal{G}$ is primitive on vertices and G is not a complete graph, then $\lambda \le 6$.*

Although this conjecture has been proved when $\mathcal{G}_v^{N(v)}$ is almost simple (see [10, Cor. 3] and [11]), it is still open for affine 2-transitive groups $\mathcal{G}_v^{N(v)}$.

If we now drop the primitivity assumption, then examples are known with arbitrarily large λ, arising from symmetric $2-(|V|, k, \lambda)$ designs $\mathcal{D}$ (where k and λ are as in Theorem 6.4). The incidence graph of $\mathcal{D}$ is the graph $G(\mathcal{D})$ with vertex-set $\mathcal{P} \cup \mathcal{B}$, where $\mathcal{P}$ is the point-set and $\mathcal{B}$ is the block-set, and where the edges are the incident point-block pairs, called the *flags* of $\mathcal{D}$. The automorphism group $\mathrm{Aut}(\mathcal{D})$ of $\mathcal{D}$ induces a group of automorphisms of $G(\mathcal{D})$, and any incidence-preserving permutation of $\mathcal{P} \cup \mathcal{B}$ that interchanges $\mathcal{P}$ and $\mathcal{B}$ (implying that $\mathcal{D}$ is self-dual) induces an automorphism of $G(\mathcal{D})$. Thus, if $\mathcal{D}$ is self-dual and $\mathrm{Aut}(\mathcal{D})$ is flag-transitive, then $(G(\mathcal{D}), \mathrm{Aut}(G(\mathcal{D})))$ is symmetric with degree k. If, in addition, the stabilizer of a block is 2-transitive on the points incident with the block, then $(G(\mathcal{D}), \mathrm{Aut}(G(\mathcal{D})))$ is 2-arc transitive and, for $v \in \mathcal{P}$, the set of vertices at distance 2 from v in $G(\mathcal{D})$ is $\mathcal{P}\backslash\{v\}$, and has size $v - 1 = k(k-1)/\lambda$. The incidence graph has girth 6 if $\mathcal{D}$ is a projective plane, and otherwise has girth 4. Note that $G(\mathcal{D})$ is bipartite and so its automorphism group is imprimitive and does not give rise to a symmetric pair addressed by Cameron's conjecture.

Cameron [8, Thms. 4.1 and 4.2] proved that, for $\lambda > k/2$, all 2-arc transitive pairs arise in this way, and he questioned whether this might be true whenever $\lambda > 6$. However, it turns out that, when $\mathcal{G}_v^{N(v)}$ is almost simple, there is a unique family of examples that do not arise from symmetric designs.

Example For $n \ge 3$ and q a prime power, the *dual orthogonal graph* $\mathrm{DO}^+(2n, q)$ has as its vertices the n-dimensional totally singular subspaces with respect to a non-degenerate quadratic form of $+$-type on a $2n$-dimensional vector space over a field of order q, with two vertices adjacent whenever their intersection has co-dimension 1 in each.

The following theorem appears in [10] and [11].

Theorem 6.5 *Let $(G, \mathcal{G})$ be 2-arc transitive of degree $k > 2$, with $\mathcal{G}_v^{N(v)}$ almost simple, and suppose that, for a vertex v, $|N_2(v)| = k(k-1)/\lambda$ with $\lambda > 6$. Then G is either the incidence graph of a self-dual flag-transitive symmetric 2-design, or a dual orthogonal graph.*

The symmetric designs arising here were described explicitly in [11]; the proof uses the classification of the almost simple 2-transitive permutation groups. It is not known whether the conclusion of Theorem 6.5 is also true in the case of 2-transitive affine groups $\mathcal{G}_v^{N(v)}$.

Cameron [8] also classified the 2-arc transitive pairs $(G, \mathcal{G})$ of degree $k > 2$ and girth 4 for which $\mathcal{G}_v^{N(v)}$ is $\mathcal{A}_k$ or $\mathcal{S}_k$. The generic examples that arise for any k are $K_{k,k}$, $K_{k+1,k+1}$ with the edges of a matching removed, or Q_k and its antipodal quotient. This classification inspired a similar classification of the odd graphs in [30], and later a study and eventual classification by Ivanov and Shpectorov of the class of Petersen and tilde geometries (see [21]).

Locally projective symmetric pairs

A family of symmetric pairs that required special attention in the above-mentioned studies of 2-arc transitive pairs is the family $(G, \mathcal{G})$, where $\mathrm{PSL}(n, q) \leq \mathcal{G}_v^{N(v)} \leq \mathrm{P\Gamma L}(n, q)$, in its natural representation of degree $(q^n - 1)/(q - 1)$ on the 1-dimensional subspaces of an n-dimensional vector space over a field of order q. Such pairs are called *locally projective*, and the following result of Weiss (which relies on the classification of finite 2-transitive permutation groups) emphasizes their special role in the study of 2-arc transitive pairs (see [21, Thm. 9.2.3]).

Theorem 6.6 *Let $(G, \mathcal{G})$ be a 2-arc transitive pair such that, for some edge vw, the pointwise stabilizer in $\mathcal{G}$ of $N(v) \cup N(w)$ is non-trivial. Then $(G, \mathcal{G})$ is locally projective.*

As mentioned above, every finite 4-arc transitive pair is locally projective. Also, when verifying Cameron's conjecture for 2-arc transitive pairs $(G, \mathcal{G})$, with $\mathcal{G}_v^{N(v)}$ unfaithful and almost simple, it is necessary to characterize an infinite family of locally projective graphs – namely, the dual orthogonal graphs (see [10, Cor. 1]).

All possibilities for the structure of a vertex-stabilizer $\mathcal{G}_v$ for a locally projective pair $(G, \mathcal{G})$ have recently been determined in a series of long and difficult papers by a number of authors; this work is described by Trofimov in [37]. A detailed account of classifications and strategies for analysing locally projective pairs can be found in [21, Ch. 9].

7. Normal quotients

The success of the classification programme for finite distance-transitive pairs $(G, \mathcal{G})$ was largely due to the fact that antipodal quotient graphs and bipartite halves of distance-transitive graphs are themselves distance-transitive. Thus, each distance-transitive pair $(G, \mathcal{G})$ is associated with at least one primitive distance-transitive pair. However, for families such as the family $\mathcal{F}_{2\mathrm{arc}}$ of 2-arc transitive pairs, quotients modulo invariant partitions may not lie in the family.

The situation for 2-arc transitive pairs is even more striking. In 1985, Babai [1, Thm. 1.5] showed that, for each finite regular graph G_0, there exists a finite 2-arc transitive pair $(G, \mathcal{G})$ for which G is a cover of G_0, relative to some vertex-partition $\mathcal{P}$. Note that $\mathcal{P}$ may not be $\mathcal{G}$-invariant, and is certainly not $\mathcal{G}$-invariant if G_0 is not symmetric. Babai's result suggested that $\mathcal{F}_{2\mathrm{arc}}$ might admit no useful organizational principle. However, it turns out that $\mathcal{F}_{2\mathrm{arc}}$ is closed under forming normal quotients, and these provide an insightful method for describing the structure of graphs in this family.

Normal partitions and normal quotients

For a transitive permutation group $\mathcal{G}$ on a set V, a partition $\mathcal{P}$ of V is $\mathcal{G}$-*normal* if $\mathcal{P}$ is the set of $\mathcal{N}$-orbits in V, for some normal subgroup $\mathcal{N}$ of $\mathcal{G}$. Since elements of $\mathcal{G}$ permute set-wise the orbits of a normal subgroup, each $\mathcal{G}$-normal partition is $\mathcal{G}$-invariant. Also, the two trivial partitions of V are both $\mathcal{G}$-normal, since they consist of the orbits either of the identity subgroup or of $\mathcal{G}$ itself. If the only $\mathcal{G}$-normal partitions are the trivial ones, or (equivalently) if every non-trivial normal subgroup of $\mathcal{G}$ is transitive, then $\mathcal{G}$ is *quasiprimitive* on V. It follows that every primitive permutation group is quasiprimitive, but there are many quasiprimitive permutation groups that are not primitive; for example, any transitive permutation representation of a non-abelian simple group, for which a point stabilizer is a non-maximal subgroup, determines a quasiprimitive permutation group that is not primitive.

A quotient $(G_{\mathcal{P}}, \mathcal{G}^{\mathcal{P}})$ of a symmetric pair $(G, \mathcal{G})$ modulo a $\mathcal{G}$-normal partition $\mathcal{P}$ is called a *normal quotient*. If $\mathcal{P}$ is $\mathcal{G}$-normal, and if $|\mathcal{P}| > 1$ and the only $\mathcal{G}$-normal partition refined by $\mathcal{P}$ is $\{V\}$, then $\mathcal{G}^{\mathcal{P}}$ is quasiprimitive. Thus, each symmetric pair $(G, \mathcal{G})$ has a quasiprimitive normal quotient. However, if G is bipartite, then the quasiprimitive normal quotient may be simply $(K_2, \mathcal{S}_2)$, providing very little information. For bipartite graphs G, the appropriate normal quotients $(G_{\mathcal{P}}, \mathcal{G}^{\mathcal{P}})$ to examine are those where the only $\mathcal{G}$-normal partitions refined by $\mathcal{P}$ are $\{V\}$ and the bipartition. Such normal quotients, and also the permutation groups $\mathcal{G}^{\mathcal{P}}$, are called *bi-quasiprimitive*. Major questions that arise when investigating a family of symmetric pairs are whether the family is closed under forming normal quotients, and if so, whether the quasiprimitive and bi-quasiprimitive pairs in the family can be analysed successfully. The relationship between a symmetric pair in the family and its normal quotients is also important.

We saw in Section 6 that a symmetric pair $(G, \mathcal{G})$ is 2-arc transitive if and only if the group $\mathcal{G}_v^{N(v)}$ is 2-transitive. Several other families of interest have defining properties given in terms of $\mathcal{G}_v^{N(v)}$: these include the family $\mathcal{F}_{\mathrm{sym}}$ of *all* symmetric pairs, where the property is that $\mathcal{G}_v^{N(v)}$ is transitive, and the families $\mathcal{F}_{\mathrm{prim}}$ and $\mathcal{F}_{\mathrm{qp}}$

of *locally primitive* and *locally quasiprimitive* symmetric pairs which comprise those $(G, \mathcal{G})$ for which $\mathcal{G}_v^{N(v)}$ is respectively primitive and quasiprimitive.

We address the above questions for these families. A symmetric pair $(G, \mathcal{G})$ is a *multicover* of its quotient $(G_{\mathcal{P}}, \mathcal{G}^{\mathcal{P}})$ if there is a constant ℓ such that, for each edge pp' of $G_{\mathcal{P}}$, each vertex of p is adjacent to ℓ vertices of p'; we also say that $(G, \mathcal{G})$ is an *ℓ-multicover* of $(G_{\mathcal{P}}, \mathcal{G}^{\mathcal{P}})$ – the 1-multicovers are therefore covers. When $\mathcal{P}$ is $\mathcal{G}$-normal, we say that $(G, \mathcal{G})$ is a *normal multicover* of $(G_{\mathcal{P}}, \mathcal{G}^{\mathcal{P}})$, or a *normal cover* if $\ell = 1$. The following theorem can be found in [31, Sec. 4].

Theorem 7.1 *Let $(G, \mathcal{G})$ be a finite connected symmetric pair of degree k in $\mathcal{F}$, where $\mathcal{F} = \mathcal{F}_{\text{sym}}, \mathcal{F}_{\text{qp}}, \mathcal{F}_{\text{prim}}$ or $\mathcal{F}_{\text{2arc}}$, and let $\mathcal{G}$ have an intransitive non-trivial normal subgroup $\mathcal{N}$ with orbit-set $\mathcal{P}$ in V. Then $(G_{\mathcal{P}}, \mathcal{G}^{\mathcal{P}})$ is a connected symmetric pair, and either*
G is bipartite and $\mathcal{P}$ is the bipartition;
or
$(G_{\mathcal{P}}, \mathcal{G}^{\mathcal{P}}) \in \mathcal{F}$, $(G, \mathcal{G})$ is an ℓ-multicover of $(G_{\mathcal{P}}, \mathcal{G}^{\mathcal{P}})$, and $G_{\mathcal{P}}$ has degree k/ℓ; further, $\ell = 1$ if $\mathcal{F} = \mathcal{F}_{\text{prim}}$ or $\mathcal{F}_{\text{2arc}}$.

The general question of constructing normal multicovers of pairs in $\mathcal{F}_{\text{qp}}$ and $\mathcal{F}_{\text{prim}}$ was addressed in [24].

Quasiprimitive groups

In order to analyse the finite quasiprimitive and bi-quasiprimitive pairs in one of the families of Theorem 7.1, we need to understand the structure of finite quasiprimitive and bi-quasiprimitive permutation groups. It turns out that finite quasiprimitive permutation groups possess many of the properties of finite primitive groups, in that similar bounds hold for their orders, minimal degrees and base sizes (see [34]). In addition, they admit a description similar to that given by Theorem 3.1 for finite primitive groups. We give a version of this result that has proved helpful in applications involving automorphism groups of vertex-transitive graphs (see [31, Sec. 5]).

Theorem 7.2 *If $\mathcal{G}$ is a quasiprimitive permutation group on a finite set V, then $\mathcal{G}$ is of one of the types in Table 1; $\mathcal{H}$ denotes a non-abelian finite simple group.*

Although some information may be inferred from Theorem 7.2 about finite bi-quasiprimitive groups, possibilities for their structure are not as well understood as in the quasiprimitive case (but see [24]).

Table 1. *Quasiprimitive types*

Type	Name	Description
HA	holomorph of an abelian group	the unique minimal normal subgroup is elementary abelian
HS	holomorph of a non-abelian simple group	there are two minimal normal subgroups, both regular, non-abelian and simple
HC	holomorph of a composite non-abelian group	there are two minimal normal subgroups, both regular, non-abelian and not simple
AS	almost simple	$\mathcal{H} \leq \mathcal{G} \leq \mathrm{Aut}(\mathcal{H})$
TW	twisted wreath product	the unique minimal normal subgroup $\mathcal{H}^k$ is regular, non-abelian and not simple
SD	simple diagonal	the point stabilizer in the unique minimal normal subgroup $\mathcal{H}^k$ is a diagonal subgroup
CD	compound diagonal	the point stabilizer in the unique minimal normal subgroup $\mathcal{H}^k$ is a product of diagonal subgroups
PA	product action type	the point stabilizer in the unique minimal normal subgroup $\mathcal{H}^k$ is a sub-direct subgroup of $\mathcal{K}^k$, for some proper subgroup $\mathcal{K}$ of $\mathcal{H}$

Quasiprimitive and bi-quasiprimitive symmetric pairs

The first application of Theorem 7.2 was an analysis of the possible types of quasiprimitive groups $\mathcal{G}$ involved in 2-arc transitive quasiprimitive pairs $(G, \mathcal{G})$. It turns out that, of the eight possible types, only four can occur, and for one of these types, Ivanov and Praeger classified all the examples (see [31, Sec. 6]).

Theorem 7.3 *If* $(G, \mathcal{G})$ *is a 2-arc transitive quasiprimitive pair, then* $\mathcal{G}$ *has type* HA, AS, TW *or* PA, *and all pairs with* $\mathcal{G}$ *of type* HA *are known.*

Examples are known for each of the types HA, AS, TW, PA, and classifications have been obtained of the 2-arc transitive quasiprimitive pairs $(G, \mathcal{G})$ for some families of low-rank Lie type almost simple groups $\mathcal{G}$. An extensive analysis of the situation when $\mathcal{G} = \mathcal{A}_n$ or $\mathcal{S}_n$ shows that $(G, \mathcal{G})$ can be described explicitly, unless a vertex-stabilizer $\mathcal{G}_v$ is itself almost simple, primitive in its action on $\{1, 2, \ldots, n\}$, and faithful in its action on $N(v)$; this suggests that, for 'large' almost simple groups $\mathcal{G}$, a complete listing of the 2-arc transitive pairs $(G, \mathcal{G})$ may be infeasible (see [31, Sec. 6]).

By Theorem 7.1, all non-bipartite 2-arc transitive pairs $(G, \mathcal{G})$ are normal covers of at least one quasiprimitive 2-arc transitive pair. It would thus be interesting to know the extent to which the 2-arc transitive normal covers of the known quasiprimitive 2-arc transitive pairs can be described or classified. The situation for bi-quasiprimitive 2-arc transitive pairs is less well understood, and a better understanding depends to some extent on a better understanding of bi-quasi-primitive groups. Currently, the best general analysis is for locally-quasiprimitive pairs $(G, \mathcal{G})$, in [24, Thm. 1.4]. The bi-quasiprimitive 2-arc transitive pairs $(G, \mathcal{G})$, in the special case where $\mathcal{G}$ is a group of affine transformations, were classified by Ivanov and Praeger. Studying s-arc transitive pairs $(G, \mathcal{G})$ by considering their normal quotients has also yielded interesting information in the case where G has odd order, as described in Li [23].

Theorem 7.4 *If $(G, \mathcal{G})$ is an s-arc transitive pair, and if G has odd order, then $s \leq 3$.*

It may be possible to complete the classification of quasiprimitive 2-arc transitive and 3-arc transitive pairs of odd order by using the classification of primitive permutation groups of odd degree (see [9, Thm. 5.5]).

In 1968, Sims conjectured, in group-theoretic language, that for a finite primitive symmetric pair $(G, \mathcal{G})$, the order of a vertex-stabilizer $\mathcal{G}_v$ must be bounded by some function of the degree of G. Weiss and the author conjectured that the same should be true for locally-primitive and locally-quasiprimitive symmetric pairs (see [24, Sec. 6]).

Bounded stabilizer conjecture *There exists an integer function f such that, if $(G, \mathcal{G})$ is a locally-quasiprimitive symmetric pair of degree k, then $|\mathcal{G}_v| \leq f(k)$.*

Sims' conjecture was proved in 1983, using Theorem 3.1 together with detailed information about finite simple groups (see [9, Thm. 5.10]). However, the Bounded stabilizer conjecture is still open, both for locally-primitive and locally-quasiprimitive pairs. Conder, Li and the author showed that, in the locally-primitive case, this conjecture is true for non-bipartite graphs if and only if it holds in the special case when the group $\mathcal{G}$ is almost simple (see [24, Sec. 6]). The proof used Theorems 7.1 and 7.2.

8. Finding automorphism groups

It is notoriously difficult to determine the full automorphism group of a graph G, even if G is symmetric. In some cases, *ad hoc* combinatorial arguments can be used to find a set of configurations (for example, a set of cycles of G) that must be left invariant by the automorphism group. The fact that $\mathrm{Aut}(G)$ is contained in the setwise stabilizer in $\mathrm{Sym}(V)$ of these configurations may facilitate the determination of $\mathrm{Aut}(G)$.

A more systematic study for some families of symmetric graphs uses group theory; this has been possible for the class of primitive symmetric pairs. Suppose that $(G, \mathcal{G})$ is primitive, so that $\mathcal{G}$ is a primitive subgroup of $\mathrm{Sym}(V)$. The group $\mathcal{G}$ may be much smaller than $\mathrm{Aut}(G)$; for example, for the complete graph $G = K_8$, all the groups in the following subgroup chain are primitive on V and symmetric on G:

$$\mathrm{PSL}(2,7) < \mathrm{AGL}(3,2) < \mathcal{A}_8 < \mathrm{Aut}(G) = \mathcal{S}_8.$$

A general determination of the subgroups of $\mathrm{Aut}(G)$ containing a given vertex primitive subgroup $\mathcal{G}$ requires a comprehensive knowledge of the subgroups of $\mathrm{Sym}(V)$ containing $\mathcal{G}$. This latter information is available (see [31, Sec. 7]), and has been applied to determine the primitive permutation groups that can act symmetrically on a graph and have a different socle from that of the full automorphism group, as shown in [26].

Theorem 8.1 *Suppose that $(G, \mathcal{G})$ is a primitive symmetric pair. Then either $\mathcal{G}$ and* $\mathrm{Aut}(G)$ *have the same socle, or there exists $\mathcal{H}$ such that $\mathcal{G} < \mathcal{H} < \mathrm{Aut}(G)$ and $\mathcal{H} = \mathrm{soc}(\mathcal{H})\mathcal{G}$, and $\mathcal{G}$ and $\mathcal{H}$ can be found explicitly.*

However, as we have seen, for many families of symmetric pairs the primitive members, although interesting, do not give a good guide to the structure of typical members of the family. Some of these families are closed under forming normal quotients, and typical graphs in these families are multicovers of their normal quotients. It follows that the quasiprimitive and bi-quasiprimitive members are of special interest, and so results like Theorem 8.1 for such graphs would be of great value. To achieve such results, the first step would be to determine the subgroups of $\mathrm{Sym}(V)$ that contain a given quasiprimitive or bi-quasiprimitive group $\mathcal{G}$. The next step would be to apply this to study the subgroups of $\mathrm{Aut}(G)$ that contain $\mathcal{G}$. The situation is more complicated than the primitive case, since such subgroups need not have the same properties. For example, there are quasiprimitive 2-arc transitive pairs $(G, \mathcal{G})$ for which $\mathrm{Aut}(G)$ is not quasiprimitive (see [31, Sec. 7]).

Given a finite quasiprimitive subgroup $\mathcal{G}$ of $\mathrm{Sym}(V)$, most primitive subgroups $\mathcal{H}$ containing $\mathcal{G}$ are determined in [2]. The cases that are not treated completely are where $\mathcal{G}$ and $\mathcal{H}$ are both almost simple, and where $\mathcal{H}$ has product action type PA and $\mathcal{G}$ has type TW or PA: dealing with these outstanding cases is a theme of ongoing research of the author and others. These results are then applied in [33] to determine the imprimitive quasiprimitive groups $\mathcal{G} < \mathcal{H} < \mathrm{Sym}(V)$ for which $\mathcal{G}$ and $\mathcal{H}$ have different quasiprimitive types or socles. Again, the cases that cannot be treated completely are those where $\mathcal{G}$ and $\mathcal{H}$ are both almost simple, or where $\mathcal{H}$ is of product action type PA.

The most important unsolved problem in this area is the determination of imprimitive quasiprimitive groups $\mathcal{G} < \mathcal{H} < \mathrm{Sym}(V)$ for which $\mathcal{G}$ and $\mathcal{H}$ are both almost simple. One of the more interesting challenges thrown up by this work is to make a thorough investigation of Cartesian decompositions of the vertex-sets that are preserved by quasiprimitive automorphism groups of symmetric graphs.

9. A geometric approach

The structure of distance-transitive graphs presents us with a tantalizing challenge to find a similar methodology that might provide a way to analyse other classes of symmetric graphs. We have seen that the class of connected symmetric pairs $(G, \mathcal{G})$ is closed under forming quotients, relative to $\mathcal{G}$-invariant vertex-partitions $\mathcal{P}$; moreover, if $\mathcal{P}$ is $\mathcal{G}$-normal and $|\mathcal{P}| > 2$, then $(G, \mathcal{G})$ is a normal multicover of $(G_{\mathcal{P}}, \mathcal{G}^{\mathcal{P}})$. However, in general, if $\mathcal{P}$ is not $\mathcal{G}$-normal, then G is not a multicover of $G_{\mathcal{P}}$ and it is unclear how much information about G can be obtained from $G_{\mathcal{P}}$. Also, the induced subgraph on a part $p \in \mathcal{P}$ is a null graph, and this provides no additional insight into the structure of G.

In [16] it was suggested that, in addition to the quotient $G_{\mathcal{P}}$, two other combinatorial configurations could be used to describe a symmetric pair $(G, \mathcal{G})$. The first is the bipartite subgraph $G(p, p')$ induced on the union $p \cup p'$ of two adjacent parts $p, p' \in \mathcal{P}$ of $G_{\mathcal{P}}$. Since $(G_{\mathcal{P}}, G^{\mathcal{P}})$ is symmetric, this subgraph is symmetric and is independent (up to equivalence) of the choice of p and p'. The second configuration is the incidence structure $\mathcal{D}_{\mathcal{P}}(p)$ induced by the adjacency relation on the pair $(p, N_{\mathcal{P}}(p))$, where $N_{\mathcal{P}}(p)$ is the set of parts adjacent to p in $G_{\mathcal{P}}$: a vertex $v \in p$ is incident with $p' \in N_{\mathcal{P}}(p)$ precisely when v is adjacent to some vertex in p'. An initial study of how these configurations might aid an understanding of symmetric graphs was made in [16], and a more extensive analysis in the case where $G_{\mathcal{P}}$ is a complete graph has been carried out in [17] and [18]. This approach raises a fundamental question:

Under what conditions is G determined by the triple $(G_{\mathcal{P}}, \mathcal{D}_{\mathcal{P}}(p), G(p, p'))$?

The incidence structure $\mathcal{D}_{\mathcal{P}}(p)$ may be interpreted as a design with point-set p and block-set $N_{\mathcal{P}}(p)$. This suggests certain other questions that might not otherwise be considered. For example, it is natural to study the case where $\mathcal{D}_{\mathcal{P}}(p)$ has 'no repeated blocks' – that is, where the subsets of p incident with parts p' and p'' in $N_{\mathcal{P}}(p)$ are distinct when $p' \neq p''$. The role of this special case is particularly striking if exactly $|p| - 1$ vertices of p are joined to some vertex of p' – that is, the blocks of $\mathcal{D}_{\mathcal{P}}(p)$ have size $|p| - 1$. An interesting family of examples is as follows.

Example Let $H = (V_H, E_H)$ be a finite connected regular graph of degree $k \geq 3$, and let Δ be a 'self-paired' set of 3-arcs of H: if $(u, v, w, x) \in \Delta$ then $(x, w, v, u) \in \Delta$. The *3-arc graph* $\mathrm{Arc}_\Delta(H)$ is the graph whose vertices are the arcs of H, and where (v, w) and (v', w') are adjacent if and only if $(w, v, v', w') \in \Delta$.

If $\mathcal{G}$ is a subgroup of $\mathrm{Aut}(H)$, then $(\mathrm{Arc}_\Delta(H), \mathcal{G})$ is a symmetric pair if and only if $(H, \mathcal{G})$ is a symmetric pair and Δ is a $\mathcal{G}$-orbit on the 3-arcs of H; in this case, $\mathcal{P} = \{p(v) : v \in V_H\}$, where $p(v) = \{(v, w) : w \in N(v)\}$, is a $\mathcal{G}$-invariant partition of the arcs of H with the property that $\mathcal{D}_\mathcal{P}(p(v))$ has no repeated blocks, and has block size $|p(v)| - 1$. It turns out that, for $|p| \geq 3$, the 3-arc graphs are the only examples with this property (see [25]).

Theorem 9.1 *Suppose that* $(G, \mathcal{G})$ *is a finite symmetric pair and that* $\mathcal{P}$ *is a* $\mathcal{G}$-*invariant vertex partition such that, for* $p \in \mathcal{P}$, $\ell = |p| \geq 3$ *and* $\mathcal{D}_\mathcal{P}(p)$ *has block size* $\ell - 1$. *Then* $\mathcal{D}_\mathcal{P}(p)$ *has no repeated blocks if and only if* $(G_\mathcal{P}, \mathcal{G}^\mathcal{P})$ *is 2-arc transitive, and in this case,* $G \cong \mathrm{Arc}_\Delta(G_\mathcal{P})$ *for some self-paired* $\mathcal{G}^\mathcal{P}$-*orbit* Δ *on the 3-arcs of* $G_\mathcal{P}$.

Theorem 9.2 *For* $(G, \mathcal{G})$ *and* $\mathcal{P}$ *as in Theorem 9.1, the following properties are equivalent:* $(G_\mathcal{P}, \mathcal{G}^\mathcal{P})$ *is 3-arc transitive,* $G(p, p') \cong K_{\ell-1,\ell-1}$, *and* Δ *is the set of all the 3-arcs of* $G_\mathcal{P}$.

Given our experience to date, one strategy for studying imprimitive symmetric pairs $(G, \mathcal{G})$ might be first to investigate proper normal quotients of order at least 3. If there are no such quotients, then $(G, \mathcal{G})$ is quasiprimitive or bi-quasiprimitive, and in this case we might study a triple $(G_\mathcal{P}, \mathcal{D}_\mathcal{P}(p), G(p, p'))$, where $\mathcal{P}$ is minimal. Here we have the additional information that $(G, \mathcal{G})$ is quasiprimitive or bi-quasiprimitive, and also that $\mathcal{G}_v^{N(v)}$ is primitive.

10. Related families of graphs

Some classes of finite graphs possess interesting symmetry properties, but are not quite symmetric. Although not the subject of this chapter, they have attracted attention in the literature, and we mention them here to set the chapter in a slightly wider context.

A connected graph G that is edge-transitive need not be vertex-transitive. In this case, G is bipartite and $\mathrm{Aut}(G)$ has two orbits on vertices forming the bipartition. If such a graph is regular, then it is called *semi-symmetric* and the first examples were constructed by Folkman [15]. Alternatively if, for each vertex v, $(\mathrm{Aut}(G))_v$ is transitive on the set of s-arcs starting with v, then G is *locally s-arc transitive*. By studying the possible amalgams of locally s-arc transitive graphs in which each vertex has degree at least 3, Stellmacher has shown that $s \leq 9$. Also, a systematic study of locally s-arc transitive graphs with $s \geq 2$, using a similar approach to Section 7, can be found in [19].

Similarly, a connected graph G that is edge-transitive and vertex-transitive need not be symmetric. If it is not, then it has even degree and $\mathrm{Aut}(G)$ has two orbits on arcs. A pair $(G, \mathcal{G})$ is called *half-arc transitive* if $\mathcal{G}$ is transitive on vertices and on edges, but not on arcs; examples of half-arc transitive pairs $(G, \mathrm{Aut}(G))$, for all even degrees, were constructed by Bouwer [6]. For a given transitive subgroup $\mathcal{G}$ of $\mathrm{Sym}(V)$, the half-arc transitive pairs $(G, \mathcal{G})$ with vertex-set V are in one-one correspondence with the orbital digraphs $G(O)$, for non-self-paired $\mathcal{G}$-orbitals O. However, for a given half-arc transitive pair $(G, \mathcal{G})$, the pair $(G, \mathrm{Aut}(G))$ may be symmetric, and deciding whether this is true may be difficult. Half-arc transitive pairs $(G, \mathcal{G})$ of degree 4 for which the stabilizers $\mathcal{G}_v$ have order 2 arise naturally when studying the embeddings of graphs in surfaces (see [27]).

Infinite symmetric graphs of finite degree have already occurred in this chapter as coset graphs of universal covers of some amalgams in Section 3. The general study of such graphs is a separate and fruitful area of combinatorics, and a good introduction may be found in [38].

References

1. L. Babai, Automorphism groups, isomorphism, reconstruction, *Handbook of Combinatorics* (eds. R. Graham, M. Grötschel and L. Lovász), Elsevier (1995), 1447–1540.
2. R. W. Baddeley and C. E. Praeger, On primitive overgroups of quasiprimitive permutation groups, *J. Algebra* **263** (2003), 294–344.
3. N. Biggs, *Algebraic Graph Theory*, Cambridge Tracts in Mathematics **67**, Cambridge Univ. Press, 1974.
4. N. L. Biggs, Presentations for cubic graphs, *Computational Group Theory (Durham, 1982)*, Academic Press (1984), 57–63.
5. N. L. Biggs, *Mathematical Reviews*, 88c:05063.
6. I. Z. Bouwer, On edge but not vertex transitive regular graphs, *J. Combin. Theory (B)* **12** (1972), 32–40.
7. A. E. Brouwer, A. M. Cohen and A. Neumaier, *Distance-Regular Graphs*, Springer-Verlag, 1989.
8. P. J. Cameron, Suborbits in transitive permutation groups, *Combinatorics (Proc. NATO Advanced Study Inst., Breukelen, 1974), Part 3: Combinatorial Group Theory*, Math. Centre Tracts **57**, Math. Centrum (1974), 98–129.
9. P. J. Cameron, Permutation groups, *Handbook of Combinatorics, Vol. 1* (eds. R. Graham, M. Grötschel and L. Lovász), Elsevier (1995), 611–645.
10. P. J. Cameron and C. E. Praeger, Graphs and permutation groups with projective subconstituents, *J. London Math. Soc. (2)* **25** (1982), 62–74.
11. P. J. Cameron and C. E. Praeger, On 2-arc transitive graphs of girth 4, *J. Combin. Theory (B)* **35** (1983), 1–11.

12. M. Conder, An infinite family of 5-arc-transitive cubic graphs, *Ars Combin.* **25(A)** (1988), 95–108.
13. M. Conder, An infinite family of 4-arc-transitive graphs each with girth 12, *Bull. London Math. Soc.* **21** (1989), 375–380.
14. M. Conder and P. Lorimer, Automorphism groups of symmetric graphs of valency 3, *J. Combin. Theory (B)* **47** (1989), 60–72.
15. J. Folkman, Regular line-symmetric graphs, *J. Combin. Theory* **3** (1967), 215–232.
16. A. Gardiner and C. E. Praeger, A geometrical approach to imprimitive graphs, *Proc. London Math. Soc. (3)* **71** (1998), 524–546.
17. A. Gardiner and C. E. Praeger, Topological covers of complete graphs, *Math. Proc. Cambridge Philos. Soc.* **123** (1998), 549–559.
18. A. Gardiner, C. E. Praeger and S. Zhou, Cross ratio graphs, *J. London Math. Soc.* **64** (2001), 257–272.
19. M. Giudici, C. H. Li and C. E. Praeger, *Analysing finite locally s-arc transitive graphs, Trans. Amer. Math. Soc.* **356** (2004), 291–317.
20. M. Heydemann, Cayley graphs and interconnection networks, *Graph Symmetry (Montreal, PQ, 1996)* (eds. G. Hahn and G. Sabidussi), Kluwer (1997), 167–224.
21. A. A. Ivanov, *Geometry of Sporadic Groups I, Petersen and Tilde Geometries*, Cambridge Univ. Press (1999).
22. C. H. Li, The finite vertex-primitive and vertex-biprimitive s-transitive graphs with $s \geq 4$, *Trans. Amer. Math. Soc.* **353** (2001), 3511–3529.
23. C. H. Li, On finite s-arc transitive graphs of odd order, *J. Combin. Theory (B)* **81** (2001), 307–317.
24. C. H. Li, C. E. Praeger, A. Venkatesh and S. Zhou, Finite locally-quasiprimitive graphs, *Discrete Math.* **246** (2002), 197–218.
25. C. H. Li, C. E. Praeger and S. Zhou, A class of finite symmetric graphs with 2-arc transitive quotients, *Math. Proc. Cambridge Philos. Soc.* **129** (2000), 19–34.
26. M. W. Liebeck, C. E. Praeger and J. Saxl, Primitive permutation groups with a common suborbit, and edge-transitive graphs, *Proc. London Math. Soc.* **84** (2002), 405–438.
27. D. Marusic and R. Nedela, Maps and half-transitive graphs of valency 4, *European J. Combin.* **19** (1998), 345–354.
28. M. J. Morton, Classification of 4- and 5-arc-transitive cubic graphs of small girth, *J. Austral. Math. Soc. (A)* **50** (1991), 138–149.
29. A. Pasini, *Diagram Geometries*, Oxford Univ. Press, 1994.
30. C. E. Praeger, Primitive permutation groups and a characterization of the odd graphs, *J. Combin. Theory (B)* **31** (1981), 117–142.
31. C. E. Praeger, Finite quasiprimitive graphs, *Surveys in Combinatorics, 1997 (London)* (ed. R. A. Bailey), Cambridge Univ. Press (1997), 65–85.
32. C. E. Praeger, Finite transitive permutation groups and finite vertex-transitive graphs, *Graph Symmetry (Montreal, PQ, 1996)* (eds. G. Hahn and G. Sabidussi), Kluwer (1997), 277–318.
33. C. E. Praeger, Quotients and inclusions of finite quasiprimitive permutation groups, *J. Algebra* **269** (2003), 329–346.
34. A. Shalev and C. E. Praeger, Bounds on finite quasiprimitive permutation groups, *J. Austral. Math. Soc.* **71** (2001), 243–258.
35. C. C. Sims. Graphs and finite permutation groups. *Math. Z.* **95** (1967), 76–86.
36. G. Stroth and R. Weiss, A new construction of the group Ru, *Quart. J. Math. (2)* **41** (1990), 237–243.

37. V. I. Trofimov, Vertex stabilisers of locally projective groups of automorphisms of graphs. A summary, Groups, Combinatorics and Geometry: Durham 2001 (eds. A. A. Ivanov, M. W. Liebeck and J. Saxl), World Scientific, 2003.
38. M. E. Watkins, Ends and automorphisms of infinite graphs, *Graph Symmetry (Montreal, PQ, 1996)* (eds., G. Hahn and G. Sabidussi), Kluwer (1997), 379–414.
39. R. Weiss, A characterization and another construction of Janko's group J_3, *Trans. Amer. Math. Soc.* **298** (1986), 621–633.
40. R. Weiss, Presentations for (G, s)-transitive graphs of small valency, *Math. Proc. Cambridge Philos. Soc.* **101** (1987), 7–20.
41. H. Wielandt, *Finite Permutation Groups*, Academic Press, 1964.

8

Strongly regular graphs

PETER J. CAMERON

Strongly regular graphs form an important class of graphs which lie somewhere between the highly structured and the apparently random. This chapter gives an introduction to these graphs with pointers to more detailed surveys of particular topics.

1. An example

Consider the Petersen graph, shown in Fig. 1.

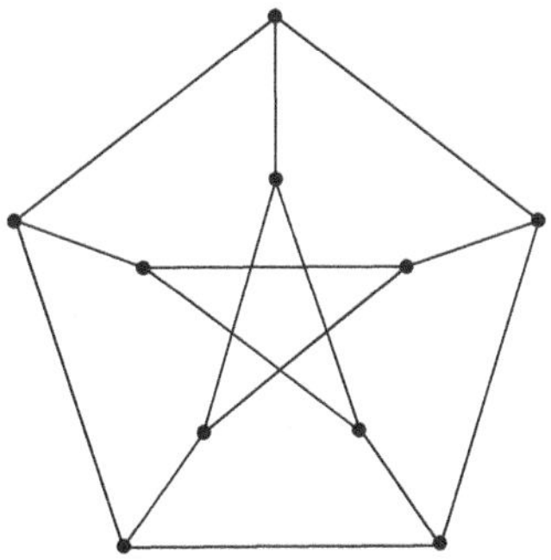

Fig. 1.

This graph has far too many remarkable properties for even a brief survey here. (It is the subject of an entire book by Holton and Sheehan [17].) We focus on a few of its properties: it has ten vertices, is regular of degree 3, and has diameter 2 and girth 5.

These properties are not all independent: simple counting arguments show that a trivalent graph with diameter 2 has at most ten vertices, with equality if and only it has girth 5; dually, a trivalent graph with girth 5 has at least ten vertices, with equality if and only it has diameter 2.

The conditions 'diameter 2 and girth 5' can be rewritten thus: any two adjacent vertices have no common neighbours and any two non-adjacent vertices have exactly one common neighbour. Replacing the particular numbers $(10, 3, 0, 1)$ by general parameters, we reach the definition of a strongly regular graph.

A *strongly regular graph* with parameters (n, k, λ, μ) is a graph on n vertices which is regular of degree k and has the following properties:

- any two adjacent vertices have exactly λ common neighbours;
- any two non-adjacent vertices have exactly μ common neighbours.

So the Petersen graph is a strongly regular graph with parameters $(10, 3, 0, 1)$.

The complete and null graphs are vacuously strongly regular, the parameters μ and λ (respectively) being undefined for them. We often exclude these trivial cases.

The four parameters are not independent. For any given vertex v, we count in two ways the ordered pairs (x, y) of adjacent vertices for which x is adjacent to v, but y is not. We obtain the following result.

Theorem 1.1 *The parameters* (n, k, λ, μ) *of a strongly regular graph satisfy the equation*

$$k(k - \lambda - 1) = (n - k - 1)\mu.$$

Later in this chapter we see that the parameters also satisfy various algebraic conditions and inequalities. However, no complete characterization of the parameter sets of strongly regular graphs is known.

This section concludes with three further basic facts about strongly regular graphs.

Theorem 1.2

(a) The complement of a strongly regular graph is strongly regular.

(b) A strongly regular graph is disconnected if and only if it is isomorphic to mK_r *(the disjoint union of* m *copies of* K_r*), for some positive integers* m *and* r*; this occurs if and only if* $\mu = 0$.

(c) Every connected strongly regular graph has diameter 2.

2. Regularity conditions

We can put the definition of a strongly regular graph into a more general context as follows. For each non-negative integer t, let $C(t)$ be the following graph property.

Let S_1 and S_2 be sets of at most t vertices. If the induced sub-graphs on S_1 and S_2 are isomorphic, then the number of vertices joined to every vertex in S_1 is equal to the number joined to every vertex in S_2.

A graph satisfying $C(t)$ is sometimes called *t-tuple regular*. In such a graph, let $\lambda(S)$ denote the number of common neighbours of a set of vertices inducing a subgraph isomorphic to S, where S runs over all isomorphism types of graphs with at most t vertices.

The conditions $C(t)$ become stronger as t increases. $C(0)$ is vacuous, and $\lambda(\emptyset)$ is just the number of vertices of the graph G. A graph satisfies $C(1)$ if and only if it is regular: for each vertex v, $\lambda(v)$ is the degree. A graph satisfies $C(2)$ if and only if it is strongly regular: λ (edge) and λ (non-edge) are the above parameters λ and μ.

In fact, the hierarchy is finite [7], as the following result shows.

Theorem 2.1 *A graph that satisfies $C(5)$ also satisfies $C(t)$ for all non-negative integers t. The only such graphs are nK_r and its complement for all n and $r \geq 1$, the 5-cycle C_5, and the 3×3 square lattice $L(K_{3,3})$.*

There are only two known examples (up to complementation) of graphs that satisfy $C(4)$ but not $C(5)$: the *Schläfli graph* on 27 vertices and the *McLaughlin graph* on 275 vertices. Infinitely many additional graphs satisfying $C(3)$ are known; all of them, except for $L(K_{n,n})$ (for $n \geq 4$) and finitely many others, are associated with geometric objects such as quadrics in projective spaces and extremal generalized quadrangles.

On the other hand, there is no shortage of graphs that satisfy $C(0)$ or $C(1)$. The number of graphs on n vertices is asymptotic to $2^{n(n-1)/2}/n!$, while the number of k-regular graphs is asymptotically equal to $c_k n^{n(k-2)/2}/n!$, for $2 < k = o(\sqrt{n})$. (Estimates exist also for $k \sim cn$; see Wormald [32].) For both graphs and regular graphs, there are well-developed theories of random objects, including the assertion that almost all of them have no non-trivial automorphisms – this explains the $n!$ in the denominators of the asymptotic formulas).

Strongly regular graphs stand on the cusp between the random and the highly structured. For example, there is a unique strongly regular graph with parameters $(36, 10, 4, 2)$, but a computation by McKay and Spence [20] showed that the number of strongly regular graphs with parameters $(36, 15, 6, 6)$ is 32548. This pattern continues: there is a unique strongly regular graph with parameters $(m^2, 2(m-1), m-2, 2)$, but more than exponentially many strongly regular graphs with parameters $(m^2, 3(m-1), m, 6)$, as we will see. This suggests that no general asymptotic results are possible, and that, depending on the parameters, strongly regular graphs can behave in either a highly structured or an apparently random manner.

Another role of strongly regular graphs is as examples for graph isomorphism testing algorithms. The global uniformity ensured by the definition makes it harder to find a canonical labelling, while the super-exponential number of graphs means that they cannot be processed as exceptions. Paley graphs and other strongly regular (and similar) graphs have also been used as models of 'pseudo-random graphs' (see Thomason [28]).

Recently, Fon-Der-Flaass [14] has observed that an old construction of Wallis [29] gives rise to more than exponentially many strongly regular graphs with various parameter sets, which we discuss below. He also used these graphs to establish the following result about the universality of strongly regular graphs.

Theorem 2.2 *Each graph on n vertices is an induced subgraph of a strongly regular graph on at most $4n^2$ vertices. This is within a constant factor of being best possible.*

It is not known whether such a universality result holds for graphs satisfying $C(3)$.

3. Parameter conditions

The parameters of strongly regular graphs satisfy a number of restrictions, the most important of which are described here.

Theorem 3.1 *Let G be a strongly regular graph with parameters (v, k, λ, μ). Then the numbers*

$$f \text{ and } g = \frac{1}{2}\left(v - 1 \pm \frac{(v-1)(\mu-\lambda) - 2k}{\sqrt{(\mu-\lambda)^2 + 4(k-\mu)}}\right)$$

are non-negative integers.

Proof Let $\mathbf{A}$ be the adjacency matrix of G. The fact that G is strongly regular shows that

$$\mathbf{A}^2 = k\mathbf{I} + \lambda\mathbf{A} + \mu(\mathbf{J} - \mathbf{I} - \mathbf{A}),$$

where $\mathbf{J}$ is the all-1 matrix. The all-1 vector $\mathbf{j}$ is an eigenvector of $\mathbf{A}$ with eigenvalue 1. Any other eigenvector of $\mathbf{A}$ is orthogonal to $\mathbf{j}$, so the corresponding eigenvalue satifies the quadratic equation

$$x^2 = k + \lambda x + \mu(-1 - x).$$

From this, we can calculate the two eigenvalues r and s and (using the fact that the trace of $\mathbf{A}$ is 0) their multiplicities f and g, obtaining the given expressions. ■

On the basis of this theorem, we can classify strongly regular graphs into two types:

Type I (or conference) graphs: for these graphs, $(v-1)(\mu-\lambda)-2k=0$. This implies that $\lambda=\mu-1$, $k=2\mu$ and $n=4\mu+1$. (They are precisely the strongly regular graphs with the same parameters as their complements.) It is known that they exist if and only if v is the sum of two squares.

Type II graphs: for these graphs, $(\mu-\lambda)^2+4(k-\mu)$ is a perfect square d^2, where d divides $(v-1)(\mu-\lambda)-2k$ and the quotient is congruent to $v-1$ (mod 2).

Examples of conference graphs include the *Paley graphs* $P(q)$: the vertex-set of $P(q)$ is the finite field GF(q), where q is a prime power congruent to 1 (mod 4), and v and w are adjacent if and only if $v-w$ is a non-zero square in GF(q) (see Paley [23]).

The 'non-principal' eigenvalues r and s of a Type II strongly regular graph are integers with opposite signs. The parameters may be conveniently expressed in terms of the eigenvalues as follows:

$$\lambda=k+r+s+rs, \qquad \mu=k+rs.$$

Of the other conditions satisfied by the parameters of a strongly regular graph, the most important are the *Kreĭn conditions*, first proved by Scott [24] using a result of Kreĭn [18] from harmonic analysis. They state that

$$(r+1)(k+r+2rs)\leq(k+r)(s+1)^2$$

and

$$(s+1)(k+s+2rs)\leq(k+s)(r+1)^2.$$

The first bound is attained by a graph if and only if the second is attained by its complement. As we shall see, the two inequalities are associated with the geometry of the two non-trivial eigenspaces of the adjacency matrix.

Some parameter sets satisfy all known necessary conditions. We mention a few of these here.

The pseudo-Latin square $PL_r(n)$, with $1\leq r\leq n$, has parameters $v=n^2$, $k=r(n-1)$, $\lambda=r^2-3r+m$, $\mu=r(r-1)$; the significance of the name will appear in the next section.

The negative Latin square $NL_r(n)$, obtained by replacing r and n by their negatives in the formulas just given. Since this gives $\lambda=r^2+3r-n$, we must have $n\geq r(r+3)$; equality holds if and only if the Kreĭn bound is attained.

Smith graphs, whose somewhat involved parameters will not be given here (see [9, p. 111]). These parameters always attain the Kreĭn bound.

These parameters arise in the following theorem of Cameron *et al.* [8]:

Theorem 3.2 *Let G be a graph satisfying C(3). Then either G is the pentagon, or its parameters are of pseudo-Latin square, negative Latin square, or Smith type.*

4. Geometric graphs

The notion of a partial geometry was introduced by Bose [3] as a tool for studying strongly regular graphs. Subsequently, partial geometries have been studied in their own right, and the concept has been extended in various ways, not all of which are related to strongly regular graphs. This section focuses only on the connections.

A *partial geometry* with parameters (s, t, α) is an incidence structure of points and lines satisfying the following axioms:

- each line contains $s + 1$ points, and each point lies on $t + 1$ lines;
- two lines meet in at most one point, and two points lie on at most one line;
- if the point p does not lie on the line L, then there are precisely α incident pairs (q, M), where q is a point of L and M is a line through p.

Note that Bose used slightly different parameters: he wrote R, K, T for what we call $s + 1, t + 1, \alpha$.

The *dual* of a partial geometry with parameters (s, t, α) is obtained by interchanging the names 'point' and 'line' for the two types of object, and dualizing the incidence relation. It is a partial geometry with parameters (t, s, α).

The *point graph* of a partial geometry is the graph whose vertices are the points of the geometry, with adjacency being defined by collinearity. The *line graph* is the point graph of the dual geometry: its vertices are the lines, and adjacency is given by concurrence.

Theorem 4.1 *The point graph of a partial geometry with parameters* (s, t, α) *is a strongly regular graph with parameters* $(n, s(t + 1), s - 1 + t(\alpha - 1), (t + 1)\alpha)$, *where* $n = (s + 1)(st + \alpha)/\alpha$.

The proof of this result is straightforward.

Motivated by this, we say that a strongly regular graph G is *geometric* if it is the point graph of a partial geometry, and that G is *pseudo-geometric* if its parameters have the form given in Theorem 4.1, for some positive integers s, t and α. Sometimes we append the triple (s, t, α) to the term 'geometric' or 'pseudo-geometric'.

Not every pseudo-geometric graph is geometric. Indeed, a pseudo-geometric (s, t, α)-graph is geometric if and only if there is a collection L of $(s + 1)$-cliques with the property that each edge lies in just one clique of L. If some edge lies in no $(s+1)$-clique, then the graph is clearly not geometric, but if there are 'too many' cliques, it is often not clear whether a suitable collection can be selected.

The major result of Bose [3] can now be stated:

Theorem 4.2 (Bose's theorem) *Suppose that s, t and α are positive integers satisfying*

$$s > \tfrac{1}{2}(t+2)(t-1+\alpha(t^2+1)).$$

Then any pseudo-geometric (s,t,α)-graph is geometric, and each edge of such a graph lies in a unique $(s+1)$-clique.

In order to see the power of this theorem, we look at partial geometries a little more closely. We divide them into six types.

Linear spaces: the case $\alpha = s+1$. A point not on a line L is collinear with every point of L. It follows that any two points lie on a unique line. Such structures are also known as *2-designs, pairwise balanced designs* or *Steiner systems*. The point graph of a linear space is just a complete graph, and is of no interest.

We note in passing the asymptotic existence theorem of Richard Wilson [30]. A necessary condition for a linear space with n points, and with $s+1$ points on each line, is that s divides $n-1$ and $s+1$ divides $n(n-1)$. In terms of s and t, this is the single condition that $s+1$ divides $t(t+1)$. Wilson showed the existence of a function $f(s)$ such that a linear space exists if $t > f(s)$ and the necessary condition is satisfied.

Dual linear spaces: the case $\alpha = t+1$. Here, the geometries are the duals of those in the preceding case, but the graphs (the line graphs of linear spaces) are much more interesting.

We examine two special cases. A linear space with two points on each line is a complete graph K_n, and its line graph (the point graph of the dual) is the line graph $L(K_n)$. We have $s=1$, $\alpha=2$ and $t=n-2$, and the inequality in Bose's theorem reduces to $n>8$. We have the following result.

Corollary 4.3 *The graph $L(K_n)$ is strongly regular and, for $n>8$, it is the unique strongly regular graph with its parameters.*

The conclusion actually holds for all $n \neq 8$. For $n=8$, Chang [10] showed that there are exactly four strongly regular graphs with parameters (28, 12, 6, 4). In the literature on strongly regular graphs, $L(K_n)$ is called the *triangular graph $T(n)$*.

A linear space with three points on each line is a *Steiner triple system*. The fact that a Steiner triple system with n points exists for all $n \equiv 1$ or 3 (mod 6) goes back to Kirkman in 1847. More recently, Wilson [30] showed that the number $v(n)$

of Steiner triple systems on an admissible number n of points satisfies

$$v(n) \geq \exp(n^2 \log n/6 - cn^2).$$

Moreover, a Steiner triple system of order $n > 15$ can be recovered uniquely from its line graph. Hence there are super-exponentally many strongly regular graphs with parameters $(n, 3s, s+3, 9)$, for $n = (s+1)(2s+3)/3$ and $s \equiv 0$ or $2 \pmod 3$.

Transversal designs: the case $s = \alpha$. In this case, it can be shown that there is a partition of the points into subsets traditionally called 'groups' (although there is no connection with the algebraic notion): each line is a transversal for the family of groups, and any two points in different groups lie on a line. Thus, the point graph is complete multipartite, and the partite sets are the groups.

Dual transversal designs: the case $t = \alpha$. For $t = 2$, the graph is the line graph of the complete bipartite graph $K_{n,n}$. Bose's theorem then gives us the following result.

Corollary 4.4 *The graph $L(K_{n,n})$ is strongly regular, and for $n > 4$ is the unique strongly regular graph with its parameters.*

This holds for all $n \neq 4$. Shrikhande showed that there are just two strongly regular graphs with parameters $(16, 6, 2, 2)$ – namely, $L(K_{4,4})$ and one other, now called the *Shrikhande graph* and defined below. In the literature on strongly regular graphs, $L(K_{n,n})$ is called the *square lattice graph $L_2(n)$*.

If $t > 2$, for reasons which will become clear, we use new parameters n and r, where $n = s + 1$ and $r = t + 1$. The 'groups' dualize to become a partition of the lines into 'parallel classes', each parallel class forming a partition of the points. There are n lines in each parallel class, with n points on each, so the total number of points is n^2. The geometry is called a *net of order n and degree r*.

Select two parallel classes $\{V_1, V_2, \ldots, V_n\}$ and $\{H_1, H_2, \ldots, H_n\}$. Then the points can be represented as an $n \times n$ grid, where the lines V_i run vertically and the H_j run horizontally, and the unique point on V_i and H_j can be labelled p_{ij}.

Now let $\{L_1, L_2, \ldots, L_n\}$ be another parallel class of lines and construct an $n \times n$ array Λ with *ij*-entry l if $p_{ij} \in L_l$: then Λ is a Latin square of order n. By reversing the construction, we find that any Latin square of order n gives rise to a net of order n and degree 3. Since the number of Latin squares of order k is asymptotically equal to $\exp(n^2 \log n - 2n^2)$, we obtain super-exponentially many strongly regular graphs with these parameters.

There are just two non-isomorphic Latin squares of order 4: the Cayley tables of the Klein group and the cyclic group of order 4. They give rise to two non-isomorphic strongly regular graphs with parameters $(16, 9, 4, 6)$, whose complements are $L_2(4)$ and the Shrikhande graph.

If $r > 3$, we have $r - 2$ additional parallel classes, giving rise to $r - 2$ Latin squares. It is easily checked that these Latin squares are *mutually orthogonal*: given any two squares Λ and Λ' and any given entries l, l', there is a unique cell in which Λ and Λ' have entries l and l'. Conversely, a set of $r - 2$ mutually orthogonal Latin squares of order n gives a net of order n and degree r, and hence a strongly regular graph.

For this reason, the point graph of a net of order n and degree r is called a *Latin square graph*, denoted by $L_r(n)$. A pseudo-geometric graph with the parameters of $L_r(n)$ is the same as the pseudo-Latin square graph $PL_r(n)$ defined earlier – hence the name.

If n is a prime power, then there exists a set of mutually orthogonal Latin squares of order n and of the maximum possible size $n - 1$; the corresponding net is an *affine plane* of order n. Choosing all subsets of $r - 2$ of these squares, where $r \sim cn$ for $0 < c < 1$, we again obtain many strongly regular graphs with the same parameters, but with only a fractional exponential number in this case.

Generalized quadrangles: the case $\alpha = 1$. In this case the geometry is trivially recoverable from its point graph, since an edge lies in a unique maximal clique. There are 'classical' generalized quadrangles, related to the classical groups (the symplectic, unitary and orthogonal groups), in much the same way as projective planes are related to the projective groups PGL$(3, q)$, and also non-classical examples, including some with non-classical parameters. Van Maldegham [21] has surveyed these geometries.

Fon-Der-Flaass [14] has pointed out that some of Wallis's graphs [29] and variants of them have pseudo-geometric parameters corresponding to generalized quadrangles with $s = q + 1$ and $t = q - 1$, or with $s = t = q$, or with $s = q - 1$ and $t = q + 1$, where q is a prime power. So there are super-exponentially many graphs for these parameter sets.

The rest: with $1 < \alpha < \min\{s, t\}$. Here some examples are known, but there is much less theory.

We conclude this section with a reference to the work of Neumaier [22], which improves Bose's classical results. From Neumaier's work, we quote two of his most notable results.

Theorem 4.5 *A strongly regular graph with parameters* (v, k, λ, μ) *and eigenvalues* k, r, s *with* $s < -1$ *and*

$$r > \tfrac{1}{2}s(s + 1)(\mu + 1) - 1$$

is the point graph of a dual linear space or a dual transversal design.

The inequality reduces to Bose's inequality in the pseudo-geometric case, but Neumaier's result applies without this assumption.

Theorem 4.6 *For each negative integer m, there is a finite set $\mathcal{L}(m)$ of strongly regular graphs with the following property: if G is any connected strongly regular graph whose adjacency matrix has eigenvalue m, then G is a complete multipartite graph with block size $-m$, or the point graph of a dual linear space or a dual transversal design with $t+1=-m$, or a member of the list $\mathcal{L}(m)$.*

For $m=-1$ the result is trivial. For $m=-2$ it was proved by Seidel [25] that the set $\mathcal{L}(-2)$ contains only the Petersen, Clebsch, Schläfli, Shrikhande, and three Chang graphs. However, all but twelve of the 32,548 graphs with parameters (36, 15, 6, 6), mentioned earlier, belong to $\mathcal{L}(-3)$: only twelve come from Latin squares.

5. Eigenvalues and their geometry

Let G be a strongly regular graph with vertex-set $V=\{v_1, v_2, \ldots, v_n\}$ and adjacency matrix $\mathbf{A}$. As we have seen, $\mathbf{A}$ has just three distinct eigenvalues k, r and s, with multiplicities 1, f and g, respectively (so that $1+f+g=n$); the eigenvector associated to the eigenvalue k is the all-1 vector. Thus,

$$\mathbf{A} = k\mathbf{E}_0 + r\mathbf{E}_1 + s\mathbf{E}_2,$$

where $\mathbf{E}_0$, $\mathbf{E}_1$, and $\mathbf{E}_2$ are the orthogonal projections of $\mathbf{R}^n$ onto the three eigenspaces V_0, V_1 and V_2 of $\mathbf{A}$.

We now fix attention on one of the non-trivial eigenspaces (V_1, say) and consider the projections of the vertices (the basis vectors of $\mathbf{R}^n$) onto V_1. Thus, let $\mathbf{x}_i = \mathbf{v}_i\mathbf{E}_1$, for $i=1,2,\ldots,n$. The basic property of these vectors is the following.

Theorem 5.1 *There are real numbers α, β and γ, expressible in terms of the parameters of G, such that the inner products of the vectors $\mathbf{x}_i$ are given by*

$$\langle \mathbf{x}_i, \mathbf{x}_j \rangle = \begin{cases} \alpha, & \text{if } v_i = v_j; \\ \beta, & \text{if } v_i \sim v_j; \\ \gamma, & \text{if } v_i \neq v_j \text{ and } v_i \not\sim v_j. \end{cases}$$

Moreover, if G is connected and not complete multipartite, then $\mathbf{x}_i \neq \mathbf{x}_j$ for $i \neq j$.

In particular, if G is connected and not complete multipartite (as we assume for the rest of this section), then the vectors $\mathbf{x}_1, \mathbf{x}_2, \ldots, \mathbf{x}_n$ lie on a sphere of radius $\sqrt{\alpha}$ in $\mathbf{R}^f$, and the angular distances between them take one of two possible values: $\arccos \beta/\alpha$ (for adjacent vertices) and $\arccos \gamma/\alpha$ (for non-adjacent vertices). It is convenient to re-scale the vectors by $1/\sqrt{\alpha}$ so that they lie on the unit sphere.

Delsarte *et al.* [11] have proved the following result:

Theorem 5.2 *The cardinality n of a two-distance set on the unit sphere in* $\mathbf{R}^f$ *satisfies*

$$n \leq \binom{f+2}{2} - 1.$$

This result can be translated into an inequality on the parameters of a strongly regular graph, which is the so-called *absolute bound.* The same authors also gave a *special bound* that depends on the values of α, β and γ – that is, on the actual distances realised by the set; it does not apply for all values of the parameters, but **x** is sometimes more powerful than the absolute bound.

A set $X = \{\mathbf{x}_1, \mathbf{x}_2, \ldots, \mathbf{x}_n\}$ of vectors lying on the unit sphere $\Omega = S^{f-1}$ in Euclidean space $\mathbf{R}^f$ is a *spherical t-design* if, for any polynomial function F of degree at most t, we have

$$\frac{1}{n}\sum_{i=1}^{n} F(\mathbf{x}_i) = \frac{1}{\text{vol}\,\Omega}\int_{\Omega} F(\mathbf{x})\text{d}\mathbf{x}.$$

In other words, the finite set 'approximates the sphere up to degree t'. For small t, there is a mechanical interpretation. Place unit masses at the points of X. Then X is a spherical 1-design if and only if the centre of mass is at the origin, and is a spherical 2-design if, in addition, the inertia ellipsoid is a sphere – that is, the moments of inertia are all equal and the products of inertia are 0.

Theorem 5.3 *Let G be a connected strongly regular graph that is not complete multipartite, and let X be the projection of the vertex-set of G onto a non-trivial eigenspace, re-scaled to lie on the unit sphere. Then*

(a) X is a spherical 2*-design.*
(b) X is a spherical 3*-design if and only if the Kreĭn bound corresponding to this eigenspace is attained – this implies that G satisfies* $C(3)$.
(c) X is a spherical 4*-design if and only if the absolute bound is attained – this implies that G satisfies* $C(4)$.
(d) X is never a spherical 5*-design.*

We saw that the parameters of a graph satisfying $C(3)$ (apart from the pentagon C_5) are either of pseudo-Latin square or negative Latin square type, or of Smith type. Such a graph gives a spherical 3-design if and only if it attains the Kreĭn bound. All Smith graphs, and also the pentagon, attain this bound, but no graph of pseudo-Latin square type does. For graphs $NL_r(n)$ of negative Latin square type, a 3-design arises if and only if $n = r(r+3)$. Only two such graphs are known, the

Clebsch graph on 16 vertices ($r = 1$) and the Higman-Sims graph on 100 vertices ($r = 2$).

We obtain a spherical 4-design in the smaller eigenspace if and only if the graph G is C_5 or a so-called 'extremal Smith graph'. Two examples of extremal Smith graphs are known, the Schläfli graph on 27 vertices and the McLaughlin graph on 275 vertices.

Information on the geometry of eigenspaces for more general classes of graphs is given in Chapter 3; see also Godsil [15].

6. Rank 3 graphs

Looking again at the picture of the Petersen graph with which we began, we see that it has five-fold symmetry, and indeed has the symmetry of a regular pentagon (the dihedral group of order 10). In fact, there is more symmetry that is not visible in the diagram. This graph has a well-known representation as the complement of the line graph of K_5 – that is, the vertices can be labelled with the 2-element subsets of $\{1, 2, 3, 4, 5\}$ in such a way that two vertices are adjacent if and only if their labels are disjoint. Now the symmetric group S_5, in its induced action on the vertex labels, acts as a group of automorphisms of the graph. It is not hard to show that S_5 is the full automorphism group. Moreover, S_5 acts transitively on the set of adjacent pairs of vertices and on the set of non-adjacent pairs of vertices.

A graph G is a *rank* 3 *graph* if it admits a group $\mathcal{G}$ of automorphisms with the property that $\mathcal{G}$ acts transitively on the set of vertices, on the set of ordered pairs of adjacent vertices, and on the set of ordered pairs of non-adjacent vertices. The term comes from permutation group theory, where the *rank* of a transitive permutation group $\mathcal{G}$ on a set Ω is the number of orbits of $\mathcal{G}$ on the set of ordered pairs of elements of Ω. In the case of a rank 3 graph, with $\Omega = V(G)$, the three orbits are $\{(v, v) : v \in V(G)\}$, $\{(v, w) : v \sim w\}$ and $\{(v, w) : v \neq w, v \nsim w\}$.

Theorem 6.1

(a) Every rank 3 *graph is strongly regular.*

(b) Let $\mathcal{G}$ be a transitive permutation group with rank 3 *and even order. Then there is a rank* 3 *graph G admitting $\mathcal{G}$ as a group of automorphisms.*

Proof

(a) This follows from the fact that the number of neighbours of a vertex (or common neighbours of a pair of vertices) is the same as the number of (common) neighbours of any image under an automorphism.

(b) The group $\mathcal{G}$ has just two orbits O_1 and O_2 on ordered pairs of *distinct* elements of Ω. Now, for any orbit O, the set $O^* = \{(w, v) : (v, w) \in O\}$ is

also an orbit, so either $O_1^* = O_1$ or $O_1^* = O_2$. However, since $\mathcal{G}$ has even order, it contains an element g of order 2 that interchanges two points v, w of Ω. If $(v, w) \in O_i$, then $O_i^* = O_i$, so the first alternative holds.

Now let the graph G have vertex-set Ω, with $v \sim w$ whenever $(v, w) \in O_1$. Our argument shows that the graph is undirected: clearly it admits $\mathcal{G}$ as a rank 3 group of automorphisms. ■

This is a special case of the general construction of $\mathcal{G}$-invariant graphs in Chapter 5. Note that by this construction the orbits O_1 and O_2 give rise to complementary strongly regular graphs.

A major result in permutation group theory, which relies heavily on the classification of finite simple groups, is the determination of all rank 3 permutation groups. We outline the argument here.

Let $\mathcal{G}$ be a rank 3 permutation group on Ω. Recall that $\mathcal{G}$ is *imprimitive* if $\mathcal{G}$ preserves a non-trivial equivalence relation, and *primitive* otherwise. Now, if $\mathcal{G}$ is imprimitive, and $\equiv$ is the equivalence relation preserved by $\mathcal{G}$, then the sets

$$\{(v, w) : v \equiv w, v \neq w\} \quad \text{and} \quad \{(v, w) : v \not\equiv w\}$$

are $\mathcal{G}$-invariant, and so must be the two $\mathcal{G}$-orbits on pairs of distinct points. The corresponding graphs are disjoint unions of complete graphs and their complements, so we may assume that $\mathcal{G}$ is primitive.

The basic analysis of such a group is done by considering the *socle* of $\mathcal{G}$, the product of its minimal normal subgroups. It follows from the O'Nan-Scott theorem that one of three possibilities must occur for the socle $\mathcal{N}$ of $\mathcal{G}$ (see Chapter 7, Section 3):

(a) $\mathcal{N}$ is an elementary abelian group and acts regularly;
(b) $\mathcal{N}$ is a non-abelian simple group;
(c) $\mathcal{N}$ is the direct product of two isomorphic non-abelian simple groups.

In case (a), because its action is regular, $\mathcal{N}$ can be identified with the set of points permuted, and is the additive group of a vector space V over the field $\mathbf{F}_p$, for some prime p. The subgroup $\mathcal{H}$ that fixes the origin is a group of linear transformations of V, with two orbits X_1 and X_2. In our case, the orbits satisfy $X_1 = -X_1$ and $X_2 = -X_2$, and the complementary graphs G_1 and G_2 have vertex-set V and satisfy $v \sim w$ in G_i if and only if $v - w \in X_i$. So the classification in this case is reduced to finding groups of matrices over $\mathbf{F}_p$ with just two orbits (each closed under negation) on non-zero vectors. Examples include the following:

- The multiplicative group of the non-zero squares in $\mathbf{F}_q$, where $q \equiv 1 \pmod 4$: the orbits are the sets of squares and non-squares in $\mathbf{F}_q$, and both graphs (which happen to be isomorphic) are the Paley graph $P(q)$.

- The *orthogonal group* preserving a non-degenerate quadratic form Q over $\mathbf{F}_2$: the orbits are the sets of non-zero vectors $\mathbf{v}$ satisfying $Q(\mathbf{v}) = 0$ or 1. Such forms can be defined on spaces of even dimension, and there are just two inequivalent forms; for example, in dimension 4, the quadratic forms

$$x_1x_2 + x_3x_4 \quad \text{and} \quad x_1x_2 + x_3^2 + x_3x_4 + x_4^2$$

  give the graphs $L_2(4)$ (and its complement) and the *Clebsch graph* (and its complement), respectively. In general, these graphs occur among Thomason's pseudo-random graphs [28]. They are of pseudo- or negative Latin square type, and satisfy $C(3)$.

The complete list of linear groups with two orbits on non-zero vectors was determined by Liebeck [19].

In case (b), where the socle $\mathcal{N}$ of $\mathcal{G}$ is non-abelian and simple, the classification of finite simple groups shows that it must be an alternating group, a group of Lie type, or one of the twenty-six sporadic groups. Moreover, the O'Nan-Scott theorem gives the extra information that $\mathcal{G}$ lies between $\mathcal{N}$ and its automorphism group, where $\mathcal{N}$ is embedded in $\text{Aut}(\mathcal{N})$ as the group of inner automorphisms.

The combined efforts of a number of mathematicians, including Bannai, Kantor, Liebler, Liebeck and Saxl, have determined all rank 3 actions of 'almost simple' groups – that is, groups lying between a simple group and its automorphism group. Examples include the following.

- The *symmetric group* S_n (for $n \geq 5$) acts on the set of 2-element subsets of $\{1, 2, \ldots, n\}$ giving the *triangular graph* $T(n)$ and its complement.
- The *projective group* $\text{PGL}(n, q)$ (for $n \geq 4$) has a rank 3 action on the set of lines of the projective space: the orbits are the sets of intersecting pairs and skew pairs of lines.
- A *classical group* (one preserving a polarity of a projective space) acts on the set of self-polar points of the projective space: these form the *polar space* associated with the polarity; the action has rank 3, except in a few low-dimensional cases where it is doubly transitive. For a few cases involving small fields, the action on the non-self-polar points also has rank 3.
- There are also various 'sporadic' examples, such as $\text{PSU}(3, 5^2)$ acting on the vertices of the Hoffman-Singleton graph, or the Higman-Sims group acting on the vertices of the Higman-Sims graph.

Several of the sporadic simple groups were first constructed as groups of automorphisms of strongly regular graphs: these were the Hall-Janko, Higman-Sims, McLaughlin, Suzuki, Fischer and Rudvalis groups.

In case (c) of the O'Nan-Scott theorem, the socle $\mathcal{N}$ of $\mathcal{G}$ is the direct product of two isomorphic simple groups. The analysis leading to this case shows that the rank 3 graphs that arise are the lattice graphs $L_2(n)$ and their complements.

7. Related classes of graphs

There are many generalizations and variants of strongly regular graphs. In this section we introduce a few of these: distance-regular graphs, association schemes, walk-regular graphs, edge-regular graphs, Deza graphs and strong graphs.

A connected graph G of diameter d is *distance-regular* if there are constants c_i, a_i and b_i, for $0 \le i \le d$, such that, if u and v are vertices at distance i, then the numbers of vertices w such that $w \sim v$ and w is at distance $i-1$, i and $i+1$ from u are c_i, a_i and b_i, respectively. The numbers c_i, a_i and b_i are the *parameters* of the graph; note that c_0, a_0 and b_d are 0.

Every distance-regular graph is regular with degree $b_0 = k$. Also,

$$c_i + a_i + b_i = k \quad \text{for all } i, \quad \text{and} \quad c_1 = 1;$$

thus there are $2d - 3$ 'independent' parameters. A distance-regular graph of diameter 2 is the same thing as a connected strongly regular graph and then $\lambda = a_1$ and $\mu = c_2$.

A connected graph G is *distance-transitive* if there is a group $\mathcal{G}$ of automorphisms of G such that, for any two pairs (v_1, w_1) and (v_2, w_2) of vertices satisfying $d(v_1, w_1) = d(v_2, w_2)$, there is an automorphism $g \in \mathcal{G}$ that maps v_1 to v_2 and w_1 to w_2. Any distance-transitive graph is distance-regular, and a distance-transitive graph of diameter 2 is the same thing as a connected rank 3 graph.

The determination of all distance-transitive graphs is not yet complete: this class of graphs is discussed further in Chapter 9. Further information about distance-regular and distance-transitive graphs can be found in Brouwer, Cohen and Neumaier [6].

Many distance-regular graphs are not distance-transitive, but such graphs become less common as the diameter increases. However, as shown by Egawa [12], there exist distance-regular graphs of arbitrarily large diameter that are not distance-transitive.

The adjacency matrix of a regular connected graph of diameter d has at least $d+1$ distinct eigenvalues, one of which is the degree, and distance-regular graphs attain this bound. As we have seen, a regular connected graph has precisely three eigenvalues if and only if it is strongly regular. However, for $d \ge 3$, there are regular connected graphs of diameter d with d eigenvalues that are not distance-regular.

The first examples were constructed by Bridges and Mena [4], but the study of this interesting class of graphs has not yet progressed much beyond the collection of examples.

Looking more closely at distance-regular graphs, we can show that there are constants p_{ij}^k, for $0 \leq i, j, k \leq d$, with the property that, given vertices u and v with $d(u, v) = k$, the number of vertices w such that $d(u, w) = i$ and $d(w, v) = j$ is precisely p_{ij}^k. We can generalize this as follows.

Suppose that the ordered pairs of points of a set Ω are partitioned into $s + 1$ classes $C_0, C_1, \ldots, C_s$ with the following properties:

the diagonal $\{(v, v) : v \in \Omega\}$ is a single class C_0;

each class C_i is symmetric (that is, if $(u, v) \in C_i$ then $(v, u) \in C_i$);

given $i, j, k \in \{1, 2, \ldots, s\}$ and $(u, v) \in C_i$, the number of w such that $(u, w) \in C_j$ and $(w, v) \in C_k$ depends only on i, j and k, and not on (u, v).

Such a structure is called an *association scheme*. Thus, any distance-regular graph gives rise to an association scheme. More about association schemes can be found in Bannai and Ito [2], Godsil [15] and Bailey [1]. A still more general concept is a *coherent configuration*, where the relations are not required to be symmetric.

Association schemes were originally used in experimental design by Bose and his school. Suppose that an experiment is being performed on a number of experimental units that are divided into b blocks of size k – for example, k fields on each of b farms, or k patients in each of b hospitals. We want to apply a number v of different treatments in such a way that no treatment occurs more than once in the same block. It is clearly a good idea to arrange that any two treatments occur together in a block the same number of times, if possible; such a design is called *balanced*. However, Fisher showed that, if $k < v$, this is not possible unless $v \leq b$: thus, to test more treatments, we must relax the condition of balance. Bose observed that the best approach is to have an association scheme on the set of treatments, and to arrange that the number of times that two treatments u and v occur together in a block depends only on which associate class C_i contains (u, v). Such a design is called *partially balanced*; see [1] for more information.

Indeed, this is a case where the applications preceded the theory, and the generalization preceded the special case. Partially balanced designs were defined by Bose and Nair in 1939. During the 1950s, association schemes became of interest in their own right, but it was not until Bose's 1963 paper [3] that the term 'strongly regular graph' was introduced.

A graph G is *walk-regular* if, for each non-negative integer i and vertex v, the number of closed walks of length i starting at v depends only on i, and not on v; equivalently, a graph is walk-regular if the characteristic polynomials of all of its vertex-deleted subgraphs are the same. The class of walk-regular graphs includes both the vertex-transitive graphs and the distance-regular graphs, and is contained in the class of regular graphs; see Godsil [15] for more about these graphs.

Recall that a strongly regular graph is defined by three conditions:

(a) each vertex has k neighbours;

(b) any two adjacent vertices have λ common neighbours;

(c) any two non-adjacent vertices have μ common neighbours.

We can weaken the definition by requiring only two of these conditions to hold. A graph satisfying (a) and (b) is *edge-regular*, and a graph satisfying (b) and (c) is a *Deza graph* (see [13]). The class of graphs satisfying (a) and (c) has not been studied, except in special cases.

More systematically, recall that a graph is t-tuple regular (that is, it satisfies the property $C(t)$) if the number of common neighbours of a set S with at most t vertices depends only on the isomorphism type of the induced subgraph on S. Let us say that a graph satisfies property $R(t)$ if this condition holds for sets S with $|S| = t$; thus, a Deza graph satisfies $R(2)$, but not necessarily $R(1)$. It seems that no systematic study has been made of the possible sets of integers t for which $R(t)$ can hold in a graph.

A variant of Deza graphs was introduced earlier by Seidel, who defined a *strong graph* to be one with the property that, for any two vertices u and v, the number of vertices joined to just one of the two depends only on whether or not u and v are joined. Using a modified adjacency matrix $\mathbf{B}$ with 0 on the diagonal, -1 for adjacent vertices and $+1$ for non-adjacent vertices, we find that

$$(\mathbf{B} - \rho_1\mathbf{I})(\mathbf{B} - \rho_2\mathbf{I}) = (n - 1 + \rho_1\rho_2)\mathbf{J},$$

for some integers ρ_1 and ρ_2. It follows that, if $n - 1 + \rho_1\rho_2 \neq 0$, then the graph is regular and thus strongly regular. In the remaining case, when $n - 1 + \rho_1\rho_2 = 0$, the graph need not be regular; such special strong graphs are closely connected with regular two-graphs (see below).

The operation σ_X of *switching* a graph G with respect to a set X of vertices is defined as follows: edges between X and its complement are 'switched' to non-edges, and non-edges are switched to edges; adjacencies within X or outside X remain unaltered. Switching with respect to all subsets generates an equivalence relation on the class of all graphs on a fixed vertex-set V. It is easy to see that, if $\mathcal{T}$ is the set of 3-subsets of V that contain an odd number of edges of G, then $\mathcal{T}$ is unaltered by switching. Moreover, a set $\mathcal{T}$ of triples arises from a graph in this way if and only if each 4-set contains an even number of members of $\mathcal{T}$; such a set is called a *two-graph*. Thus, there is a bijection between the set of two-graphs on V and the set of switching equivalence classes on V.

Switching a graph G has the effect of pre- and post-multiplying the $(0, -1, +1)$ adjacency matrix of G (defined above) by a diagonal matrix with entries 1 and -1. The matrix equation

$$(\mathbf{B} - \rho_1\mathbf{I})(\mathbf{B} - \rho_2\mathbf{I}) = \mathbf{0},$$

satisfied by special strong graphs, is unaffected by this, so if a graph satisfies this equation, then so do all graphs in its switching class. In this case the corresponding two-graph is called *regular*. Regular two-graphs are also characterized by the property that any two vertices in V lie in a constant number of triples in $\mathcal{T}$.

There are many connections between regular two-graphs, strongly regular graphs, sets of equiangular lines in Euclidean space, doubly transitive permutation groups, antipodal distance-regular graphs of diameter 3, and several other topics. We refer to Seidel's surveys [26] and [27].

References

1. R. A. Bailey, *Association Schemes: Designed Experiments, Algebra and Combinatorics*, Cambridge Univ. Press, 2004.
2. E. Bannai and T. Ito, *Algebraic Combinatorics I: Association Schemes*, Benjamin, 1984.
3. R. C. Bose, Strongly regular graphs, partial geometries, and partially balanced designs, *Pacific J. Math.* **13** (1963), 389–419.
4. W. G. Bridges and R. A. Mena, Multiplicative cones – a family of three eigenvalue graphs, *Aequat. Math.* **22** (1981), 208–214.
5. A. E. Brouwer, Strongly regular graphs, *The CRC Handbook of Combinatorial Designs* (eds. C. J. Colbourn and J. H. Dinitz), CRC Press (1996), 667–685.
6. A. E. Brouwer, A. M. Cohen and A. Neumaier, *Distance-regular Graphs*, Springer, 1989.
7. P. J. Cameron, 6-transitive graphs, *J. Combin. Theory* (*B*) **28** (1980), 168–179.
8. P. J. Cameron, J.-M. Goethals and J. J. Seidel, Strongly regular graphs with strongly regular subconstituents, *J. Algebra* **55** (1978), 257–280.
9. P. J. Cameron and J. H. van Lint, *Graphs, Codes, Designs and their Links*, Cambridge Univ. Press, 1991.
10. L.-C. Chang, The uniqueness and non-uniqueness of triangular association schemes, *Science Record* **3** (1959), 604–613.
11. P. Delsarte, J.-M. Goethals and J. J. Seidel, Spherical codes and designs, *Geometriae Dedicata* **6** (1977), 363–388.
12. Y. Egawa, Association schemes of quadratic forms, *J. Combin. Theory* (*A*) **38** (1985), 1–14.
13. M. Erickson, S. Fernando, W. H. Haemers, D. Hardy and J. Hemmeter, Deza graphs: A generalization of strongly regular graphs, *J. Combin. Design* **7** (1999), 395–405.
14. D. G. Fon-Der-Flaass, New prolific constructions of strongly regular graphs, *Adv. Geom.* **2** (2002), 301–306.
15. C. D. Godsil, *Algebraic Combinatorics*, Chapman & Hall/CRC Press, 1993.

16. A. J. Hoffman, On the uniqueness of the triangular association scheme, *Ann. Math. Stat.* **31** (1960), 492–497.
17. D. A. Holton and J. Sheehan, *The Petersen graph*, Austral. Math. Soc. Lecture Series **7**, Cambridge Univ. Press, 1993.
18. M. G. Kreĭn, Hermitian-positive kernels, II, *Amer. Math. Soc. Transl.* (2) **34** (1963), 109–164.
19. M. W. Liebeck, The affine permutation groups of rank 3, *Proc. London Math. Soc.* (3) **54** (1987), 477–516.
20. B. D. McKay and E. Spence, see `http:/gauss.maths.gla.ac.uk/~ted/srgraphs.html`.
21. H. Van Maldeghem, *Generalized Polygons*, Birkhäuser, 1998.
22. A. Neumaier, Strongly regular graphs with least eigenvalue $-m$, *Arch. Math.* **33** (1979), 392–400.
23. R. E. A. C. Paley, On orthogonal matrices, *J. Math. Phys.* **12** (1933), 311–320.
24. L. L. Scott, Jr., A condition on Higman's parameters, *Notices Amer. Math. Soc.* **20** (1973), A-97.
25. J. J. Seidel, Strongly regular graphs with $(-1, 1, 0)$ adjacency matrix having eigenvalue 3, *Linear Alg. Appl.* **1** (1968), 281–298.
26. J. J. Seidel, A survey of two-graphs, *Proc. Internat. Coll. Teorie Combinatorie* (Roma 1973), Accad. Naz. Lincei (1977), 481–511.
27. J. J. Seidel and D. E. Taylor, Two-graphs: a second survey, *Algebraic methods in graph theory, II*, Conf. Szeged 1978, Colloq. Math. János Bolyai **25** (1981), 689–711.
28. A. Thomason, Random graphs, strongly regular graphs and pseudorandom graphs, *Surveys in Combinatorics 1987* (ed. C. A. Whitehead), London Math. Soc. Lecture Note Series **123**, Cambridge Univ. Press (1987), 173–195.
29. W. D. Wallis, Construction of strongly regular graphs using affine designs, *Bull. Austral. Math. Soc.* **4** (1971), 41–49.
30. R. M. Wilson, Non-isomorphic Steiner triple systems, *Math. Z.* **135** (1974), 303–313.
31. R. M. Wilson, An existence theory for pairwise balanced designs, III: Proof of the existence conjectures, *J. Combin. Theory* (*A*) **18** (1975), 72–79.
32. N. C. Wormald, Models of random regular graphs, *Surveys in Combinatorics, 1999* (eds. J. D. Lamb and D. A. Preece), London Math. Soc. Lecture Notes Series **267**, Cambridge Univ. Press (1999), 239–298.

9

Distance-transitive graphs

ARJEH M. COHEN

In this chapter we investigate the classification of distance-transitive graphs: these are graphs whose automorphism groups are transitive on each of the sets of pairs of vertices at distance i, for $i = 0, 1, \ldots$. *We provide an introduction into the field. By use of the classification of finite simple groups, it seems possible to find all distance-transitive graphs. Priority is given to the classification of the so-called primitive ones. We give an account of the achievements in this direction.*

1. Introduction

An automorphism of a graph reflects the symmetry of the graph. A high degree of symmetry can be related to both aesthetics and efficiency. It is often the apparent symmetry of a graph when pictured in the plane or in space that makes viewing it a pleasant activity. If the graph has a high degree of symmetry, a set of permutations of the automorphism group of the graph can be used to describe the graph rather succinctly: we just list a few edges, and assert that the others can be obtained as images of those given under successive applications of the generating permutations. Since groups can be generated by relatively small sets of permutations – for any finite simple group, two generators suffice! – this representation of a graph can be much more efficient than listing all the edges. An extreme example is the complete graph on the vertex-set $\{1, 2, \ldots, n\}$, that can be described by saying that $\{1, 2\}$ is an edge and that the permutations $(1\ 2)$ and $(1\ 2\ 3\ \ldots\ n)$ are automorphisms.

In this chapter, we are interested in the case where the automorphism group of the graph is as large as possible. There are various interpretations of largeness, most of which use the transitivity of the group on sets of vertices sharing (possibly) common substructures; for instance, we might require that the group be transitive on the set of all paths in the graph of a certain length. Here we focus on one particular interpretation of the group being large.

2. Distance-transitivity

Let G be a connected graph of diameter d, and let $\mathcal{G}$ be a group of automorphisms of G. Recall that $d(x, y)$ is the distance between the vertices x and y of G. Then the group $\mathcal{G}$ is *distance-transitive* on G if it is transitive on each of the relations $G_i = \{(x, y) \in G \times G : d(x, y) = i\}$, where $i = 0, 1, \ldots, d$. A graph is *distance-transitive* if its automorphism group is distance-transitive on it.

As an example, if $G = K_n$ is the complete graph on $\{1, 2, \ldots, n\}$ and $\mathcal{G}$ is the symmetric group $\mathcal{S}_n$, then $\mathcal{G}$ is clearly transitive on the vertex-set, and hence on $G_0 = \{(x, x) : x \in G\}$. Moreover, the diameter of G is 1. Let $(x, y) \in G_1$; then $x \neq y$. It is readily seen that the pair (x, y) can be mapped to the pair $(1, 2)$ by means of an element of S_n – for instance, if $x \neq 1, 2$ and $y \neq 1, 2$, then the permutation $(1, x)(2, y)$ works. So S_n is transitive on G_1, and hence is distance-transitive.

A bipartite graph $G = K_{n,n}$ with parts of size n is also distance-transitive. Any permutation that interchanges the two parts is an automorphism, and so is any permutation that preserves both parts. The diameter is 2, and so we have to check the transitivity of the automorphism group on the three sets G_i, for $i = 0, 1, 2$. If $i = 0$, the transitivity assertion means the transitivity on the vertex-set, which is immediate from the presence of the above-mentioned automorphisms. If $i = 1$, then we have a pair (x, y) with x in one part and y in the other. But then we can move this pair to any other pair of this kind by permutations that preserve both parts. Finally, if $i = 2$, then we have two points in the same part. They can be moved simultaneously to the other part, or to any other pair in the part to which they belong. It follows that the automorphism group of $K_{n,n}$ is distance-transitive.

The diameters of the distance-transitive graphs introduced so far have been restricted to 1 and 2. The following example gives distance-transitive graphs with arbitrarily large diameters; recall that the stabilizer of a vertex x in a group $\mathcal{G}$ is the subgroup $\mathcal{G}_x := \{g \in \mathcal{G} : xg = x\}$ of $\mathcal{G}$.

Example 2.1 Let d and n be positive integers, with $d \leq n/2$. The *Johnson graph J(n, d)* has as its vertex-set the d-subsets of $X = \{1, 2, \ldots, n\}$, with adjacency defined by $x \sim y$ if and only if $|x \cap y| = d - 1$. Its diameter is d. The group $\mathcal{G} = \mathcal{S}_n$ is a distance-transitive group of automorphisms on this graph, with vertex-stabilizer $\mathcal{S}_d \times \mathcal{S}_{n-d}$.

For $d = 1$, we recover the complete graphs. For $d = 2$, the complement of a distance-transitive graph, if connected, is again distance-transitive. The graph $J(4, 2)$ is the complement of three disjoint copies of K_2, and $J(5, 2)$ is the complement of the well-known Petersen graph of degree 3 (Fig. 1).

Let us take a closer look at distance-transitive groups of automorphisms of graphs of diameter 1; here the graph is a complete graph K_n. A permutation group on a set V is called *doubly transitive* (or *2-transitive*) if it is transitive on the set $\{(x, y) \in V \times V : x \neq y\}$ of all pairs from V. The following theorem shows that $\mathcal{S}_n$ need not be the only distance-transitive automorphism group on K_n.

Theorem 2.1 *A group is a distance-transitive group of automorphisms of K_n if and only if it is doubly transitive on its vertex-set.*

If $n \geq 4$, then $\mathcal{A}_n$ is a proper and doubly transitive subgroup of the automorphism group $\mathcal{S}_n$ of K_n. There are many more examples. The doubly transitive permutation groups have been determined by means of the classification of finite

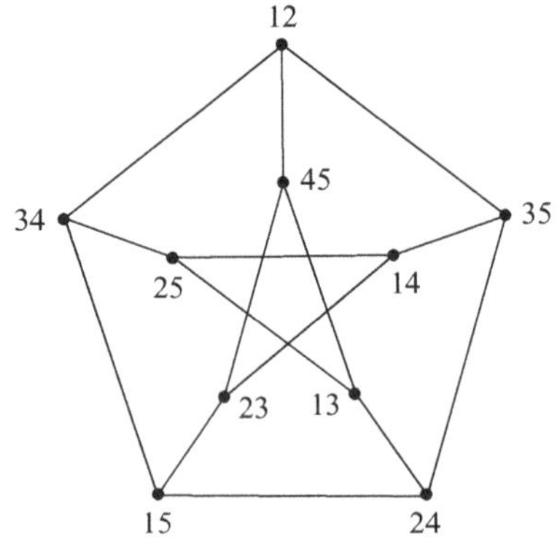

Fig. 1.

simple groups (see Cameron [15]), so the distance-transitive groups of automorphisms of graphs with diameter $d = 1$ are all known.

In this smallest diameter case, we encounter some of the properties that will recur for all d. One of them is the phenomenon that it is easier to classify distance-transitive graphs than to classify distance-transitive groups. If G admits a distance-transitive group, then Aut(G) acts distance-transitively; but when G is distance-transitive, it is harder to establish which subgroups of Aut(G) are still distance-transitive on G.

Another property valid for all diameters d is that, when $\mathcal{G}$ acts distance-transitively on G, it has *permutation rank* $d + 1$ on $V(G)$ – that is, the number of orbits of $\mathcal{G}$ on $V(G) \times V(G)$ is $d + 1$; here, each $G_i (i = 0, 1, \ldots, d)$ is an orbit. As we have seen, this obvious fact is both necessary and sufficient for the case $d = 1$. It does not suffice for higher d, but it is close to being sufficient for $d = 2$.

We will consider graphs G with vertex-set V for which $\mathcal{G}$ is a group of automorphisms of G; such a graph is a *$\mathcal{G}$-invariant graph structure* on V. The next example shows that a $\mathcal{G}$-invariant graph structure on a vertex-set V constructed from a transitive permutation representation of $\mathcal{G}$ on V need not be connected.

Example 2.2 Consider the permutation group $\mathcal{G}$ on the set $V = \{1, 2, 3, 4, 5, 6\}$ generated by the three permutations

$$a = (1\ 6\ 3\ 2\ 5\ 4), \quad b = (1\ 5\ 3)(2\ 4\ 6), \quad c = (2\ 6)(3\ 5).$$

Clearly, $\mathcal{G}$ is transitive on V. The stabilizer of 1 in $\mathcal{G}$ is the subgroup $\mathcal{H}$ generated by $a^2b = (2\ 6\ 4)$ and c; this can be verified by using the algorithms explained in Chapter 10. Besides $\{1\}$, the subgroup $\mathcal{H}$ of $\mathcal{G}$ has orbits $K = \{2, 6, 4\}$ and $L = \{3, 5\}$ on V. Since $\mathcal{G}$ is transitive on V, its permutation rank equals the number of $\mathcal{H}$-orbits on V, which is 3.

If G is a connected $\mathcal{G}$-invariant graph structure on V, vertex 1 must have a neighbour. Since $\mathcal{H}$ fixes 1 and consists of automorphisms of G, the presence of a vertex x in the neighbourhood of 1 implies that all vertices in the $\mathcal{H}$-orbit of x are adjacent to 1. Thus, apart from the complete graph on V, there are two possibilities: either K or L is the set of all neighbours of 1.

If we let K be the neighbours of 1, then the whole graph G is determined by the fact that two vertices are adjacent whenever they are in the same $\mathcal{G}$-orbit as $(1\ x)$, for some (and hence all) $x \in K$. We find the other graph by taking L instead of K. The two graphs are shown in Fig. 2. The graph defined by L is the disjoint union of two 3-cycles, $\{1, 3, 5\}$ and $\{2, 4, 6\}$. It turns out that the diameter of a $\mathcal{G}$-invariant graph structure is not necessarily $e - 1$, if e is the permutation rank. The graph defined by K is the complement of the graph defined by L; it is the complete bipartite graph $K_{3,3}$ on V, whose parts are $\{1, 3, 5\}$

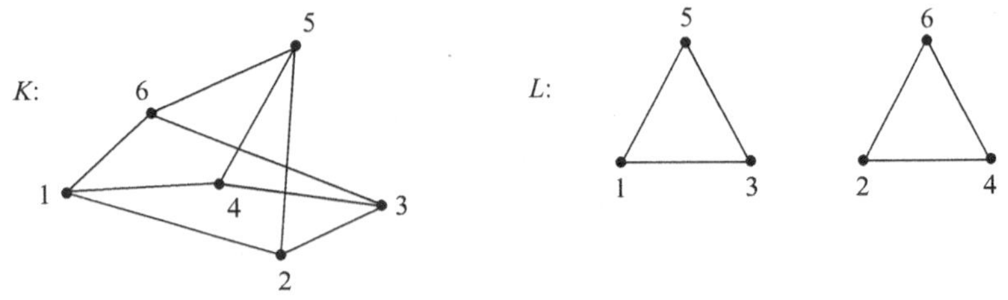

Fig. 2.

and $\{2, 4, 6\}$. The diameter of $K_{3,3}$ is 2, which is one less than the permutation rank.

The construction of the graph G from the group $\mathcal{G}$ using the subgroup orbit K works well, because the $\mathcal{G}$-orbit of a pair $(1, k)$ with $k \in K$ is *self-paired* – that is, it also contains $(k, 1)$. Self-pairedness is necessary for an orbit on pairs to define the adjacency of a graph structure, since otherwise the graph definition would lead to a directed graph.

The group $\mathcal{G}$ has size 36, whereas the full automorphism group of $K_{3,3}$ has order $2 \cdot 6^2 = 72$ (see Chapter 5, Theorem 1.1), so this is another example of a proper distance-transitive subgroup of the full automorphism group of a graph.

In conclusion, when we build distance-transitive graphs on a vertex set V by use of a permutation group on V, we need criteria for self-pairedness of orbits on pairs and for connectedness of the graph in terms of the group. This will be the topic of the next section.

3. Graphs from groups

Applying the definition of distance-transitivity with $i = 0$ and $i = 1$, respectively, we find that a distance-transitive group $\mathcal{G}$ of automorphisms on a graph G is transitive on both the vertex-set $V(G)$ and on the edge-set $E(G)$ of G. It can therefore be described in terms of a subgroup $\mathcal{H}$ of $\mathcal{G}$ and an element r of $\mathcal{G}$. To see this, take $\mathcal{H} = \mathcal{G}_v$, the stabilizer in $\mathcal{G}$ of a vertex v of G, and choose $r \in \mathcal{G}$ in such a way that vr is adjacent to v. Then G is isomorphic (with equivalence of the $\mathcal{G}$-actions) to the graph $G(\mathcal{G}, \mathcal{H}, r)$, whose vertex-set is $\mathcal{G}/\mathcal{H}$ (the set of cosets of $\mathcal{H}$ in $\mathcal{G}$) and whose adjacency $\sim$ is given by $\mathcal{H}x \sim \mathcal{H}y$ when $y \in \mathcal{H}r\mathcal{H}x$.

Theorem 3.1 *Let $\mathcal{G}$ be a group of automorphisms of a graph G. If $\mathcal{G}$ is transitive on the vertex-set and the edge-set of G and v is a vertex, then G is isomorphic to $G(\mathcal{G}, \mathcal{H}, r)$ where $\mathcal{H} = \mathcal{G}_v$ and $r \in \mathcal{G}$ is chosen so that $\{v, vr\}$ is an edge of G.*

In fact, the isomorphism between G and $G(\mathcal{G}, \mathcal{H}, r)$ can even be chosen to preserve the $\mathcal{G}$-actions on the vertex-sets.

In the identification of $V = \{1, 2, \ldots, 6\}$ of Example 2.2 with cosets of $\mathcal{H}$ in $\mathcal{G}$, the coset $\mathcal{H}r$ with $r = a^2$ corresponds to 3, a member of L. The subgroup of $\mathcal{G}$ generated by $\mathcal{H}$ and r does not coincide with $\mathcal{G}$, since it leaves invariant the subset $\{1, 3, 5\}$ of V, whereas $\mathcal{G}$ is transitive on V. This implies that the graph on V built using L cannot be connected: the connected component containing 1 does not reach an even-labelled vertex.

Suppose that we are given a permutation group $\mathcal{G}$ on a vertex-set V, a subgroup $\mathcal{H}$ of $\mathcal{G}$, and an element $r \in \mathcal{G}$. Then the above-defined construction for $G(\mathcal{G}, \mathcal{H}, r)$ leads to a directed graph that need not be a graph: it is a graph if and only $\mathcal{H}x \sim \mathcal{H}y$ implies that $\mathcal{H}y \sim \mathcal{H}x$, for all $x, y \in G$. This is readily seen to be equivalent to self-pairedness of the $\mathcal{G}$-orbit on $(\mathcal{G}/\mathcal{H}) \times (\mathcal{G}/\mathcal{H})$ containing $(\mathcal{H}, \mathcal{H}r)$; in turn, this is equivalent to $r^{-1} \in \mathcal{H}r\mathcal{H}$. In Example 2.2, taking $r = a$, we find that $r^{-1} = (a^2b)r(a^2b) \in \mathcal{H}r\mathcal{H}$, so the $\mathcal{G}$-orbit containing $(1\ 1a) = (1\ 6)$ is self-paired and $G(\mathcal{G}, \mathcal{H}, a)$ is indeed a graph. In conclusion, we have the following criteria for connectedness of $G(\mathcal{G}, \mathcal{H}, r)$ (disregarding the directions of the edges) and for self-pairedness of the adjacency orbit.

Theorem 3.2 *Let $\mathcal{G}$ be a group with subgroup $\mathcal{H}$, and let $r \in \mathcal{G}$. Let G be the directed graph whose vertices are the cosets $\mathcal{H}g$, for $g \in \mathcal{G}$, and in which two vertices $\mathcal{H}x, \mathcal{H}y$ are adjacent if and only if $y \in \mathcal{H}r\mathcal{H}x$. Then:*

- *G is connected if and only if $\langle \mathcal{H}, r\rangle = \mathcal{G}$;*
- *G is undirected if and only if $r^{-1} \in \mathcal{H}r\mathcal{H}$.*

Apart from the anomalies illustrated by the above example, the picture is satisfactory for $d = 2$ in that transitive groups with permutation rank 3 give rise to distance-transitive group actions on graphs with diameter 2.

Theorem 3.3 *Let $\mathcal{G}$ act transitively on a set V. Then $\mathcal{G}$ is distance-transitive on a $\mathcal{G}$-invariant graph structure on V with diameter 2 if and only if it is transitive of rank 3 on V, with self-paired orbits on pairs.*

In view of Theorem 3.2, the only part of the proof that needs attention is the existence of a connected $\mathcal{G}$-invariant graph structure in the 'if' part. If $\mathcal{G}$ has permutation rank 3 on a subgroup $\mathcal{H}$, then there are vertices $r, s \in G$ such that

$$\mathcal{G} = \mathcal{H} \cup \mathcal{H}r\mathcal{H} \cup \mathcal{H}s\mathcal{H}.$$

Suppose that all orbits on pairs are self-paired. Then, by Theorem 3.2, both $G(\mathcal{G}, \mathcal{H}, r)$ and $G(\mathcal{G}, \mathcal{H}, s)$ are well-defined graphs. Since they are complements of each other, at least one of them is connected.

As a consequence of the classification of finite simple groups, all rank 3 permutation groups are known (see [13, p. 229]). We next show that a rank 4 permutation group does not necessarily lead to a graph on which it acts distance-transitively.

Example 3.1 Consider the Mathieu group M_{11} on 11 letters, the smallest sporadic simple group. It can be realized as the permutation group on $\Omega = \{1, 2, \ldots, 11\}$ generated by the permutations

$$(1\ 2\ 3\ 4\ 5\ 6\ 7\ 8\ 9\ 10\ 11) \quad \text{and} \quad (3\ 7\ 11\ 8)(4\ 10\ 5\ 6).$$

The M_{11}-orbit V of the 6-subset $v = \{1, 11, 6, 9, 3, 7\}$ of Ω has 66 elements. We shall analyse the M_{11}-invariant graph structures on V. The stabilizer $\mathcal{H}$ in M_{11} of v has size 120 (in accordance with $|M_{11}| = 66 \times 120 = 7920$), by the famous result of Lagrange that the cardinality of an orbit is the quotient of the group order and the stabilizer order. The subgroup $\mathcal{H}$ is transitive on v and on $\Omega \backslash v$, and is easily identified with the symmetric group $\mathcal{S}_5$.

Let V_i be the set of elements $w \in V$ for which $|v \cap w| = i$. Then

$$|V_1| = 0, \quad |V_2| = 15, \quad |V_3| = 20, \quad |V_4| = 30, \quad |V_5| = 0.$$

Moreover, $\mathcal{H}$ is transitive on each of V_2, V_3 and V_4, and so M_{11} has a transitive permutation representation on V of permutation rank 4. Since all $\mathcal{H}$-orbit sizes are distinct, each orbit on pairs must be self-paired. We can therefore define three non-trivial M_{11}-invariant graph structures on V, according as V_2, V_3 or V_4 is chosen as the set of neighbours of v. Denote the resulting graphs by G_2, G_3 and G_4. In Fig. 3 we record some basic properties of these graphs. On the left, each of the diagrams has a small circle, representing the vertex v of G. The larger circles represent the other $\mathcal{H}$-orbits on V. The number inside each circle is the size of the orbit.

The numbers next to a circle are to be thought of as being attached to a 'loop' on the circle. The remaining numbers are attached to arrows from one circle to another. The number f attached to an arrow from circle i to circle j indicates that there are precisely f vertices of the $\mathcal{H}$-orbit in circle j that are adjacent to a fixed vertex in circle i. We can reformulate this as

$$f = |G_j(v) \cap G_1(w)| \quad \text{for} \quad w \in G_i(v).$$

As suggested by the definition, this number does not depend on the choice of the vertex w in circle i. For example, the number 10 above the circle containing the

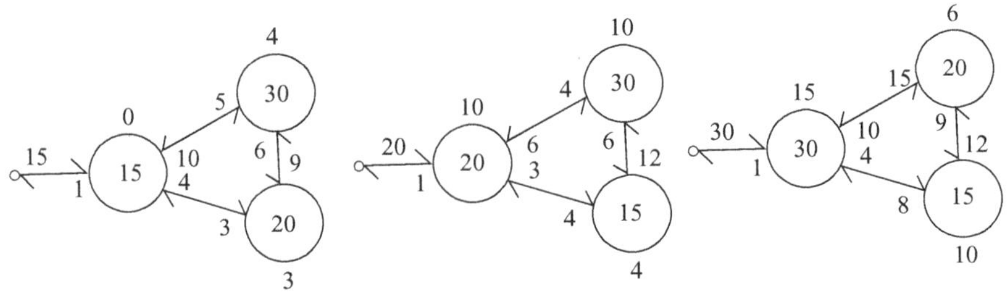

Fig. 3.

number 20 in the middle diagram indicates that a neighbour of v is adjacent to ten other neighbours of v.

Several important properties of G can be read off directly from these so-called *collapsed adjacency diagrams*: the terminology is very close to Soicher's terminology (see Chapter 10) in the software GRAPE for the analysis of graphs with group action in the software system GAP; compare with the *double coset diagram* in [13]. One of the key properties of a distance-transitive graph is the fact that, if G is such a graph, then the collapsed adjacency diagram of G is a path. For example, the collapsed adjacency diagram of the graph G of Example 2.2 is as shown in Fig. 4. It is a path of length 2, in accordance with the fact that the graph is distance-transitive with diameter 2.

The graphs of Example 3.1 are connected and have diameter 2, but they are not paths. Hence, there is no M_{11}-invariant graph structure on V of which M_{11} is a distance-transitive group of automorphisms.

Example 3.2 The *Hamming graph* $H(d, q)$ has vertex-set $V = \mathbf{Z}_q^d$, where two vertices are adjacent if they differ in exactly one coordinate. The wreath product $\mathcal{S}_q \wr \mathcal{S}_d$ is a semi-direct product of d copies of the symmetric group $\mathcal{S}_q$ (each acting on a single component) and $\mathcal{S}_d$ (permuting the coordinates of the vertices), and is a distance-transitive group of automorphisms of $H(d, q)$. The diameter of $H(d,q)$ is d, and its collapsed adjacency diagram is shown in Fig. 5.

Example 3.3 Let V be the vector space of all homogeneous polynomials over $\mathbf{F}_3$ in two variables x, y of degree 2; then V has dimension 3. Define a graph G on V by decreeing that two members of V are adjacent if and only if they differ by an irreducible polynomial. Since a polynomial f in V is irreducible if and only if $-f$ is, this defines a graph on V. The neighbours of 0 in G are $x^2 + y^2, x^2 + xy - y^2, x^2 - xy - y^2$, and their negatives; thus, $G(0)$ has 6 vertices. We shall write G_i for the graph on V whose edges are those pairs of vertices

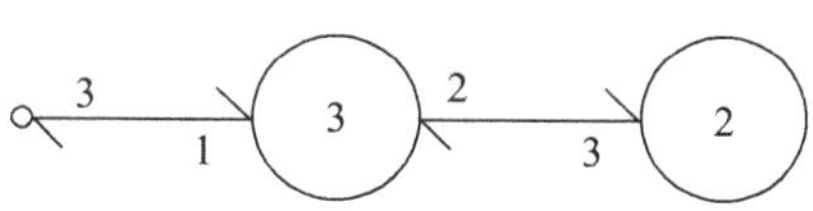

Fig. 4.

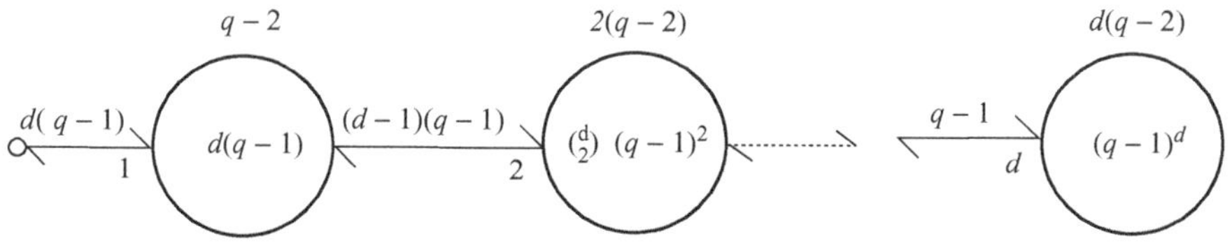

Fig. 5.

$x, y \in V$ with $d(x, y) = i$, where the distance $d(x, y)$ is taken in G. Forming all differences of irreducible polynomials, we see that $G_2(0)$ consists of all polynomials in V that are products of two distinct linear factors, so $|G_2(0)| = 12$. Similarly, all distinct triples of elements of $G_1(0)$ yield squares of linear terms and their negatives, so $|G_3(0)| = 8$. Since $1 + 6 + 12 + 8 = 27 = |V|$, we have accounted for all members of V.

Let us now consider the group $\mathcal{G}$ of automorphisms of G. The group of translations of V preserves G and is transitive on V, so each vertex has degree 6. The stabilizer $\mathcal{G}_0$ of 0 in $\mathcal{G}(= \mathrm{Aut}(G))$ contains the linear transformations induced from a projective linear transformation in x and y. More explicitly, if the latter is determined by $x \mapsto ax + by$ and $y \mapsto cx + dy$, with $ad - bc = \pm 1$, then the linear transformation of V determined by $x^2 \mapsto a^2x^2 - abxy + b^2y^2$, $xy \mapsto acx^2 + (ad + bc)xy + bdy^2$ and $y^2 \mapsto c^2x^2 - cdxy + d^2y^2$, belongs to $\mathcal{G}_0$. Moreover, homotheties in x and y (that is, linear transformations with $b = c = 0$ and $a = d$) act trivially, so $\mathrm{PGL}(2, 3)$ acts, and the homotheties on V belong to $\mathcal{G}_0$. So $\mathcal{G}$ contains the semidirect product of the additive group V and the linear group $\mathrm{PGL}(2, 3) \times \langle -1 \rangle$. It follows directly that $\mathcal{G}_0$ is transitive on $G_i(0)$, for $i = 0, 1, 2$, and hence that $\mathcal{G}$ is distance-transitive on G. The collapsed adjacency diagram of G is shown in Fig. 6. In fact, G is isomorphic with the Hamming graph $H(3, 3)$, and so its full automorphism group $\mathcal{G}$ is the wreath product $\mathcal{S}_3 \wr \mathcal{S}_3$, of size 6^4. By an order argument, this is isomorphic with the group $3^3 \rtimes (\mathrm{PGL}(2, 3) \times \langle -1 \rangle)$.

4. Combinatorial properties

Distance-transitivity is naturally associated with the presence of a group of automorphisms. Nevertheless, we can derive some properties of distance-transitive graphs that are purely combinatorial. Here are the most prominent of these properties.

Theorem 4.1 *If G is a distance-transitive graph with diameter d, then there are non-negative integers*

$$a_1, a_2, \ldots, a_d, \quad b_0, b_1, \ldots, b_{d-1}, \quad c_1, c_2, \ldots, c_d, \quad k_1, k_2, \ldots, k_d$$

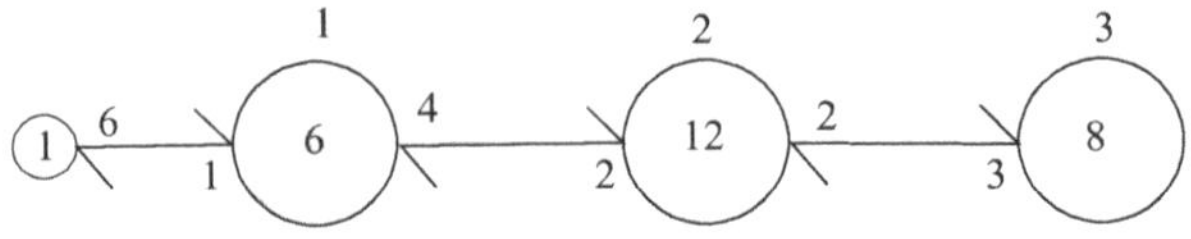

Fig. 6.

such that the collapsed adjacency diagram of G is

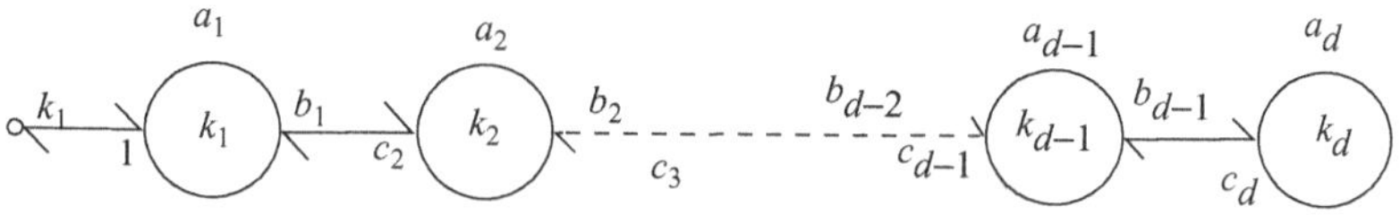

These numbers satisfy the following relations:

- $b_0 = k_1, c_1 = 1$;
- $a_i + b_i + c_i = k_1$;
- $k_{i-1}b_{i-1} = k_i c_i$, *for* $i = 1, 2, \ldots, d$.

In particular, they are all determined by the array

$$(b_0, b_1, \ldots, b_{d-1}; c_1, c_2, \ldots, c_d).$$

The array mentioned at the end of this theorem is called an *intersection array*. A graph is *distance-regular* if its collapsed adjacency diagram can be drawn in this manner, with one 'circle' of vertices for each distance from a fixed vertex v, and with the same intersection array for each vertex. Recall from Example 3.3 that G_h is the graph on the vertex-set of G in which two vertices are adjacent if and only if they are at distance h in Γ. If G is a graph for which the cardinality $|G_i(v) \cap G_j(w)|$ does not depend on the choice of the edge $\{v, w\}$ of G_h, then G is distance-regular. In this case, one usually writes p_{ij}^h for $|G_i(v) \cap G_j(w)|$.

Such a graph is regular; we usually denote its degree by k, so $k = k_1$. If G is distance-regular, then the numbers p_{ij}^h are uniquely determined by the numbers $b_i = p_{i+1,1}^i$ and $c_{i+1} = p_{i-1,1}^i$, for $i = 0, 1, \ldots, d-1$. For example, $k = p_{1,1}^0 = b_0$, $a_i = k - b_i - c_i$, $p_{i,j}^h = 0$ if $i + j < h$, and $p_{i-2,2}^i = c_i c_{i-1}/c_2$.

The Petersen graph (see Example 2.1) is distance-regular with intersection array (3, 2; 1, 1); its diagram is shown in Fig. 7.

These combinatorial properties hold in graphs that admit distance-transitive groups of automorphisms.

Theorem 4.2 *Each distance-transitive graph is distance-regular.*

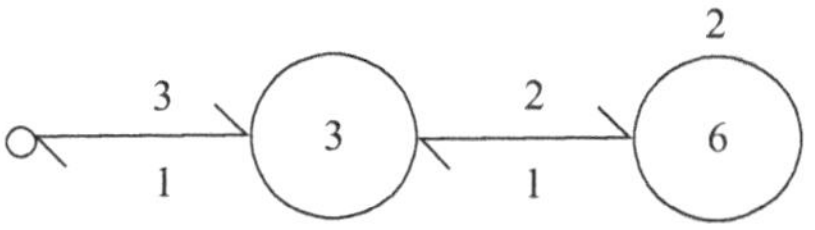

Fig. 7.

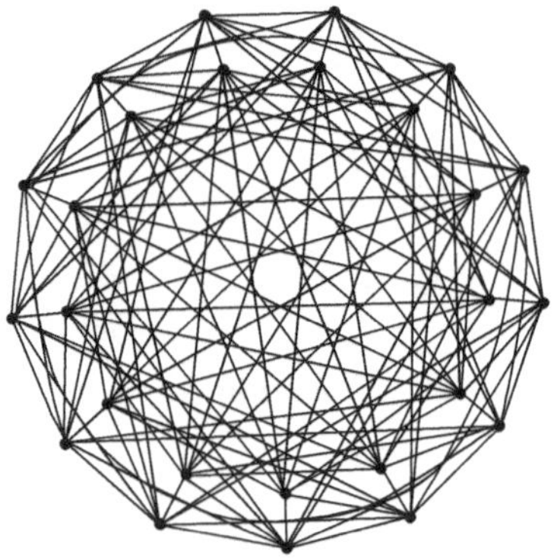

Fig. 8.

The result is easily proved, but the converse statement is false. Figure 8 shows a counter-example due to Adels'son-Vel'skii, Veĭsfeiler, Leman and Faradžev (see [13, p. 136]). It is a distance-regular graph on 26 vertices with intersection array $(10, 6; 1, 4)$. It has an automorphism of order 13 that can be visualized as a rotation of the picture, but has no distance-transitive group.

For an algebraic description of this graph, we write $V = \{a_i, b_i : i = 1, 2, \ldots, 13\}$, and define adjacency $\sim$ on V by

$$a_i \sim a_j \Leftrightarrow |i - j| \in \{1, 3, 4\},$$
$$b_i \sim b_j \Leftrightarrow |i - j| \in \{2, 5, 6\},$$
$$a_i \sim b_j \Leftrightarrow i - j \in \{0, 1, 3, 9\}.$$

For each diameter larger than 1, distance-regular graphs that are not distance-transitive are known (see [13, p. 290]).

The intersection array of the Johnson graph $J(n, d)$ is

$$\big(d(n-d), (d-1)(n-d-1), \ldots, (n-2d+1); 1, 4, 9, \ldots, d^2\big).$$

The group $\mathcal{G} = \mathcal{S}_n$ acts distance-transitively on this graph, with vertex-stabilizer $\mathcal{S}_d \times \mathcal{S}_{n-d}$. This illustrates the remarkable fact that a single permutation representation of a group $\mathcal{G}$ on V may correspond to more than one group-invariant distance-transitive graph structure on V. To see this, let $n = 2d + 1$ and define the *odd graph* O_{d+1} to be the graph whose vertex-set consists of the d-subsets of $X = \{1, 2, \ldots, n\}$, and where adjacency is defined by $x \sim y$ if and only if $x \cap y = \emptyset$ – that is, the vertices x, y are at distance d in $J(2d + 1, d)$. Then O_{d+1} is a distance-transitive graph with diameter d and with intersection array

$$(d+1, d, d, d-1, d-1, \ldots, \lceil (d+1)/2 \rceil; 1, 1, 2, 2, 3, \ldots, \lfloor (d+1)/2 \rfloor).$$

For $d = 2$, the intersection array is $(3, 2; 1, 1)$, and we recover the Petersen graph. The permutation action of $\mathcal{S}_{2d+1}$ on the vertex-set of O_{d+1} is the same as that on $J(2d + 1, d)$.

5. Imprimitivity

Although we are primarily interested in distance-transitive graphs, some notions can be treated entirely within the wider context of distance-regular graphs. An example of such a notion is *imprimitivity*, with which we deal next. For a permutation group $\mathcal{G}$, it means that the stabilizer of a vertex is not a maximal subgroup of $\mathcal{G}$.

Recall that G_i denotes the relation 'being at distance i' on the vertex-set of G; thus, $(v, w) \in G_i$ and $v \in G_i(w)$ both mean that v is at distance i from w in G. The following beautiful characterization of distance-regular graphs that are imprimitive in a combinatorial sense is due to Smith (see [13]).

Theorem 5.1 *Let G be a connected distance-regular graph with diameter d and intersection array $(b_0, \ldots, b_{d-1}; c_1, \ldots, c_d)$, and let I be a minimal proper subset of $\{0, 1, \ldots, d\}$ for which $\bigcup_{i \in I} G_i$ is an equivalence relation with $I \neq \{0\}$. Then at least one of the following properties holds:*

- *G is antipodal – that is, for each vertex v, the distance between any two vertices from $G_d(v)$ is d, and $b_i = c_{d-i}$ for all $i \neq \lfloor d/2 \rfloor$; in this case, $I = \{0, d\}$;*
- *G is bipartite – that is, $a_1 = a_2 = \cdots = a_d = 0$, $c_i = k - b_i$ for $i < d$, and $c_d = k$; in this case, $I = \{0, 1, \ldots, d\} \cap 2\mathbf{Z}$.*

If $\mathcal{G}$ acts distance-transitively on G, and if the stabilizer $\mathcal{H}$ in $\mathcal{G}$ of the vertex v is contained in a proper subgroup $\mathcal{K}$ of $\mathcal{G}$, then the $\mathcal{K}$-orbit of v is a union of $\mathcal{H}$-orbits and is thus of the form $\bigcup_{i \in I} G_i(v)$, for some subset I of $\{0, 1, \ldots, d\}$. If $I \neq \{0\}$, then $\mathcal{K}$ strictly contains $\mathcal{H}$. Note that I is a proper subset, since $\mathcal{K}$ is a proper subgroup of $\mathcal{G}$ containing $\mathcal{H}$. Thus, $\bigcup_{i \in I} G_i$ is an equivalence relation on $V(G)$, and so the above theorem applies.

Conversely, if I is a proper subset of $\{0, 1, \ldots, d\}$ for which $\bigcup_{i \in I} G_i$ is an equivalence relation on $V(G)$, then the stabilizer of the subset $\bigcup_{i \in I} G_i(v)$ is a proper subgroup of $\mathcal{G}$ containing $\mathcal{H}$. It coincides with $\mathcal{H}$ if and only if $I = \{0\}$. Consequently, we have the following result.

Corollary 5.2 *Let $\mathcal{G}$ be a group acting distance-transitively on a connected graph G. Then $\mathcal{G}$ is imprimitive if and only if G is antipodal or bipartite.*

The classification problem of distance-transitive graphs can be reduced to *primitive* (that is, non-imprimitive) distance-transitive graphs, by the following beautiful result and Theorem 5.4 below.

Theorem 5.3 *Suppose that $\mathcal{G}$ acts distance-transitively on the connected graph G with diameter d.*
- *If G is antipodal, then $\mathcal{G}$ acts distance-transitively on the graph whose vertices are the equivalence classes of $G_0 \cup G_d$, and in which two vertices are adjacent whenever they contain adjacent vertices in G.*
- *If G is bipartite, then $\mathcal{G}$ acts distance-transitively on each of the two graphs obtained from G by taking one of the bipartite classes, and letting two vertices be adjacent wherever they are at distance 2 in G.*

For imprimitive distance-regular graphs, these two constructions lead to distance-regular graphs. They are called the *antipodal quotient* in the antipodal case, and the *halved graphs* in the bipartite case.

An example of an antipodal quotient comes from the Johnson graph $J(2d, d)$. The complement in $\{1, 2, \ldots, 2d\}$ of a vertex v is the unique vertex of $J(2d, d)$ at distance d from v. Thus, $G_0 \cup G_d$ is an equivalence relation, and $J(2d, d)$ is antipodal; the antipodal quotients of $J(4, 2)$ and $J(6, 3)$ are the complete graphs K_3 and K_{10}, respectively, The Hamming graph $H(d, q)$ is antipodal if and only if $q = 2$, in which case it is also bipartite.

To each imprimitive distance-regular graph G, we can apply a finite sequence of halvings and antipodal quotients, until we arrive at a primitive distance-regular graph Δ. We say that G is *associated* with Δ.

Theorem 5.4 *For each primitive distance-regular graph of degree $k > 2$, the number of associated imprimitive distance-regular graphs is finite.*

The proof of this result is not complicated. As a consequence, the classifications of distance-regular graphs and of distance-transitive graphs can be thought of as two-stage procedures: first find all primitive examples, and then, for each individual primitive example, determine all the associated imprimitive examples. In what follows, we focus on the first stage for distance-transitive graphs. Much has been done regarding the second stage; see, for example, [13], [25], [37], [17] and [21].

In [21], many interesting examples of antipodal covers of complete graphs can be found. Complete bipartite graphs $K_{k,k}$ with parts of equal size k are distance-transitive with intersection array $(k, k-1; 1, k)$; antipodal covers of $K_{k,k}$ are studied in [25] and [17].

6. Bounds

The intersection array provides a very useful set of numerical invariants for the classification of distance-regular graphs; for example, it specifies all the eigenvalues of the adjacency matrix and their multiplicities. We present a selection of the many inequalities satisfied by the numbers in the array. As usual, for a graph G, a vertex v and a distance i, we write k_i for the cardinality of $G_i(v)$, the set of vertices at distance i from v.

Theorem 6.1 *Let G be a distance-regular graph with diameter d and degree k. Then its intersection array $(b_0, b_1, \ldots, b_{d-1}; c_1, c_2, \ldots, c_d)$ satisfies the following conditions:*

- $1 = c_1 \leq c_2 \leq \cdots \leq c_d$;
- $k = b_0 \geq b_1 \geq \cdots \geq b_{d-1}$;
- *if* $i + j \leq d$, *then* $b_i \geq c_j$;
- $c_2 \geq k - 2b_1$;
- *if G contains a 4-cycle, then* $c_{i+2} - b_{i+2} \geq c_{i+1} - b_{i+1} + a_1 + 2$, *for* $i \leq d - 3$;
- *there exist i and j, with $i < j$, such that*

$$k = k_1 < k_2 < \cdots < k_{i-1} < k_i = k_{i+1} = \cdots = k_j > k_{j+1} > \cdots > k_d;$$

- *for these i and j, if* $\alpha = k_i/k_{i-1}$ *and* $\beta = k_{i+j}/k_{i+j+1}$, *then*

$$|V| \leq \left(\frac{\alpha}{\alpha - 1} + \frac{\beta}{\beta - 1} + j - 1\right) k_i.$$

If $\mathcal{G}$ is a permutation group on V that leaves invariant a distance-transitive graph structure on V, then often the only parameters besides $|V|$, and sometimes d, that we can evaluate among those relevant for the intersection array are the numbers k_i – and frequently just a few of them. In this light, the above inqualities for the k_i are important when finding the graphs on which $\mathcal{G}$ acts distance-transitively.

Recall the definition of a *group character*: this is the function that assigns to a group element the trace of a matrix of the element in a given linear representation of the group. A group character χ is *irreducible* if the representation is irreducible, and *real* if the representation can be defined over the reals.

Theorem 6.2 *Suppose that $\mathcal{G}$ is distance-transitive on a graph G with diameter d, with stabilizer $\mathcal{H}$. Then*

$$|\mathcal{H}| \geq \sqrt{|\mathcal{G}|/(d+1)} \geq \sqrt{|\mathcal{G}|/(r+1)},$$

where r is the number of irreducible real characters of $\mathcal{G}$. If, moreover, G contains a 4-cycle, then $|\mathcal{H}| \geq \sqrt{|\mathcal{G}|/5}$.

The latter part follows directly from the last two parts of Theorem 6.1. The proof of the first part rests on the powerful observation below. A group character χ is *multiplicity-free* if no irreducible character occurs more than once in the decomposition of χ as a sum of irreducibles. We often say that the representation itself is *multiplicity-free* if its character is.

Theorem 6.3 *The permutation character of a distance-transitive group action on a graph is multiplicity-free.*

The proof of the theorem follows from the fact that a permutation representation is multiplicity-free if all of its orbits on pairs are self-paired. The idea of studying multiplicity-free permutation representations stems from Saxl (see [13, p. 230]). Baddeley [3] proved an O'Nan-Scott type theorem for multiplicity-free permutation representations, which essentially reduces the study of primitive permutation groups with only self-paired orbits to almost simple groups.

An outstanding question is whether there are only a finite number of distance-regular graphs of a fixed degree $k > 2$ (see [14]). If, in addition to the degree, the so-called 'geometric girth' is also fixed, then the diameter is known to be bounded (see [13, p. 184]). This result was used by Weiss to obtain a bound for distance-transitive graphs of degree $k > 2$: the diameter does not exceed $(k^6)!2^{2k}$ (see [13, Cor. 7.3.2]). This implies immediately the following result.

Theorem 6.4 *There are only a finite number of distance-transitive graphs of any given degree $k > 2$.*

The condition $k > 2$ is necessary in view of polygons that can have arbitrarily large diameter. For $k \leq 13$, the explicit list of distance-transitive graphs is known – see [13], for details and other references. Despite the fact that the upper bounds on the diameter are astronomical, the result indicates that there is some hope for a classification of distance-transitive graphs. The real progress in the classification of primitive distance-transitive graphs, however, is due to the classification of finite simple groups, as we shall see in the next sections.

7. Finite simple groups

In this section, we summarize the classification of finite simple groups. A *classical finite simple group* is a simple group, obtained from a group $\mathcal{H}$ of linear transformations on a finite vector space V that leave invariant a quadratic or sesquilinear form on V by taking the quotient of the commutator subgroup of $\mathcal{H}$ by its centre.

Table 1.

Name	Condition	Order
$A_{n-1}(q)$	$n \geq 2$	$q^{\frac{n(n-1)}{2}}(q^2-1)(q^3-1)\ldots(q^n-1)/\gcd(n, q-1)$
${}^2A_{n-1}(q)$	$n \geq 3$	$q^{\frac{n(n-1)}{2}}(q^2-1)(q^3+1)\ldots(q^n-(-1)^n)/\gcd(n, q+1)$
$B_n(q)$	$n \geq 2$	$q^{n^2}(q^2-1)(q^4-1)\ldots(q^{2n}-1)/\gcd(2, q-1)$
$C_n(q)$	$n \geq 3$	$q^{n^2}(q^2-1)(q^4-1)\ldots(q^{2n}-1)/\gcd(2, q-1)$
$D_n(q)$	$n \geq 4$	$q^{n^2-n}(q^2-1)(q^4-1)\ldots(q^{2n-2}-1)(q^n-1)/\gcd(4, q^n-1)$
${}^2D_n(q)$	$n \geq 4$	$q^{n^2-n}(q^2-1)(q^4-1)\ldots(q^{2n-2}-1)(q^n+1)/\gcd(4, q^n+1)$

Let q be a prime power, and let n (≥ 2) be an integer.

If the form is identically zero, and if $V = \mathbf{F}_q^n$, then the classical group is denoted by $A_{n-1}(q)$, and if the form is non-degenerate hermitian and $V = \mathbf{F}_{q^2}^n$, then the classical group is denoted by ${}^2A_{n-1}(q)$.

If the form is non-degenerate and quadratic, and if $V = \mathbf{F}_q^{2n}$, then the classical group is denoted by ${}^2D_n(q)$, if the Witt index, the dimension of a maximal linear subspace of V on which the form vanishes identically, is $n-1$, and by $D_n(q)$, if the Witt index equals n.

If the form is non-degenerate and quadratic, and if $V = \mathbf{F}_q^{2n+1}$, then the classical group is denoted by $B_n(q)$.

If the form is non-degenerate, anti-symmetric, and bilinear, and if $V = \mathbf{F}_q^{2n}$, then the classical group is denoted by $C_n(q)$.

Each classical simple group is isomorphic to one of those listed in Table 1. Except for $A_1(2)$, $A_1(3)$, ${}^2A_2(2)$, which are solvable, and for $B_2(2)$, which is isomorphic to $\mathcal{S}_6$, all of these groups are simple.

There are some further series of simple groups obtained from more elaborate constructions in multilinear algebra or Lie algebras. When q is a prime power, these are as shown in Table 2. The commutator subgroups of the groups in the table are simple, with the exception of ${}^2B_2(2)$, which is solvable. All of the other groups are isomorphic to their commutator subgroups, with the exception of $G_2(2)$ (whose commutator subgroup is isomorphic to ${}^2A_2(3)$), ${}^2G_2(3)$ (whose commutator subgroup is isomorphic to $A_1(8)$) and ${}^2F_4(2)$ (whose commutator subgroup has index 2). The simple groups described in this paragraph are the *exceptional finite simple groups*.

A group is called *of Lie type* if it is either classical or exceptional; the prime divisor of q is called the *defining characteristic* of the group.

Finally, 26 *sporadic groups* have been found to exist. Among these are the five Mathieu groups (Example 3.1 introduced the smallest of these) and the Monster group.

Table 2.

Name	Condition	Order
${}^2\mathrm{B}_2(q)$	$q = 2^{2m+1}$	$q^2(q-1)(q^2+1)$
${}^3\mathrm{D}_4(q)$		$q^{12}(q^2-1)(q^6-1)(q^8+q^4+1)$
$\mathrm{E}_6(q)$		$\gcd(3, q-1)^{-1}q^{36}\prod_{m\in\{2,5,6,8,9,12\}}(q^m-1)$
${}^2\mathrm{E}_6(q)$		$\gcd(3, q+1)^{-1}q^{36}\prod_{m\in\{2,5,6,8,9,12\}}(q^m-(-1)^m)$
$\mathrm{E}_7(q)$		$\gcd(2, q-1)^{-1}q^{63}\prod_{m\in\{2,6,8,10,12,14,18\}}(q^m-1)$
$\mathrm{E}_8(q)$		$q^{120}\prod_{m\in\{2,8,12,14,18,20,24,30\}}(q^m-1)$
$\mathrm{F}_4(q)$		$q^{24}(q^2-1)(q^6-1)(q^8-1)(q^{12}-1)$
${}^2\mathrm{F}_4(q)$	$q = 2^{2m+1}$	$q^{12}(q-1)(q^3+1)(q^4-1)(q^6+1)$
$\mathrm{G}_2(q)$		$q^6(q^2-1)(q^6-1)$
${}^2\mathrm{G}_2(q)$	$q = 3^{2m+1}$	$q^3(q-1)(q^3+1)$

The above-mentioned simple groups are all finite simple groups, according to the classification of finite simple groups (see [22] for more details).

Theorem 7.1 *Let $\mathcal{G}$ be a finite non-abelian simple group. Then $\mathcal{G}$ is isomorphic to $\mathcal{A}_n$ (for $n \geq 5$), to a classical simple group, to an exceptional simple group, or to a sporadic simple group.*

8. The first step

The rest of this chapter is devoted to the prospective classification of primitive distance-transitive graphs. Previous overviews can be found in [24] and [10].

The starting point in the classification of primitive distance-transitive graphs is the following theorem, proved by Praeger, Saxl and Yokoyama (see [4] and [13, p. 229]). In view of the determination of all permutation groups with permutation rank 3 (see [13, p. 229]), for the classification of distance-transitive graphs, we may assume that $d \geq 3$. Also, since the only connected distance-regular graphs with degree $k = 2$ and diameter d are the $2d$-gon and the $(2d+1)$-gon, we may also assume that $k \geq 3$.

Theorem 8.1 *Let G be a primitive distance-regular graph with a distance-transitive group $\mathcal{G}$ of automorphisms, and assume that $k \geq 3$ and $d \geq 3$. Then one of the following holds:*

(i) G is associated with a Hamming graph;

(ii) $\mathcal{G}$ has an elementary abelian normal subgroup which is regular on $V(G)$;

(iii) there is a simple non-abelian normal subgroup $\mathcal{N}$ of $\mathcal{G}$ for which $\mathcal{G}$ embeds canonically in Aut($\mathcal{N}$) *– that is, the centralizer $C_{\mathcal{G}}(\mathcal{N})$ of $\mathcal{N}$ in $\mathcal{G}$ is trivial.*

The basic ingredient of the proof of Theorem 8.1 is the celebrated O'Nan-Scott Theorem (see [2]), which does not use the classification of finite simple groups. In Case (i), the graph G is well known, but the possibilities for the group $\mathcal{G}$ are not completely determined – it is typically a wreath product. Case (ii) has been dealt with completely, and will be discussed in Section 9; we refer to it as the *affine case*. Case (iii), the *simple socle case*, is the hardest, since no strong general techniques have been found for ruling out subgroups of small index as stabilizers of distance-transitive actions on graphs. In Section 10, we go into more detail on this case.

Let us discuss the distinct cases in terms of subgroups. A *subnormal subgroup* of a group $\mathcal{G}$ is a subgroup $\mathcal{N}$ for which there exists a chain

$$\mathcal{N} = \mathcal{N}_1 \trianglelefteq \mathcal{N}_2 \trianglelefteq \cdots \trianglelefteq \mathcal{N}_t = \mathcal{G},$$

with the property that $\mathcal{N}_i$ is a normal subgroup of $\mathcal{N}_{i+1}$, for each i. The *generalized Fitting subgroup* $F^*(\mathcal{G})$ of $\mathcal{G}$ is the product of all minimal non-trivial subnormal subgroups of $\mathcal{G}$. In Case (ii) of the Praeger-Saxl-Yokoyama theorem, the generalized Fitting subgroup satisfies $F^*(\mathcal{G}) \cong \mathbf{F}_s^n$ for some prime power s, and in Case (iii), $F^*(\mathcal{G})$ is a (single) non-abelian simple group – the *simple socle*. To conclude this section, we give examples of graphs belonging to each of the Cases (ii) and (iii).

Case (ii): the affine case. The Hamming graphs can be thought of as the 'field-size-1' specializations of the so-called 'bilinear forms graphs'. For $n \geq d$, the *bilinear forms graph* $G = \mathrm{BF}(n, d, q)$ has as its vertex-set the $(n \times d)$-matrices over $\mathbf{F}_q$, and two vertices v and w are adjacent whenever $\mathrm{rk}(v - w) = 1$. For each vertex v, the map $x \mapsto x + v$ is an automorphism of G. In this way, the additive group $\mathcal{N}$ of the vector space $\mathbf{F}_q^{n \times d}$ is a group of automorphisms of G: it is even a regular permutation group on the vertex-set.

Let $\mathcal{H}$ be the subgroup of $\mathrm{GL}(\mathbf{F}_q^{n \times d})$ generated by all left-multiplications by invertible $(n \times n)$-matrices and all right-multiplications by invertible $(d \times d)$-matrices. Then the semidirect product $\mathcal{G} = \mathcal{N} \rtimes \mathcal{H}$ acts distance-transitively on G. For, $\mathcal{N}$ is regular on the vertex-set and so, in order to verify that $\mathcal{G}$ is distance-transitive on G, we need establish only that $G_i(0)$ is a single $\mathcal{H}$-orbit, for $0 \leq i \leq d$. A matrix lies in $G_i(0)$ if and only if it is the sum of i matrices of rank 1 and no fewer; but this is equivalent to having rank i. It is well known that the matrices of rank i form a single $\mathcal{H}$-orbit for each value of i, and so $G_i(0)$ is a single $\mathcal{H}$-orbit; in particular, the diameter of G is d. Moreover, G belongs to the affine case with socle $\mathcal{N}$; its intersection array is

$$\left(\frac{(q^n - 1)(q^d - 1)}{q - 1}, \frac{(q^n - q)(q^d - q)}{q - 1}, \ldots, \frac{(q^n - q^{d-1})(q^d - q^{d-1})}{q - 1}; \right.$$
$$\left. 1, \frac{q^3 - q}{q - 1}, \ldots, \frac{q^{2d-1} - q^{d-1}}{q - 1} \right).$$

Case (iii): the simple socle case. The Johnson graphs can be thought of as the 'field-size-1' specializations of the so-called 'Grassmann graphs'. For $d \leq n/2$, the *Grassmann graph* $G(n, d, q)$ has as its vertex-set the d-dimensional subspaces of the vector space $\mathbf{F}_q^n$ of dimension n over $\mathbb{F}_q$, and two vertices v and w are adjacent whenever $\dim(v \cap w) = d - 1$. The general linear group $\mathrm{GL}(n, q)$ acts distance-transitively on $G(n, d, q)$. The pairs of vertices at mutual distance i form a single orbit, for each value of i in the range $0 \leq i \leq d$, which coincides with $G(n, d, q)_i$; in particular, the diameter of $G(n, d, q)$ is d. The kernel of the action of $\mathrm{GL}(n, q)$ on $G(n, d, q)$ is the centre of the group. For $n \geq 3$, the quotient group $\mathrm{PGL}(n, q)$ has generalized Fitting subgroup $\mathrm{PSL}(n, q) = \mathrm{A}_{n-1}(q)$. So this is a simple socle case when $(n, q) \neq (2, 3)$ or $(2, 4)$; its intersection array is

$$\left(\frac{(q^d - 1)(q^{n+1} - q)}{(q-1)^2}, \frac{(q^d - q)(q^n - q^2)}{(q-1)^2}, \ldots, \right.$$
$$\left. \frac{(q^d - q^{d-1})(q^{n-d+2} - q^d)}{(q-1)^2}; 1, \frac{(q^2-1)^2}{(q-1)^2}, \ldots, \frac{(q^d-1)^2}{(q-1)^2} \right).$$

9. The affine case

In this section, G is a graph admitting a primitive group of automorphisms $\mathcal{G}$ in such a way that the vertex-set of G can be identified with a vector space V over the field $\mathbf{F}_s$ of order s; here, $s = r^b$ is some power of a prime r, maximal with respect to $\mathcal{G}_0 \subseteq \Gamma\mathrm{L}(V)$, where $\mathcal{G}_0$ is the stabilizer in $\mathcal{G}$ of $0 \in V$, and $\Gamma\mathrm{L}(V)$ is the group of all semi-linear transformations of V.

The assumption that G is primitive amounts to requiring that the action of $\mathcal{G}_0$ on V must be irreducible. In his thesis [4] (see also [6]), van Bon showed that there is a reduction to the case where $\mathcal{G}_0$ is quasi-simple; more precisely, he proved Theorem 9.1 below. Here, a *projective representation* of a group $\mathcal{H}$ on a vector space is a linear representation on that vector space of a group $\widehat{\mathcal{H}}$ for which the quotient $\widehat{\mathcal{H}}/Z(\widehat{\mathcal{H}})$ of $\widehat{\mathcal{H}}$ by its centre $Z(\widehat{\mathcal{H}})$ coincides with $\mathcal{H}$. For every finite non-abelian simple group $\mathcal{H}$, there is a minimal finite group $\widehat{\mathcal{H}}$ for which every projective representation of $\mathcal{H}$ comes from a linear representation of $\widehat{\mathcal{H}}$. This group satisfies $\widehat{\mathcal{H}}/Z(\widehat{\mathcal{H}}) \cong \mathcal{H}$, and is equal to its commutator subgroup. A projective or linear representation is *absolutely irreducible* if it is irreducible and remains irreducible after every finite extension of the field of scalars of the vector space.

Theorem 9.1 *Let $\mathcal{G}$ be an affine group that acts primitively and distance-transitively on a connected non-complete graph G with degree and diameter at least 3. Then, with s and V as above, one of the following must hold:*

(i) *G is associated with a Hamming graph;*
(ii) *G is a bilinear forms graph;*
(iii) $V = \mathbf{F}_s$ *is 1-dimensional, and* $\mathcal{G}_0$ *is a subgroup of* $\Gamma\mathrm{L}(\mathbf{F}_s)$*;*
(iv) *the 'generalized Fitting subgroup'* $F^*(\mathcal{G}_0/Z(\mathcal{G}_0 \cap \mathrm{GL}(V)))$ *is non-abelian and simple, and its projective representation on* V *is absolutely irreducible and can be realized over no proper subfield of* $\mathbf{F}_s$.

The importance of this theorem lies in the fact that it reduces the affine case to the study of simple groups and their projective representations. For, in Case (*iv*), the group $\mathcal{G}_0$ is – up to a finite number of possibilities – determined by projective representations of the generalized Fitting subgroup appearing there, and this must be a non-abelian finite simple group. Given such a group $\mathcal{K}$, the group $\mathcal{G}_0$ is contained in the normalizer of the image of $\mathcal{K}$ in $\Gamma\mathrm{L}(V)$ for an irreducible representation space V for $\mathcal{K}$. The number of possibilities for V leading to a $V \rtimes \mathcal{G}_0$-invariant distance-transitive graph structure on V is not yet under control. Theorem 9.3 takes care of this.

The proof of Theorem 9.1 is based on Aschbacher's analysis [1] of subgroups of classical groups. In the first two cases, the graphs are fully determined (as discussed above). The next case was dealt with by Cohen and Ivanov [18], who proved the following result.

Theorem 9.2 *Suppose that Case (iii) of Theorem 9.1 holds. Then* $s = 64$, $\mathcal{G}_0$ *is a semidirect product of the cyclic group of order* 9 *with a cyclic group of order* 3 *or* 6, *and* G *is the Hamming graph* $H(4, 3)$.

The proof uses the Hasse-Weil estimates of the number of vectors $(x_1, x_2, \ldots, x_n)$ in $\mathbf{F}_q^n$ that satisfy $x_1^m + \cdots + x_n^m = 0$.

The remainder of this section is devoted to Case (*iv*) of Theorem 9.1. The general technique here is to use the bounds of Theorem 6.2 and the following result of van Bon [4].

Theorem 9.3 *Let* $\mathcal{G}$, $s = r^b$, *and* V *be as in Theorem 9.1. Then:*

(i) $\mathcal{G}_0$ *is closed under scalar multiplication by* $\mathbf{F}_r^*$*;*
(ii) *if* $\dim_{\mathbf{F}_r} V = c$, *then* $\mathcal{G}_0$ *has at most* c *orbits in* V*;*
(iii) $|V| \leq 5|\mathcal{G}_0|$.

Part (*iii*) follows from the observation that G contains a 4-cycle, so Theorem 6.2 can be applied: $|\mathcal{G}_0| \geq \sqrt{|\mathcal{G}|/5}$.

The general idea of the approach for a fixed simple group $\mathcal{K}$ is to determine all possible projective representations of $\mathcal{K}$ on finite vector spaces V for which there exist $\mathcal{G}_0$ and G, as in Case (*iv*) of Theorem 9.1, with V as the vertex-set of G and with $\mathcal{K}$ isomorphic to $(\mathcal{G}_0 \cap \mathrm{GL}(V))/Z(\mathcal{G}_0 \cap \mathrm{GL}(V))$.

Up to relatively small factors for the central part $Z(\mathcal{G}_0 \cap \mathrm{GL}(V))$ (bounded by $|Z(\mathcal{K})|$) and for the non-linear part $\Gamma\mathrm{L}(V)/\mathrm{GL}(V)$ (bounded by b), the upper bound for $|V|$ in terms of $\mathcal{G}_0$ in Theorem 9.3(*iii*) can be translated into a slightly weaker upper bound for $|V|$ in terms of $\mathcal{K}$. Consequently, when given $\mathcal{K}$, we find a finite list of projective representations V for $\mathcal{G}_0$ which is usually short. For each $\mathcal{K}$, there is a finite list of pairs $(\mathcal{G}_0, V)$ that possibly lead to a $\mathcal{G}$-invariant structure on V for which the permutation group $\mathcal{G} = V \rtimes \mathcal{G}_0$ is a primitive distance-transitive group of automorphisms.

Although there are an infinite number of possibilities for $\mathcal{K}$, the simple groups can be dealt with series-by-series corresponding to the lines of the tables in Section 6. As we have seen in Section 3, we need to analyze $\mathcal{G}_0$-orbits on V.

One of the few results of a general nature for deciding whether a subgroup $\mathcal{G}_0$ of $\Gamma\mathrm{L}(V)$ leads to a distance-transitive graph is due to van Bon [7]. It deals with the case where $\mathbf{F}_s^*$ is contained in $\mathcal{G}_0$ and $\mathcal{K}$ leaves invariant a unitary or orthogonal form on V.

The remainder of this section follows the above subdivision of types for a non-abelian finite simple group $\mathcal{K}$.

Affine alternating groups

For $\mathcal{K} \cong \mathcal{A}_n (n \geq 5)$, the complete analysis was achieved by Liebeck and Praeger [34].

Theorem 9.4 *Suppose that $\mathcal{G}$, $s = r^b$, G and $\mathcal{K}$ are as in Theorem 9.1(iv), where $\mathcal{K} \cong \mathcal{A}_n$ for some integer $n \geq 5$. Then G is associated with a Hamming graph.*

Affine groups of Lie type in the same characteristic

We next take $\mathcal{K}$ to be a group of Lie type with defining characteristic r, the characteristic of the field over which the projective representation spaces V to be considered here are defined. The following result is due to van Bon, Cohen and Cuypers; see [8] and [11].

Theorem 9.5 *Suppose that $\mathcal{G}$, $s = r^b$, V, G, $\mathcal{K}$ are as in Theorem 9.1(iv), where $\mathcal{K} = F^*(\mathcal{G}_0 \cap \mathrm{GL}(V))/Z(\mathcal{G}_0 \cap \mathrm{GL}(V))$ is a simple group of Lie type with defining characteristic r. Then one of the following holds:*

(*i*) $\mathcal{K} \cong \mathrm{E}_6(q)$, *V is a 27-dimensional vector space over* $\mathbf{F}_q$, *and* $q = s$; *the intersection array of G is*

$$\left(\frac{(q^{12}-1)(q^9-1)}{q^4-1}, q^8(q^4+1)(q^5-1), q^{16}(q-1); 1, q^8+q^4, \frac{q^{20}-q^8}{q^4-1}\right);$$

(*ii*) $\mathcal{K} \cong \mathrm{A}_{m-1}(q)$, *and G is the bilinear forms graph* $\mathrm{BF}(m, m, q)$ *on* $\mathbf{F}_q^m$, *so* $\mathcal{G}_0$ *contains* $\mathrm{SL}(\mathbf{F}_q^m)/\langle -(-1)^m I_m\rangle$;

(*iii*) $\mathcal{K} \cong \mathrm{A}_{m-1}(q)$, *and V is the* $(m(m-1)/2)$*-dimensional vector space of alternating forms on* $\mathbf{F}_q^m$, *so* $\mathcal{G}_0$ *contains* $\mathrm{SL}(\mathbf{F}_q^m)/\langle -(-1)^m I_m\rangle$;

(*iv*) $\mathcal{K} \cong \mathrm{A}_{m-1}(q^2)$, *and V is the* $(m(m+1)/2)$*-dimensional vector space over* $\mathbf{F}_q$ *of all Hermitian forms on* $\mathbf{F}_{q^2}^m$, *so* $\mathcal{G}_0$ *contains* $\mathrm{SL}(m, q^2)/\langle -(-1)^m I_m\rangle$;

(*v*) $\mathcal{K} \cong \mathrm{A}_1(8)$, *and G is the primitive distance-regular graph* Δ *associated with the Hamming graph* $H(9, 2)$;

(*vi*) $\mathcal{K} \cong \mathrm{A}_1(8)$, *and G is the distance 2 graph* Δ_2 *of the graph* Δ *in part* (*v*).

The graphs in the conclusions exist and are distance-transitive. Those in Cases (*ii*) and (*iii*) are known as the *alternating* and *Hermitian forms graphs*, denoted by $\mathrm{AF}(m, q)$ and $\mathrm{HF}(m, q^2)$, respectively (see [13]).

We outline the proof. By the use of representation theory of the groups of Lie type, it is relatively easy to provide a list of small-dimensional projective representation spaces V for $\mathcal{K}$ over $\mathbf{F}_s$, among which all affine distance-transitive examples must occur. Here, 'small' means that the somewhat larger group $\mathcal{G}_0$ with $F^*(\mathcal{G}_0 \cap \mathrm{GL}(V)/Z(\mathcal{G}_0 \cap \mathrm{GL}(V))) \cong \mathcal{K}$ satisfies $s^m \leq 5|\mathcal{G}_0|$, in accordance with Theorem 9.3(*iii*).

An intuitive argument (based on closures of orbits in the module over the algebraic closure of $\mathbf{F}_s$) tells us that the so-called 'highest-weight orbit' of $\mathcal{G}_0$ in such a representation space V is among the smallest orbits; the notion 'highest-weight orbit' comes from the theory of groups of Lie type. In view of Theorem 6.1, the $\mathcal{G}_0$-orbit that defines adjacency in G must be among the smallest. Thus, it makes sense to exploit knowledge of the highest-weight orbit to analyse the possible distance-transitive graph structures on V.

To this end, for a $\mathcal{G}_0$-orbit $\mathcal{O}$ of vectors in V, we consider the following two properties:

($O1$) if $v \in \mathcal{O}$, then $\lambda v \in \mathcal{O}$ for all $\lambda \in \mathbf{F}_q^*$;

($O2$) for each $v, w \in \mathcal{O}$ with $w \notin \langle v\rangle$, there exists $g \in \mathcal{G}_{0,v}$ with $gw - w \in \mathcal{O}$.

The next theorem shows that the existence of such an orbit $\mathcal{O}$ often leads to a small diameter or is the adjacency orbit in G.

Theorem 9.6 *Suppose that $\mathcal{G}$, $s = r^b$, V and G are as in Theorem 9.1 (iv), and let $\mathcal{O}$ be a $\mathcal{G}_0$-orbit satisfying (O1) and (O2). Then, with $i = d(0, v)$, one of the following holds for any $v \in \mathcal{O}$:*
(i) $i = 1$;
(ii) $i = 2$ *and, if* $a_2 \neq 0$, *then there exists* $w \in \mathcal{O}$ *with* $v - w \in G_1(0)$;
(iii) $i = 3$ *or* 4, *and there exists* $w \in \mathcal{O}$ *with* $v - w \in G_2(0)$; *moreover,*

- *(a) if* $a_1 \neq 0$, *then* $i = d$;
- *(b) if there is no non-trivial* $\mathcal{G}_0$*-orbit that consists of sums of two elements of* $\mathcal{O}$ *of size smaller than* $|\mathcal{O}|$, *then* $d \leq 5$;

(iv) $i = d$, $a_d = 0$, $b_{d-1} = 1$, *and* $\mathcal{G}_0 = \mathcal{G}_{0,v}\mathcal{G}_{0,w}$, *for some* $w \in G_1(0)$.

If $\mathcal{O}$ is the smallest of all non-trivial $\mathcal{G}_0$-orbits, then either $d \leq 4$ or (*i*) or (*iv*) holds. But case (*iv*) leads to a factorization of the group $\mathcal{K}$ of Lie type into two factors, one of which (the one corresponding to $\mathcal{G}_{0,v}$) is a 'parabolic subgroup' of $\mathcal{K}$ – for our purposes, it suffices to know that a parabolic subgroup is a well-studied subgroup that occurs as the stabilizer in $\mathcal{G}$ of the 1-dimensional subspace generated by a highest-weight vector in a projective representation of $\mathcal{G}$. This situation has been studied satisfactorily, leaving a limited number of possibilities for $\mathcal{G}_{0,w}$; see [23] and [35].

Affine groups of Lie type of cross characteristic

There is an irreducible projective representation on $V = \mathbf{F}_3^6$ of the sporadic group M_{12}, discovered by Mathieu and realizable as a 5-transitive permutation group on 12 letters. The resulting semi-direct product of $\mathbf{F}_3^6$ and a group $\mathcal{G}_0$ with centre of size 2 and quotient isomorphic to M_{12} is a distance-transitive group of automorphisms on a graph of 729 vertices with intersection array (24, 22, 20; 1, 2, 12) (see [13, p. 359]). This graph may thus be expected to occur in the classification of affine distance-transitive graphs with (almost) sporadic vertex-stabilizer. Indeed it does (see Theorem 9.8, below); however it also appears here. The following result was obtained by Cohen, Magaard and Shpectorov [19].

Theorem 9.7 *Suppose that $\mathcal{G}$, $s = r^b$, V, G and $\mathcal{K}$ are as in Theorem 9.1(iv), where $\mathcal{K}$ is a simple group of Lie type whose defining characteristic is a prime p distinct from r, and suppose that $\mathcal{K}$ cannot be defined as a group of Lie type with defining characteristic r. Then $\mathcal{K} \cong \mathrm{A}_1(11)$ (so $p = 11$) and G is the graph on 729 vertices associated with M_{12} (discussed above).*

As in the other cases, the first stage in the proof of this theorem is the determination of pairs $\mathcal{K}$, $V = \mathbf{F}_s^n$ for which the bound of Theorem 9.3(*iii*) is satisfied. Here, the results of Landazuri, Seitz and Zalesskii [39] are used to obtain lower bounds for non-trivial projective representation spaces for $\mathcal{K}$, in characteristic $r \neq p$. As a consequence, a finite (but not really small) list of modules V for groups $\mathcal{G}_0$ with $\mathcal{K} = F^*(\mathcal{G}_0 \cap \mathrm{GL}(V)/Z(\mathcal{G}_0 \cap \mathrm{GL}(V)))$ is found. The next stage is an analysis of $\mathcal{G}_0$-orbits on V. One typically arrives at too many $\mathcal{G}_0$-orbits on V for G to exist, in view of Theorem 8.3(*ii*), or at one or two candidate $\mathcal{G}_0$-orbits for $G_1(0)$, in which case inspection of $G_2(0)$ often leads to more than one $\mathcal{G}_0$-orbit.

Affine sporadic groups

The successful classification of those affine distance-transitive graphs for which $\mathcal{K}$ is a simple sporadic group was achieved by van Bon, Ivanov and Sax [12].

Theorem 9.8 *Suppose that $\mathcal{G}$, s, V, G, $\mathcal{K}$ are as in Theorem 9.1(iv), where $\mathcal{K}$ is a sporadic simple group. Then G and $\mathcal{K}$ are as described in Table 3.*

All of the graphs appearing in this table have been discussed in [13]. For most of the 26 sporadic simple groups $\mathcal{K}$, the representation spaces V satisfying the bounds of Theorem 9.3 could be found by use of the Modular Atlas [27]. For the Conway groups, special arguments were needed in order to find a finite list of all modules V that might occur as regular normal subgroups in Case (*iv*) of Theorem 9.1.

10. The simple socle case

The classification of finite simple groups can be invoked to make a further subdivision of the possibilities for $F^*(\mathcal{G})$. Knowledge of the maximal subgroups of $F^*(\mathcal{G})$ of moderate index should then help to finish the classification of primitive distance-transitive graphs. The multiplicity freeness criteria (see Theorem 6.3) give a method for ruling out several cases.

In cases where the permutation representation of $\mathcal{G}$ on the vertex-set of G is equivalent to conjugation on a conjugacy class of involutions of $\mathcal{G}$, van Bon [5] has given strong restrictions. This is one of the few necessary conditions for a

Table 3.

$\lvert V\rvert$	Array	Name	$\mathcal{K}$
3^6	(26, 24, 19; 1, 3, 8)	ternary Golay	M_{12}
2^{10}	(22, 21, 20; 1, 2, 6)	truncated Golay	M_{22}
2^{11}	(23, 22, 21; 1, 2, 3)	perfect Golay	M_{23}
2^{12}	(24, 23, 22, 21; 1, 2, 3, 24)	extended Golay	M_{24}

permutation representation to come from a distance-transitive action that rules out many examples that survive the multiplicity freeness criteria.

Alternating groups

The case $F^*(\mathcal{G}) \cong \mathcal{A}_n (n \geq 5)$ is dealt with by Ivanov as well as by Liebeck, Praeger and Saxl; the multiplicity-free permutation representations of the alternating and symmetric groups were found by Saxl (see [13, p. 230]).

Theorem 10.1 *Let $\mathcal{G}$ be a primitive distance-transitive group on the graph G, where $d \geq 3$ and $k \geq 3$. If $F^*(\mathcal{G}) \cong \mathcal{A}_n$ for some $n \geq 5$, then G is associated with a Johnson graph $J(n, d)$ or (if $n = 2d + 1$ is odd) an odd graph O_{d+1}.*

Groups of Lie type

This is the hardest subproblem that remains in the classification of primitive distance-transitive graphs. The parabolic subgroups for which the permutation representation is distance-transitive are well understood; roughly speaking, they correspond to the (possibly imprimitive) distance-transitive permutation representations of the corresponding Weyl groups – that is, 'field-size-1' specializations of the groups of Lie type (see [13]).

The classification is expected to follow the pattern of $\mathrm{PSL}(n, q)$, which was treated by van Bon and Cohen in [9]. The proof uses results of Inglis, Liebeck and Saxl (see [13, p. 230]) to exclude various permutation representations that are not multiplicity free.

Theorem 10.2 *Let $\mathcal{G}$ be a group with $\mathrm{PSL}(n, q) \trianglelefteq \mathcal{G} \leq \mathrm{Aut}(\mathrm{PSL}(n, q))$, for $n \geq 2$ and $(n, q) \neq (2, 2), (2, 3)$. If G is a graph with diameter $d \geq 3$ on which $\mathcal{G}$ acts primitively and distance-transitively, then G is either a Grassmann graph or is as listed in Table 4.*

See [13] for details on these graphs.

For those $F^*(\mathcal{G})$ that are of classical Lie type, major progress has been made by van Bon, Inglis and Saxl. They used Aschbacher's list of subgroups of classical groups (see, for example, [1] or [29]) to distinguish between cases, and they used known bounds on subgroups to limit the possibilities for vertex-stabilizers when $F^*(\mathcal{G})$ is given. Lawther has analysed several series of pairs of a group of Lie type and a maximal subgroup with respect to multiplicity freeness (see [30], [31] and [32]).

For those $F^*(\mathcal{G})$ that are of exceptional Lie type, Cohen, Liebeck and Saxl have obtained partial results. As before, the idea of the proof is to use results of Liebeck

Table 4.

$\lvert V \rvert$	(n, q)	Array	Name
28	(2, 7)	(3, 2, 2, 1; 1, 1, 1, 2)	Coxeter
36	(2, 9)	(5, 4, 2; 1, 1, 4)	Sylvester
45	(2, 9)	(4, 2, 2, 2; 1, 1, 1, 2)	generalized 8-gon (2, 1)
68	(2, 16)	(12, 10, 3; 1, 3, 8)	Doro
102	(2, 17)	(3, 2, 2, 2, 1, 1, 1; 1, 1, 1, 1, 1, 3)	Biggs-Smith
57	(2, 19)	(6, 5, 2; 1, 1, 3)	Perkel
65	(2, 25)	(10, 6, 4; 1, 2, 5)	Hall
$q^3 + 2q^2 + 2q + 1$	$(3, q)$	$(2q, q, q; 1, 1, 2)$	generalized 6-gon $(q, 1)$
280	(3, 4)	(9, 8, 6, 3; 1, 1, 3, 8)	$HF(3, 4)_3$
56	(4, 2)	(15, 8, 3; 1, 4, 9)	$J(8, 3)$

and Saxl [36] to determine the maximal groups of relatively small index, and to prune the series that fail to be multiplicity-free (see [30] and [32]). A case-by-case study for each of the remaining permutation groups then needs to be conducted. There exists a short list of hard cases that have not yet been dealt with.

Sporadic groups

For $F^*(\mathcal{G})$ sporadic, the possible graphs G have been determined by Ivanov, Linton, Lux, Saxl and Soicher [26]. The proof was based on a computer search for all possible multiplicity-free permutation characters and subsequent calculation of the collapsed adjacency diagrams.

Theorem 10.3 *If $\mathcal{G}$ is a primitive distance-transitive group of automorphisms of a graph G, with $d \geq 3$ and a sporadic simple generalized Fitting subgroup $\mathcal{K}$, then G and $\mathcal{K}$ are as described in Table 5.*

Details of all six graphs can be found in [13].

Table 5.

$\lvert V \rvert$	$\mathcal{K}$	Array	Name
266	J_1	(11, 10, 6, 1; 1, 1, 5, 11)	Livingstone graph
315	J_2	(10, 8, 8, 2; 1, 1, 4, 5)	near octagon
759	M_{24}	(30, 28, 24; 1, 3, 15)	Witt graph
506	M_{23}	(15, 14, 12; 1, 1, 9)	truncated from Witt
330	M_{22}	(7, 6, 4, 4; 1, 1, 1, 6)	doubly truncated Witt
22880	Suz	(280, 243, 144, 10; 1, 8, 90, 280)	Patterson graph

11. Conclusion

We conclude by providing references to three topics that we have not touched upon:

- in view of their high degree of symmetry, it should be no surprise that there have been some applications of distance-regular graphs: (see [20], [28] and [38]);
- infinite distance-transitive graphs are treated in [16] and [13];
- there is a 'directed' version of distance-transitive graphs due to Lam; see [13, p. 232] and [33] for results on these.

References

1. M. Aschbacher, On the maximal subgroups of the finite classical groups, *Invent. Math.* **76** (1984), 469–514.
2. M. Aschbacher and L. L. Scott, Maximal subgroups of finite groups, *J. Algebra* **92** (1985), 44–80.
3. R. W. Baddeley, Multiplicity-free and self-paired primitive permutation groups, *J. Algebra* **162** (1993), 482–530.
4. J. van Bon, *Affine distance-transitive groups*, Ph.D. thesis, University of Utrecht, 1990.
5. J. van Bon, On distance-transitive graphs and involutions, *Graphs Combin.* **7** (1991), 377–394.
6. J. van Bon, Affine distance-transitive groups, *Proc. London Math. Soc.* **67** (1993), 1–52.
7. J. van Bon, Affine distance-transitive graphs with quadratic forms, *Math. Proc. Cambridge Philos. Soc.* **112** (1992), 507–517.
8. J. van Bon and A. M. Cohen, Affine distance transitive graphs and exceptional Chevalley groups, *Proc. London Math. Soc.* **83** (2001), 51–70.
9. J. van Bon and A. M. Cohen, Linear groups and distance-transitive graphs, *Europ. J. Combin.* **10** (1989), 394–411.
10. J. van Bon and A. M. Cohen, Prospective classification of distance-transitive graphs, *Proc. Combinatorics 1988 conference at Ravello* (eds. A. Barlotti *et al.*), Mediterranean Press (1991), 25–38.
11. J. van Bon, A. M. Cohen and H. Cuypers, Affine distance-transitive graphs and classical groups, preprint TU/e, Eindhoven, 2002.
12. J. van Bon, A. A. Ivanov and J. Saxl, Affine distance-transitive graphs with sporadic stabilizer. *Europ. J. Combin.* **20** (1999), 163–177.
13. A. E. Brouwer, A. M. Cohen and A. Neumaier, Distance-regular graphs, *Ergebnisse der Math. u.i. Grenzgebiete* 3. Folge Band 18, Springer-Verlag, 1989.
14. A. E. Brouwer and J. H. Koolen, The distance-regular graphs of valency four, *J. Alg. Combin.* **10** (1999), 5–24.
15. P. J. Cameron, Finite permutation groups and finite simple groups, *Bull. London Math. Soc.* **13** (1981), 1–22.
16. P. J. Cameron, A census of infinite distance-transitive graphs. *Discrete Math.* **192** (1998), 11–26.

17. I. V. Chuvaeva and D. V. Pasechnik, Distance-transitive graphs of type $q \cdot K_{q,q}$ and projective planes, *Europ. J. Combin.* **11** (1990), 341–346.
18. A. M. Cohen and A. A. Ivanov, Affine distance-transitive groups of dimension one, *Europ. J. Combin.* **21** (2000), 191–195.
19. A. M. Cohen, K. Magaard and S. Shpectorov, Affine distance-transitive graphs: the cross characteristic case, *Europ. J. Combin.* **20** (1999), 351–373.
20. J. R. Driscoll, D. M. Healy, Jr. and D. N. Rockmore, Fast discrete polynomial transforms with applications to data analysis for distance-transitive graphs, *SIAM J. Comput.* **26** (1997), 1066–1099.
21. C. D. Godsil, R. A. Liebler and C. E. Praeger, Antipodal distance-transitive covers of complete graphs, *Europ. J. Combin.* **19** (1998), 455–478.
22. D. Gorenstein, R. Lyons and R. Solomon, The classification of the finite simple groups, *Mathematical Surveys and Monographs* **40**, Amer. Math. Soc., 1994.
23. C. Hering, M. W. Liebeck and J. Saxl, The factorizations of the finite exceptional groups of Lie type, *J. Algebra* **106** (1987), 517–527.
24. A. A. Ivanov, Distance-transitive graphs and their classification, *Investigations in Algebraic Theory of Combinatorial Objects* (eds. I. A. Faradẓev, A. A. Ivanov, M. H. Klin and A. J. Woldar), Math. and its Appl. **84**, Kluwer (1994), 283–378.
25. A. A. Ivanov, R. A. Liebler, T. Penttila and C. E. Praeger, Antipodal distance-transitive covers of complete bipartite graphs, *Europ. J. Combin.* **18** (1997), 11–33.
26. A. A. Ivanov, S. A. Linton, K. Lux, J. Saxl and L. H. Soicher, Distance-transitive representations of the sporadic groups, *Comm. Algebra* **23** (1995), 3379–3427.
27. C. Jansen, K. Lux, R. Parker and R. Wilson, *An atlas of Brauer characters*, London Mathematical Society Monographs, New Series **11**, Oxford Univ. Press, 1995.
28. J. S. Jwo and T. C. Tuan, On transmitting delay in a distance-transitive strongly antipodal graph, *Inform. Process. Lett.* **51** (1994), 233–235.
29. P. Kleidman and M. Liebeck, *The subgroup structure of the finite classical groups*, Cambridge Univ. Press, 1990.
30. R. Lawther, Some (almost) multiplicity-free coset actions, *Groups, Combinatorics and Geometry* (eds. M. W. Liebeck and J. Saxl), London Math. Soc. Notes **165**, Cambridge Univ. Press (1992), 292–310.
31. R. Lawther, Folding actions, *Bull. London Math. Soc.* **25** (1993), 132–144.
32. R. Lawther, The action of $F_4(q)$ on cosets of $B_4(q)$, *J. Algebra* **212** (1999), 79–118.
33. D. A. Leonard and K. Nomura, The girth of a directed distance-regular graph, *J. Combin. Theory (B)* **58** (1993), 34–39.
34. M. W. Liebeck and C. E. Praeger, Affine distance-transitive groups with alternating or symmetric point stabilizer, *Europ. J. Combin.* **13** (1992), 489–501.
35. M. W. Liebeck, C. E. Praeger and J. Saxl, The maximal factorizations of the finite simple groups and their automorphism groups, *Memoirs Ameri. Math. Soc.* **432**, 1990.
36. M. W. Liebeck and J. Saxl, On the orders of maximal subgroups of the finite exceptional groups of Lie type, *Proc. London Math. Soc.* **55** (1987), 299–330.
37. R. A. Liebler, The classification of distance-transitive graphs of type $q \cdot K_{q,q}$, *Europ. J. Combin.* **12** (1991), 125–128.
38. J. Rifà and J. Pujol, Completely transitive codes and distance-transitive graphs, *Applied algebra, algebraic algorithms and error-correcting codes (New Orleans, LA, 1991)*, Lecture Notes in Computer Science **539**, Springer (1991), 360–367.
39. G. Seitz and A. Zalesskii, On the minimal degrees of projective representations of finite Chevalley groups, **II**, *J. Algebra* **158** (1993), 233–243.

10

Computing with graphs and groups

LEONARD H. SOICHER

In this chapter we discuss the computational study of graphs with groups acting on them, and demonstrate various ways in which computational group-theoretic methods are used in the study of graphs and groups. We place particular emphasis on the ideas and methods behind the GRAPE *and* nauty *computer packages.*

1. Introduction

The study of graphs with groups acting on them is the study of $\mathcal{G}$-graphs. A *$\mathcal{G}$-graph* $G = (G, \mathcal{G}, \phi)$ consists of a graph or digraph G, a group $\mathcal{G}$, and a homomorphism $\phi : \mathcal{G} \to \mathrm{Aut}(G)$. (The study of graphs without groups acting on them is just the special case of $\mathcal{G}$-graphs for which $\mathcal{G}$ is the trivial group.) $\mathcal{G}$-graphs arise naturally in many areas, most obviously in the study of graphs related to permutation groups, but also in the study of finite geometries and designs.

An example of a $\mathcal{G}$-graph (where $\mathcal{G}$ is the symmetric group $\mathcal{S}_n$) is the *Johnson graph* $J(n, k)$, defined to be the graph whose vertex-set consists of all k-subsets of $\{1, 2, \ldots, n\}$, with a vertex v adjacent to a vertex w exactly when $|v \cap w| = k - 1$.

Now $J(n,k) = (J(n,k), \mathcal{S}_n, \phi)$ is an $\mathcal{S}_n$-graph: if $v = \{i_1, i_2, \ldots, i_k\}$ is a vertex of $J(n,k)$ and $x \in \mathcal{S}_n$, then $v\phi(x) = \{i_1x, i_2x, \ldots, i_kx\}$.

The GAP package GRAPE (see [13] and [38]) is designed for computing with $\mathcal{G}$-graphs, and in particular makes use of the group $\mathcal{G}$ acting on a $\mathcal{G}$-graph G in order to construct, store, and compute with G efficiently. The nauty package (see [26]) contains the most powerful programs available for computing automorphism groups of graphs and testing graph isomorphism; nauty is available as a stand-alone package, or may be used from within GRAPE or MAGMA (see [27]).

There are many computer systems that are useful for studying graphs and groups which we do not discuss here. These include the algebra system MAGMA, which supports computation with graphs and includes the nauty package, the CoCo package [11] for computing with coherent configurations (see [5] and [12]), and Kocay's package Groups & Graphs [21] (for Macintosh and Windows), which includes a graphical user interface to compute with graphs, geometric configurations, combinatorial designs, and their automorphism groups. Also worth mentioning are the algorithms of Rees and the author for computing fundamental groups and covers of combinatorial cell complexes, and which have been implemented for simplicial complexes (see [32]).

This chapter is organized as follows. We first describe some basic algorithms for permutation groups that are used in the study of $\mathcal{G}$-graphs. We then discuss the efficient storage and construction of $\mathcal{G}$-graphs, and how to use a modified form of breadth-first search to determine efficiently many properties of a $\mathcal{G}$-graph. Following that, we concentrate on the methods used by nauty. Next, we discuss the application of computational methods to the study of vertex-transitive graphs. Then, after a brief discussion of the coset enumeration procedure, we describe some applications of coset enumeration to the study of symmetric graphs.

Throughout this chapter, a digraph is allowed to have loops, but no multiple arcs, and by the *adjacency set* or *neighbourhood* $G(v)$ of a vertex v in a graph or digraph G we mean the set of all vertices w such that vw is an edge or arc of G.

2. Permutation group algorithms

In this section, we describe some basic permutation group algorithms that are important in the study of $\mathcal{G}$-graphs; more detailed elementary expositions can be found in [4] and [9], and a comprehensive treatment of the state of the art in permutation group algorithms is the book by Seress [33]. Throughout this section, V is a finite set of size n, and we are computing with the permutation group $\mathcal{G}$ given by a generating set X of permutations of V. The image of $v \in V$ under the permutation x of V is denoted by vx.

Orbits and Schreier vectors

Let $v \in V$. The *orbit* of v under $\mathcal{G}$ is the set $v\mathcal{G} = \{vg : g \in \mathcal{G}\}$. The calculation of all of the orbits $\{v\mathcal{G} : v \in V\}$ of $\mathcal{G} = \langle X \rangle$ is equivalent to determining the connected components of the 'Schreier graph' with respect to X, for $\mathcal{G}$ acting on V. This *Schreier graph* $S(V, X)$ is a digraph with vertex-set V, where (v, w) is an arc exactly when $vx = w$ for some $x \in X$. A set of rooted directed spanning trees of the connected components of $S(V, X)$ describes a set R of orbit representatives (the roots) and their $\mathcal{G}$-orbits (the vertices in each tree). These trees are called *Schreier trees*, and we always direct a Schreier tree so that the direction of each arc is away from the root. A set of Schreier trees can be compactly encoded using a 'Schreier vector', usually implemented as an array. A *Schreier vector* $s = s(V, X)$, for a set R of orbit representatives and a set of Schreier trees in $S(V, X)$ that they root, is a map

$$s : V \to \{0\} \cup X,$$

for which $s(v) = 0$ if $v \in R$, and $s(v) = x \in X$ means that (vx^{-1}, v) is an arc of one of the Schreier trees (so vx^{-1} is the 'parent' of v in that tree).

The computation of a single orbit $r\mathcal{G}$, and the definition of the Schreier vector entries $s(v)$ for the vertices v in this orbit, are usually done using a breadth-first search of the connected component containing r. We first define $s(r) = 0$ and perform the breadth-first search outwards from r, finding the adjacency set of a vertex u by applying each element of X to u. Whenever we encounter a new vertex v in our breadth-first search – say by applying x to u – we define $s(v) = x$.

If we are given a Schreier vector $s = s(V, X)$ and a point $v \in V$, then we can determine a pair $(r, \mathbf{w})$ for which r is the root of the Schreier tree containing v, and $\mathbf{w}$ is a word in X such that $r\mathbf{w} = v$. This calculation proceeds as follows:

1. $\mathbf{w}$:= `emptyword`; $r := v; x := s(r)$;
2. `while` $x \neq 0$ do $\mathbf{w} := x\mathbf{w}$; $r := rx^{-1}; x := s(r)$; `end_do`;

On completion, $(r, \mathbf{w})$ is the required pair.

Some brief remarks are in order. We do not keep evaluating the inverses of elements of X, since we either make X inverse-closed, and so can access directly the inverse of a generator in X, or we compute rx^{-1} by tracing r through its cycle in x until we find the point mapped to r by x. Finally, note that our definition of a Schreier vector (for *all* the orbits of a group simultaneously) is not the usual one.

Bases and strong generating sets

A sequence $B = (b_1, b_2, \ldots, b_m)$ of elements of V is a *base* for $\mathcal{G}$ if the (pointwise) stabilizer $\mathcal{G}_B$ of B is the trivial group. A base B defines a chain

$$\mathcal{G} = \mathcal{G}^{(1)} \geq \mathcal{G}^{(2)} \geq \cdots \geq \mathcal{G}^{(m)} \geq \mathcal{G}^{(m+1)} = \{1\}$$

of subgroups of $\mathcal{G}$, for which $\mathcal{G}^{(i)}$ is the (pointwise) stabilizer of $(b_1, b_2, \ldots, b_{i-1})$. We observe that

$$|\mathcal{G}| = \prod_{i=1}^{m} |\mathcal{G}^{(i)} : \mathcal{G}^{(i+1)}| \quad \text{and} \quad |\mathcal{G}^{(i)} : \mathcal{G}^{(i+1)}| = |b_i\mathcal{G}^{(i)}|.$$

A *strong generating set* for $\mathcal{G}$, relative to B, is a generating set Y for $\mathcal{G}$ with the property that

$$\langle Y \cap \mathcal{G}^{(i)} \rangle = \mathcal{G}^{(i)}, \quad \text{for } i = 1, 2, \ldots, m+1.$$

Given a strong generating set relative to B, the orbits $b_i\mathcal{G}^{(i)}$ can easily be computed, and we can thus obtain the order of $\mathcal{G}$. A base and associated strong generating set are required by most advanced permutation group algorithms, and are very useful in $\mathcal{G}$-graph computations.

As well as introducing the fundamental concepts of base and strong generating set, Sims [34] devised an algorithm, now called the *Schreier-Sims algorithm*, to construct a base and associated strong generating set for the permutation group $\mathcal{G} = \langle X \rangle$. Modern variants of this algorithm and others, implemented in GAP and MAGMA, can be used to compute bases and strong generating sets for many permutation groups of degree 10^5 or more (see [33, Chs. 5, 8]). It is shown in [33, Ch. 5] that, given a base B for $\mathcal{G}$, a strong generating set for $\mathcal{G}$ can be computed in $O(n|B|^2|X|(\log|G|)^3)$ time. Of course, we could take B to be a sequence of length n of all of the elements of V, but in practice we often know (or can compute) a much shorter base.

Some permutation group algorithms, such as those for determining the $\mathcal{G}$-stabilizer of a set of points of V or determining the centralizer of a subgroup of $\mathcal{G}$, currently use backtrack search, and are not polynomial-time algorithms. Modern implementations of certain permutation group backtrack algorithms can be quite efficient in practice, however, even when V has size 10^5 or more (see [22]). It would be extremely interesting if polynomial-time algorithms could be found for set-stabilizer or subgroup-centralizer, especially since the problem of determining graph isomorphism is polynomial-time reducible to each of these tasks (see [33, Ch. 3]).

3. Storing and accessing a $\mathcal{G}$-graph

We now consider how *GRAPE* stores a $\mathcal{G}$-graph $G = (G, \mathcal{G}, \phi)$ in a compact way that also enables the efficient recovery of basic information about the graph or digraph G. In this implementation, we store $\phi(\mathcal{G})$ by a generating set, so we now assume that $\mathcal{G} = \phi(\mathcal{G}) \leq \mathrm{Aut}(G)$ and let $\mathcal{G}$ be given by a generating set X.

Let $V_1, V_2, \ldots, V_k$ be the orbits of $\mathcal{G}$ on the set V of vertices of G, with respective representatives $v_1, v_2, \ldots, v_k$. We store G using a record data structure, containing:

- the permutation generators X for $\mathcal{G}$;
- the list $v_1, v_2, \ldots, v_k$ of orbit representatives;
- the list $G(v_1), G(v_2), \ldots, G(v_k)$ of the adjacency sets of these $\mathcal{G}$-orbit representatives;
- a Schreier vector $s(V, X)$.

Note that, even if $\mathcal{G}$ is trivial, the above method of storing G is not significantly worse than a represention by a list of adjacency sets of the vertices.

We now describe how to use the Schreier vector $s = s(V, X)$ to determine whether (v, w) is an arc of G, for vertices v and w, and to calculate the adjacency set $G(v)$ of v. We first use s to determine a pair $(v_i, \mathbf{w})$ for which v_i is the orbit representative of $v\mathcal{G}$ and $\mathbf{w}$ is a word in X mapping v_i to v. Then (v, w) is an arc of G if and only if $w\mathbf{w}^{-1} \in G(v_i)$, and we have $G(v) = G(v_i)\mathbf{w}$.

Our method of storing a $\mathcal{G}$-graph is space-efficient at the cost of time for the recovery of the adjacency sets of vertices. It is a good idea for the Schreier trees encoded by s to be as shallow as is reasonably possible – hence the use of breadth-first search. A thorough discussion on computing shallow Schreier trees is given in [33, Ch. 4]. We do remark, however, that our method of storing $\mathcal{G}$-graphs can often save time when constructing a graph, since we need only calculate the adjacency sets of orbit representatives for $\mathcal{G}$ on V.

We discuss one final point concerning the GRAPE data structure for a $\mathcal{G}$-graph G, which has turned out to be extremely useful when doing 'real-life' calculations. Internally, the n vertices of G are represented by the integers $1, 2, \ldots, n$, but each vertex also has a 'name' which can be an object of any GAP type. When constructing a new graph (possibly from another graph), the vertices of this new graph are numbered $1, 2, \ldots, n$, but their names are chosen to reflect the mathematical nature of the vertices. Details of this naming for specific functions can be found in the GRAPE manual. (The use of this naming, which is implemented by a list, was suggested by P. J. Cameron.)

4. Constructing $\mathcal{G}$-graphs

Although not always necessary, we assume from here on that if we are computing with a $\mathcal{G}$-graph $G = (G, \mathcal{G}, \phi)$, then we have a base and associated strong generating set for $\phi(\mathcal{G})$ acting on $V = V(G)$. This allows, for example, the efficient calculation of the $\phi(\mathcal{G})$-stabilizer of an arbitrary point in V (see [33, Ch. 5]). We remark that, in many constructions of $\mathcal{G}$-graphs, our knowledge about $\mathcal{G}$ often allows us to compute this base and strong generating set much more efficiently than otherwise. For example, we often know the order of $\phi(\mathcal{G})$, and this allows the use of a very fast randomized algorithm to compute a base and strong generating set (see [33, Ch. 8]), often applicable to groups of degree 10^6 or more. We also remark that the *nauty* package [26] outputs the automorphism group of a graph as a base and associated strong generating set for that group (see Section 6).

GRAPE has many functions to construct $\mathcal{G}$-graphs, including Cayley graphs, orbital graphs and digraphs, induced subgraphs and quotient graphs. However, the most useful and general way of constructing a $\mathcal{G}$-graph in GRAPE is to use the function Graph, which behaves as follows. The input is a group $\mathcal{G}$ (which may or may not be a permutation group), a finite set V on which $\mathcal{G}$ acts (the action can be one of the standard actions in GAP or may be one supplied by the user), and a $\mathcal{G}$-invariant relation *rel* on V (given as a function of two vertices v and w and which returns TRUE or FALSE, according to whether or not (v, w) is in the relation). The output is the $\mathcal{G}$-graph G with vertex-set V, where (v, w) is an arc if and only if $rel(v, w)$.

The first step is to compute the orbits of $\mathcal{G}$ acting on V and the associated Schreier vector. Then, for each orbit representative r, we need to determine those vertices $v \in V$ such that $rel(r, v)$. In fact, we need check $rel(r, s)$ only for those s in a set of orbit representatives of the orbits on V of the $\mathcal{G}$-stabilizer $\mathcal{H}$ of r, since $rel(r, t)$ for each t in $s\mathcal{H}$ if and only if $rel(r, s)$.

5. $\mathcal{G}$-breadth-first search in a $\mathcal{G}$-graph

In this section, for ease of exposition, we assume that our $\mathcal{G}$-graph $G = (V, E)$ is a simple graph, and that $\mathcal{G}$ is a subgroup of Aut(G). We define $G_i(v)$ to be the set of all vertices at distance i from a vertex v of G.

Many properties of G can be determined by (possibly repeated) applications of breadth-first search. These include finding the connected components, diameter and girth, as well as determining various regularity properties such as whether G is distance-regular. We describe here a version of breadth-first search, which we call $\mathcal{G}$-breadth-first search, which takes into account $\mathcal{G}$ as a subgroup of Aut(G).

Let $v \in V$, and let $\mathcal{H} = \mathcal{G}_v$ be the stabilizer in $\mathcal{G}$ of v. The key observation is that if w is a vertex at distance i from v, then each vertex in the orbit $w\mathcal{H}$ is at distance i from v.

Suppose that $V_1(=\{v\}), V_2, \ldots, V_k$ are the orbits of $\mathcal{H}$ on V, with respective representatives $v_1(=v), v_2, \ldots, v_k$, and let $R = \{v_1, v_2, \ldots, v_k\}$. In a $\mathcal{G}$-breadth-first search from v, we determine the sets $R_0, R_1, \ldots$ of $\mathcal{H}$-orbit representatives, where

$$R_i = G_i(v) \cap R.$$

Given R_i, we then have $G_i(v)$ as the union of the (already computed) orbits represented by the elements of R_i.

Clearly $R_0 = \{v\}$. The basic step is to obtain R_{i+1} from R_i. We actually do more, in order to obtain more information from our $\mathcal{G}$-breadth-first search. We start by setting $R_{i+1} := \{\}$ and then, for each $r \in R_i$, do the following:

1. determine $C := G(r) \cap G_{i-1}(v)$, $A := G(r) \cap G_i(v)$ and $B := G(r) - (C \cup A)$;
2. add to R_{i+1} the representatives of the $\mathcal{H}$-orbits that intersect B non-trivially;
3. for later use, store $c_i(v, r) := |C|$, $a_i(v, r) := |A|$, and $b_i(v, r) := |B|$.

The $\mathcal{G}$-breadth-first search stops with $R_0, R_1, \ldots, R_m$, when R_{m+1} is empty but R_m is not. If required, it is trivial to recover $G_0(v), G_1(v), \ldots, G_m(v)$, the union of these sets being the vertices in the connected component containing v. Also note that m is the greatest distance $d(v)$ from v to any vertex in the connected component of v. Moreover, if G is connected, then its diameter $\mathrm{diam}(G)$ is the maximum value of $d(w)$, as w ranges over a set of representatives of the $\mathcal{G}$-orbits on V.

Let $g(v)$ be the length of a shortest cycle containing v if v is on some cycle of G, and let $g(v) = \infty$ otherwise. The numbers $c_i(v, r)$ and $a_i(v, r)$ computed above (for each $i = 1, 2, \ldots, m$ and each $r \in R_i$) can be used to determine $g(v)$, as follows.

Let t be the least value of $i \in \{1, 2, \ldots, m\}$ for which $c_i(v, r) \geq 2$ or $a_i(v, r) \geq 1$, for some $r \in R_i$, if such an i exists. If no such i exists, then $g(v) = \infty$; otherwise, if $c_t(v, r) \geq 2$ for some $r \in R_i$, then $g(v) = 2t$; if not, then $g(v) = 2t + 1$.

Note that the girth of G is the minimum value of $g(w)$ as w ranges over a set of representatives of the $\mathcal{G}$-orbits on V (a girth of ∞ means that G has no cycles).

Now define $c_0(v, v) = 0$, $a_0(v, v) = 0$ and $b_0(v, v) = |G(v)|$. Let $0 \leq i \leq m$ and $r \in R_i$. If $c_i(v, r)$ depends only on i and v (and not on the $\mathcal{H}$-orbit representative r), then we denote this quantity by $c_i(v)$ and call it a *local parameter* of G. Similarly, if $a_i(v, r)$ and $b_i(v, r)$ do not depend on r, then these too are called local parameters, and are respectively denoted by $a_i(v)$ and $b_i(v)$. Such local parameters, if and when they exist, are used in the determination of various regularity properties of G, the strongest of which is distance-regularity. Indeed, the graph G is distance-regular if and only if G is connected, $d(v)$ is the same for all vertices v (so each $d(v) = \mathrm{diam}(G)$), and for each vertex v and $i = 0, 1, \ldots, \mathrm{diam}(G)$, all local parameters $c_i(v)$, $a_i(v)$ and $b_i(v)$ exist and do not depend on v, thus giving the parameters c_i, a_i and b_i of a distance-regular graph; see Chapter 9. Of course, we need to check these conditions only for those vertices v in a set of orbit representatives of $\mathcal{G}$ on V.

Finally, we remark that $\mathcal{G}$ acts distance-transitively on G (see Chapter 9) if and only if G is connected, $\mathcal{G}$ has just one orbit on V, and, for some (and hence all) vertices v, the number of $\mathcal{G}_v$-orbits on V is $\mathrm{diam}(G) + 1$.

$\mathcal{G}$-breadth-first search is efficiently implemented in GRAPE in order to determine connected components, diameter, girth, and regularity properties of $\mathcal{G}$-graphs. Although we do not discuss it here, GRAPE includes many other functions for computing with $\mathcal{G}$-graphs, such as backtrack search functions for classifying complete subgraphs of given weight-sum in a vertex-weighted $\mathcal{G}$-graph and for classifying partial linear spaces with given point graph and parameters.

6. Automorphism groups and graph isomorphism

Automorphism groups of graphs are discussed in Chapter 5. The most advanced algorithms and programs for computing automorphism groups of graphs and for testing graph isomorphism are those of McKay, freely available as part of his *nauty* package [26] and useful for graphs with up to about 10^4 vertices; nauty also works with digraphs, although we do not consider them in this section. We shall briefly describe McKay's method of 'partition backtrack', which has been very influential in computational group theory. Indeed, partition-based backtrack methods are now used in the most advanced available algorithms for calculating set-stabilizers, centralizers and normalizers in permutation groups (see [33, Ch. 9]).

Let $G = (V, E)$ be a graph, with $V = \{1, 2, \ldots, n\}$, and let π be an ordered partition of V. Thus π is a sequence $(V_1, V_2, \ldots, V_k)$ of distinct subsets of V, for which $\{V_1, V_2, \ldots, V_k\}$ is a partition of V: the elements of V are called *cells*. (One can think of (G, π) as being a (not necessarily properly) vertex-coloured graph, with vertices v and w having the same colour if and only if they belong to the same cell of π.) Define the ordered partition $c(\pi)$ of V to be

$$(\{1, 2, \ldots, |V_1|\}, \{|V_1| + 1, |V_1| + 2, \ldots, |V_1| + |V_2|\}, \ldots,$$
$$\{n - |V_k| + 1, n - |V_k| + 2, \ldots, n\}).$$

If x is a permutation of V, define Gx to be the graph (V, Ex), where $Ex = \{\{vx, wx\} : \{v, w\} \in E\}$, and define $\pi x = (V_1x, V_2x, \ldots, V_kx)$, where $V_ix = \{vx : v \in V_i\}$. The *automorphism group* $\mathrm{Aut}(G, \pi)$ of (G, π) is the group of all permutations x of V for which $(Gx, \pi x) = (G, \pi)$. Thus, when $\pi = (V)$, the automorphism group of (G, π) is just $\mathrm{Aut}(G)$.

The main functions of the nauty package are to determine $\mathrm{Aut}(G, \pi)$, in the form of a base and associated strong generating set, and in the process, to compute the image of (G, π) under a canonical labelling map (described below) which is used for isomorphism testing.

A *canonical labelling map* is a function $\mathcal{C}$ such that, for each graph G with vertex-set $V = \{1, 2, \ldots, n\}$, each ordered partition π of V, and each permutation x of V, we have:

- $\mathcal{C}(G, \pi) = Gy$, for some permutation y of V such that $\pi y = c(\pi)$;
- $\mathcal{C}(Gx, \pi x) = \mathcal{C}(G, \pi)$.

The importance of a canonical labelling map $\mathcal{C}$ is this. Suppose that G_1 and G_2 are graphs on the same vertex-set $V = \{1, 2, \ldots, n\}$, and that π_1 and π_2 are ordered partitions of V with $c(\pi_1) = c(\pi_2)$. Then there is a permutation y of V such that $(G_1 y, \pi_1 y) = (G_2, \pi_2)$ if and only if $\mathcal{C}(G_1, \pi_1) = \mathcal{C}(G_2, \pi_2)$. In particular, G_1 is isomorphic to G_2 if and only if $\mathcal{C}(G_1, (V)) = \mathcal{C}(G_2, (V))$.

Partition backtrack

As before, G is a graph with vertex-set $V = \{1, 2, \ldots, n\}$ and $\pi = (V_1, V_2, \ldots, V_k)$ is an ordered partition of V. The cells of π of size 1 are called *singletons*, and a *discrete* partition is one in which all cells are singletons.

For each $v \in V$, let $u(v, \pi)$, denote the index of the cell in π containing v; in other words, $u(v, \pi) = i$ means that $v \in V_i$. We say that an ordered partition ν of V is a *refinement* of π if:
- each cell of ν is contained in some cell of π;
- for each $v, w \in V$: if $u(\pi, v) < u(\pi, w)$, then $u(\nu, v) < u(\nu, w)$.

A *refinement process* is a function $\mathcal{R}$ such that, for each graph G with vertex-set V, each ordered partition π of V, and each permutation x of V,

$$\mathcal{R}(G, \pi) \text{ is a refinement of } \pi, \text{ and } \mathcal{R}(G, \pi)x = \mathcal{R}(Gx, \pi x).$$

In particular, if $x \in \mathrm{Aut}(G)$, then $\mathcal{R}(G, \pi)x = \mathcal{R}(G, \pi x)$, and so x maps the ordered partition π to ν only if x maps the refinement $\mathcal{R}(G, \pi)$ to $\mathcal{R}(G, \nu)$. In particular, if $c(\mathcal{R}(G, \pi)) \neq c(\mathcal{R}(G, \nu))$, then there is no element of $\mathrm{Aut}(G)$ that maps π to ν. Note that, if $\mathcal{R}(G, \pi)$ is discrete, then $\mathrm{Aut}(G, \pi)$ is trivial.

An ordered partition $\pi = (V_1, V_2, \ldots, V_k)$ of V is *G-equitable* if there are constants d_{ij} $(1 \leq i, j \leq k)$ such that $|G(v) \cap V_j| = d_{ij}$ for each vertex $v \in V_i$. The default refinement process $\mathcal{R}$ used by nauty (for simple graphs) maps (G, π) to an ordered G-equitable partition ν, such that ν is a refinement of π and, up to the order of its cells, is the coarsest equitable partition that is a refinement of π. More precise details can be found in McKay [25].

We now outline (roughly) how the nauty procedure finds $\mathrm{Aut}(G, \pi)$. For more details, together with how a canonical labelling map is found, see [25] (see also [20], [24] and [28]).

We let $\mathcal{R}$ be the default refinement process used by nauty. The nauty procedure proceeds by a depth-first search in a search tree whose nodes are ordered G-equitable partitions of V; the root is $\mathcal{R}(G, \pi)$ and the leaves are discrete partitions.

We call the first leaf found ζ. Any other leaf λ may give rise to a new automorphism of (G, π), and we check to see whether the unique permutation of V mapping ζ to λ is in $\mathrm{Aut}(G, \pi)$. A non-leaf ν in the search tree is not discrete, and we obtain its 'children' as follows. First, we choose a non-singleton cell C of ν according to some rule – we usually just take the first non-singleton cell. Then, for each $v \in C$ (in ascending order of v), we *isolate* v – that is, we form the ordered partition $\nu \circ v$ obtained from ν by replacing C by $\{v\}, C - \{v\}$, and then we add $\mathcal{R}(G, \nu \circ v)$ as a 'child' of ν. The search tree is pruned in various ways, so as to avoid searching subtrees providing no new information (see [25]). The search is structured so as to provide a base and associated strong generating set for $\mathrm{Aut}(G, \pi)$. In determining the first leaf ζ, G-vertices $w_1, w_2, \ldots, w_m$, say, are isolated (w_i at depth i) for the formation of the ancestors of ζ (other than the root) and the formation of ζ itself. It is not difficult to see that $(w_1, w_2, \ldots, w_m)$ is a base for $\mathrm{Aut}(G, \pi)$; nauty computes a strong generating set relative to this base.

7. Computing with vertex-transitive graphs

Recall that a graph or digraph G is *vertex-transitive* if $\mathrm{Aut}(G)$ acts transitively on $V(G)$. The class of vertex-transitive graphs includes Cayley graphs and symmetric graphs (studied in Chapters 6 and 7), which further includes the class of distance-transitive graphs (studied in Chapter 9). In this section we consider the computational study of a $\mathcal{G}$-graph G, where $\mathcal{G}$ (a subgroup of $\mathrm{Aut}(G)$) acts transitively on $V(G)$. Note that the GRAPE data structure for storing a $\mathcal{G}$-graph is especially compact in this case.

Collapsed adjacency matrices

Let $\mathcal{G}$ be a transitive permutation group on a finite set V. Then $\mathcal{G}$ has a natural action on $V \times V$, defined by $(v, w)x = (vx, wx)$. The orbits of this action are called *orbitals*, and the orbits of the stabilizer $\mathcal{G}_v$ of a point $v \in V$ are called *suborbits*. It is well known that the orbitals for $\mathcal{G}$ are in one-to-one correspondence with these suborbits: this correspondence maps an orbital E to the suborbit $\{w : (v, w) \in E\}$. The *orbital digraph* for $\mathcal{G}$ associated with an orbital E is simply the digraph (V, E). If the orbital E is non-diagonal and self-paired (see Chapter 8), then we associate the *orbital graph* $(V, \{\{v, w\} : (v, w) \in E\})$ with E.

Let $v_1 \in V$, and suppose that $V_1 (= \{v_1\}), V_2, \ldots, V_k$ is an ordering of the orbits of $\mathcal{G}_{v_1}$, with respective representatives $v_1, v_2, \ldots, v_k$; k is the *rank* of $\mathcal{G}$. Let $G = (V, E)$ be a graph or digraph on which $\mathcal{G}$ acts vertex-transitively, so that E is a union of orbitals and $G(v)$ is the union of the corresponding suborbits contained in $\{V_1, V_2, \ldots, V_k\}$. For $i, j = 1, 2, \ldots, k$, define

$$a_{ij} = |G(v_i) \cap V_j|.$$

Note that a_{ij} does not depend on the choice v_i of suborbit representative, and it can easily be computed using the GRAPE data structure for the $\mathcal{G}$-graph G – in practice, for $|V|$ up to about 10^6. The $k \times k$ integer matrix $\mathbf{A} = (a_{ij})$ is the *collapsed adjacency matrix* for G, with respect to $\mathcal{G}$ and the ordering of the suborbits. This matrix (which is extremely compact when k is small) contains at least as much information as that computed in a $\mathcal{G}$-breadth first search from v.

Since $\mathcal{G}$ acts vertex-transitively, the single collapsed adjacency matrix $\mathbf{A}$ for G can be used to determine whether G is (strongly) connected, and if so, what its diameter is; whether G is a simple graph, and if so what its girth is; whether G is distance-regular, and whether $\mathcal{G}$ acts distance-transitively on G. See Praeger and Soicher [31] for more detailed information and applications of collapsed adjacency matrices, and also Chapter 9 for examples of 'collapsed adjacency diagrams'.

We remark that the collapsed adjacency matrices for the orbital digraphs for the transitive group $\mathcal{G}$ are useful in studying the coherent configuration associated with $\mathcal{G}$ (see [5] and [12]), since, with respect to a fixed ordering $V_1, V_2, \ldots, V_k$ of the suborbits as above, the collapsed adjacency matrix for an orbital digraph (V, E) for $\mathcal{G}$ is the transpose of the intersection matrix (as defined in [5, Ch. 3]) corresponding to the orbital paired with E.

Distance-transitive graphs

Chapter 9 provides an overview of the state of the classification of distance-transitive graphs. Here we discuss the application of computing, which has been used in the discovery, analysis and classification of certain distance-regular and distance-transitive graphs (see, for example, [37], [31], [23] and [18]). We also remark that computing is used in the determination of feasible intersection arrays for possible distance-regular graphs (see, for example, [3]).

Suppose that $\mathcal{G}$ acts distance-transitively on a graph G. Then G must be an orbital graph for $\mathcal{G}$ of the smallest or second-smallest vertex-degree (see [18]). Furthermore, each orbital for $\mathcal{G}$ must be *self-paired*, which is equivalent to the property that the permutation character of $\mathcal{G}$ on V is the sum of distinct complex irreducible characters, each with Frobenius-Schur indicator $+1$ (see [2, p. 64]). For these, and other reasons, it makes sense to analyse the lower degree orbital graphs of permutation representations whose character is *multiplicity-free* – that is, the sum of distinct complex irreducible characters.

Computation has been applied extensively in the classification of the graphs on which a sporadic simple group or its automorphism group acts primitively and distance-transitively (see [23] and [18]; the results of this classification are given in Chapter 9). In the process of this classification, the primitive multiplicity-free permutation characters for these sporadic groups were also determined, and for most of the corresponding permutation representations of sporadic groups, a collapsed adjacency matrix was computed for the orbital graph of least degree. This built on the work of Praeger and Soicher [31], where collapsed adjacency matrices were computed for the orbital digraphs for all permutation representations of rank at most 5 of the sporadic simple groups and their automorphism groups. The practical computational determination of permutation characters is described in some detail in Linton, Lux and Soicher [23], which also details randomized techniques for computing collapsed adjacency matrices for certain permutation representations of degree about 10^{11}, where it would be impossible to store explicit permutation generators. These techniques make use of graph algorithms as well as permutation group algorithms.

An ambitious project at Lehrstuhl D für Mathematik, RWTH (Aachen), involving T. Breuer, I. Höhler and J. Müller, has since determined collapsed adjacency matrices for all orbital digraphs for all multiplicity-free permutation representations of the sporadic simple groups and their automorphism groups, and the results are published on the world-wide web [1] (although what they call 'collapsed adjacency matrices' we would call intersection matrices).

We remark that many permutation and matrix representations of finite simple groups and related groups can be downloaded from the online ATLAS of group representations (see [41]). These group representations are very useful for constructing related $\mathcal{G}$-graphs and collapsed adjacency matrices.

8. Coset enumeration

Coset enumeration is one of the oldest and most useful methods of computational group theory (see [30], [35], [14] and their references). For this chapter, we concentrate on what coset enumeration does, and on the application of coset enumeration and related procedures to problems in graph theory.

Let $\mathcal{G} = \langle X : R \rangle$ be a finitely presented group – that is, $\mathcal{G}$ is generated by the finite set X, subject (only) to the finite set R of relators which are words in $X \cup X^{-1}$ that evaluate to the identity in $\mathcal{G}$ (where $X^{-1} = \{x^{-1} : x \in X\}$). The input to coset enumeration is (X, R, Y), where Y is a set of words in $X \cup X^{-1}$ that generates a subgroup $\mathcal{H}$ of $\mathcal{G}$.

The coset enumeration process attempts to construct a set V, with $1 \in V$, and a transitive permutation representation $\rho : \mathcal{G} \to \mathrm{Sym}(V)$ for which, in this representation, $\mathcal{H}$ is the stabilizer in $\mathcal{G}$ of the point 1. Coset enumeration does this by using a trial-and-error process for constructing the permutations $\rho(X \cup X^{-1})$.

The name *coset enumeration* comes from the fact that, if a coset enumeration is successful, then there is a one-to-one correspondence between the elements of V and the cosets of $\mathcal{H}$ in $\mathcal{G}$, with 1 corresponding to $\mathcal{H}$.

If the index of $\mathcal{H}$ in $\mathcal{G}$ is infinite, then the coset enumeration process does not terminate; if it is finite, then the process terminates, but there can be no computable general bound (in terms of the size of the input and the putative index) on the time or store required for termination (see, for example, Neubüser [30]).

There is an enormous amount of flexibility in the coset enumeration process, and many different approaches have been suggested and experimented with (see [30], [35] and [14]). Depending on the presentation and the approach used, there can be huge variations in the time and store taken. Currently, the most advanced methods are due to Havas and Ramsay, and these methods are available in their ACE package [15], also available as a GAP package [16] and within the MAGMA system [27].

There are many useful variations on coset enumeration. For example, a 'modified Todd-Coxeter' enumeration gives a presentation for $\mathcal{H}$, and the 'low-index subgroups procedure' determines (up to permutation isomorphism) all transitive representations of $\mathcal{G}$ up to some given degree k; see [30] for an excellent introduction to coset enumeration and its variations.

9. Coset enumeration for symmetric graphs

Recall that a graph G is *symmetric* if Aut(G) acts transitively on both its vertices and arcs (ordered pairs of adjacent vertices). A subgroup $\mathcal{G}$ of Aut(G) acts *symmetrically* on G if $\mathcal{G}$ acts transitively on both the vertices and arcs of G.

One common way to study connected symmetric graphs with given properties is by determining the groups that act on them symmetrically, as quotients of universal completions of appropriate amalgams (see Chapter 7 and [19] for useful overviews, and [17] for a more general geometric context). In this approach, we first use the given graph-theoretical properties to determine the possible amalgams of the form $\mathcal{A} = \{\mathcal{G}_v, \mathcal{G}_{\{v,w\}}\}$, where (v, w) is an arc in the putative graph on which $\mathcal{G}$ acts symmetrically. Since the universal completion $U(\mathcal{A})$ of such an amalgam of two groups (with neither a subgroup of the other) is infinite, we need to add further relations to $U(\mathcal{A})$ to obtain the finite groups of automorphisms we seek. Such extra relations could come from cycles in our graph or from the local graph structure.

Coset enumeration can then be used in an effort to determine the (hopefully finite) index of $\mathcal{H} = \mathcal{G}_v$ in $\mathcal{G}$, and to construct the representation of $\mathcal{G}$ acting on the set V of right cosets of $\mathcal{H}$ in $\mathcal{G}$. Given such a representation of $\mathcal{G}$ on V, we can reconstruct and study the graph G which may have the properties we seek – or we may have been able to prove theoretically that G has the required properties. This graph G is simply the orbital graph for $\mathcal{G}$ for which the orbital contains $(\mathcal{H}, \mathcal{H}g)$, where g is an element of $\mathcal{G}_{\{v,w\}} - \mathcal{H}$. Applications of this kind include those in Morton [29], where the 4-arc- and 5-arc-transitive connected cubic graphs of girth up to 11 and girth 13 are classified; see Chapter 7 for a discussion of s-arc transitivity, and Conder and Dobcsányi (see [6] and [7]), where the connected symmetric cubic graphs on up to 768 vertices are determined, using a powerful new low-index subgroups procedure, parallel computation and coset enumeration. Of course, computational studies of this kind often lead to conjectures and theoretical results. For more applications of coset enumeration to the study of symmetric graphs, and for a beautiful and natural characterization and construction of the sporadic simple group J_3, see Weiss [39] and also [40], [36] and [10].

Graphs that are locally a given graph

We now give an example where additional transitivity assumptions and local structure specification lead to an amalgam of three groups. We use the ATLAS notation [8] for group structures.

Let G and H be graphs. Then G is said to be *locally-H* if, for each vertex v of G, the induced subgraph on $G(v)$ is isomorphic to H (this situation is discussed briefly in Chapter 5). Given a graph H, coset enumeration can sometimes be used effectively to study presentations that arise in the classification of the connected graphs G that are *locally-H*, for which $\mathrm{Aut}(G)$ acts transitively on the ordered triangles of G. These presentations come from applying the amalgam method to putative stabilizers of a vertex, incident edge and triangle, contained in a fixed triangle of such a graph G.

A simple, but good, example of this application of coset enumeration is given in [10], where H is the incidence graph of the unique 2-(11, 5, 2) design, and where the ordered-triangle-transitive graphs G that are locally-H are classified. We discuss here the case where a vertex-stabilizer in the automorphism group of G is isomorphic to $\mathrm{PGL}_2(11)$. For such a graph G, it is shown that there is (essentially) only one possible amalgam $\mathcal{A} = \{\mathcal{X}, \mathcal{Y}, \mathcal{Z}\}$ of the $\mathrm{Aut}(G)$-stabilizers $\mathcal{X}, \mathcal{Y}, \mathcal{Z}$ of $x, \{x, y\}$, and $\{x, y, z\}$, contained in a triangle $\{x,y,z\}$ of G. In this amalgam, $\mathcal{X} \cong \mathrm{PGL}_2(11)$, $\mathcal{Y} \cong \mathcal{S}_5$, $\mathcal{Z} \cong (\mathcal{A}_4 \times 3) : 2$, $\mathcal{X} \cap \mathcal{Y} \cong \mathcal{A}_5$, $\mathcal{X} \cap \mathcal{Z} \cong \mathcal{S}_4$, and $\mathcal{Y} \cap \mathcal{Z} \cong \mathcal{S}_4$. Then $\mathrm{Aut}(G)$ is a homomorphic image of $U(\mathcal{A})$ which has the following presentation, determined in [10]:

$$\big\langle a, b, c, d, e : a^3 = b^2 = c^2 = d^2 = (ab)^3 = (ac)^2 = (ad)^2 = a(cd)^4 \\ = (bc)^3 = (bd)^2 = e^2 = (ae)^2 = (be)^2 = (ce)^2 = (de)^3 = 1\big\rangle.$$

For this presentation of $U(\mathcal{A})$, $\mathcal{X}$ maps onto $\langle a, b, c, d\rangle$, $\mathcal{Y}$ maps onto $\langle a, b, c, e\rangle$, and $\mathcal{Z}$ maps onto $\langle a, b, d, e\rangle$; the relations $(ac)^2 = (ad)^2 = 1$ are consequences of the others.

Applying coset enumeration, we find that $\langle a, b, c, d\rangle$ has index 432 in $U(\mathcal{A})$. It is then shown that $U(\mathcal{A}) \cong (3 \times M_{12}) : 2$, and that there are just two connected ordered-triangle-transitive locally-H graphs whose vertex-stabilizer is $\mathrm{PGL}_2(11)$, having (respectively) 432 and 144 vertices and automorphism groups isomorphic to $U(\mathcal{A})$ and $U(\mathcal{A})/\langle (bcde)^{11}\rangle \cong M_{12} : 2$, the automorphism group of the Mathieu group M_{12}.

References

1. T. Breuer and J. Müller, The character tables of endomorphism rings of multiplicity-free permutation modules of the sporadic simple groups, their automorphism groups, and their cyclic central extension groups; `http://www.math.rwth-aachen.de/~Juergen.Mueller/mferctbl/mferctbl.html`.
2. A. E. Brouwer, A. M. Cohen and A. Neumaier, *Distance-Regular Graphs*, Springer, 1989.
3. A. E. Brouwer and J. H. Koolen, The distance-regular graphs of valency four, *J. Alg. Combin.* **10** (1999), 5–24.
4. G. Butler, *Fundamental Algorithms for Permutation Groups,* Lecture Notes in Computer Science **559**, Springer, 1991.
5. P. J. Cameron, *Permutation Groups*, London Math. Soc. Student Texts **45**, Cambridge Univ. Press, 1999.
6. M. Conder and P. Dobcsányi, Trivalent symmetric graphs on up to 768 vertices, *J. Combin. Math. Combin. Comput.* **40** (2002), 41–63.
7. M. Conder and P. Dobcsányi, Applications and adaptations of the low index subgroups procedure, *Math. Comput.*, 2004.
8. J. H. Conway, R. T. Curtis, S. P. Norton, R. A. Parker and R. A. Wilson, *ATLAS of Finite Groups*, Oxford Univ. Press, 1985.
9. H. Cuypers, L. H. Soicher and H. Sterk, Working with finite groups, *Some Tapas of Computer Algebra* (eds. A. M. Cohen, H. Cuypers and H. Sterk), Springer (1999), 184–207.
10. H. Cuypers, L. H. Soicher and H. Sterk, The small Mathieu groups (Project), *Some Tapas of Computer Algebra* (eds. A. M. Cohen, H. Cuypers and H. Sterk), Springer (1999), 323–337.
11. I. A. Faradžev and M. H. Klin, Computer package for computations with coherent configurations, *Proc. ISSAC '91* (ed. S. M. Watt), Assoc. Comp. Mach. (1991), 219–223.
12. I. A. Faradžev, M. H. Klin and M. E. Muzichuk, Cellular rings and groups of automorphisms of graphs, *Investigations in Algebraic Theory of Combinatorial Objects* (eds. I. A. Faradžev, A. A. Ivanov, M. H. Klin and A. J. Woldar), Kluwer (1994), 1–152.
13. The GAP Group, *GAP – Groups, Algorithms, and Programming, Version 4.3*, Aachen, St Andrews (2002); `http://www.gap-system.org`.

14. G. Havas and C. Ramsay, Experiments in coset enumeration, *Groups and Computation III* (eds. W. M. Kantor and Á. Seress), de Gruyter (2001), 183–192.
15. G. Havas and C. Ramsay, The "Advanced Coset Enumerator" ACE; `http://www.csee.uq.edu.au/~cram/ce.html`.
16. G. Havas, C. Ramsay, G. Gamble and A. Hulpke, The ACE Package for GAP; `http://www.gap-system.org/Share/ace.html`.
17. A. A. Ivanov, *Geometry of Sporadic Groups. I: Petersen and Tilde Geometries*, Cambridge Univ. Press, 1999.
18. A. A. Ivanov, S. A. Linton, K. Lux, J. Saxl and L. H. Soicher, Distance-transitive representations of the sporadic groups, *Comm. Algebra* **23** (1995), 3379–3427.
19. A. A. Ivanov and S. V. Shpectorov, Applications of group amalgams to algebraic graph theory, *Investigations in Algebraic Theory of Combinatorial Objects* (eds. I. A. Faradžev, A. A. Ivanov, M. H. Klin and A. J. Woldar), Kluwer (1994), 417–441.
20. W. Kocay, On writing isomorphism programs, *Computational and Constructive Design Theory* (ed. W. D. Wallis), Kluwer (1996), 135–175.
21. W. Kocay, Groups & Graphs; `http://www.paddle.mb.ca/G&G/G&G.html`.
22. J. S. Leon, Partitions, refinements, and permutation group computation, *Groups and Computation II* (eds. L. Finkelstein and W. M. Kantor), DIMACS Series in Discrete Mathematics and Theoretical Computer Science **28**, Amer. Math. Soc. (1997), 123–158.
23. S. A. Linton, K. Lux and L. H. Soicher, The primitive distance-transitive representations of the Fischer groups, *Experimental Math.* **4** (1995), 235–253.
24. B. D. McKay, Computing automorphisms and canonical labellings of graphs, *Combin. Math., Proc. Int. Conf., Canberra 1977*, Lecture Notes in Mathematics **686**, Springer (1978), 223–232.
25. B. D. McKay, Practical graph isomorphism, *Congr. Numerantium* **30** (1981), 45–87.
26. B. D. McKay, nauty; `http://cs.anu.edu.au/people/bdm/nauty`.
27. The Magma Computational Algebra System, Computer Algebra Group, School of Mathematics and Statistics, University of Sydney; `http://magma.maths.usyd.edu.au`.
28. T. Miyazaki, The complexity of McKay's canonical labelling algorithm, *Groups and Computation II* (eds. L. Finkelstein and W. M. Kantor), DIMACS Series in Discrete Mathematics and Theoretical Computer Science **28**, Amer. Math. Soc. (1997), 239–256.
29. M. J. Morton, Classification of 4- and 5-arc-transitive cubic graphs of small girth, *J. Austral. Math. Soc. (A)* **50** (1991), 138–149, and Corrigendum, **52** (1992), 419–420.
30. J. Neubüser, An elementary introduction to coset table methods in computational group theory, *Groups – St. Andrews 1981* (eds. C. M. Campbell and E. F. Robertson), London Math. Soc. Lecture Note Series **71**, Cambridge Univ. Press (1982), 1–45.
31. C. E. Praeger and L. H. Soicher, *Low Rank Representations and Graphs for Sporadic Groups*, Australian Math. Soc. Lecture Series **8**, Cambridge Univ. Press, 1997.
32. S. Rees and L. H. Soicher, An algorithmic approach to fundamental groups and covers of combinatorial cell complexes, *J. Symbolic Comp.* **29** (2000), 59–77; GAP program available at `http://www.maths.qmul.ac.uk/~leonard/fundamental`.
33. Á. Seress, *Permutation Group Algorithms*, Cambridge Univ. Press, 2003.
34. C. C. Sims, Computation with permutation groups, *Proc. Second Symposium on Symbolic and Algebraic Manipulation* (ed. S. R. Petrick), Assoc. Comp. Mach. (1971), 23–28.

35. C. C. Sims, *Computation with Finitely Presented Groups*, Cambridge Univ. Press, 1994.
36. L. H. Soicher, On simplicial complexes related to the Suzuki sequence graphs, *Groups, Combinatorics and Geometry* (eds. M. W. Liebeck and J. Saxl), London Math. Soc. Lecture Note Series **165**, Cambridge Univ. Press (1992), 240–248.
37. L. H. Soicher, Three new distance-regular graphs, *Europ. J. Comb.* **14** (1993), 501–505.
38. L. H. Soicher, The GRAPE Package for GAP; `http://www.gap-system.org/Share/grape.html`.
39. R. Weiss, A characterization and another construction of Janko's group J_3, *Trans. Amer. Math. Soc.* **298** (1986), 621–633.
40. R. Weiss, Presentations for (G, s)-transitive graphs of small valency, *Math. Proc. Camb. Philos. Soc.* **101** (1987), 7–20.
41. R. A. Wilson *et al.*, ATLAS of finite group representations; `http://www.mat.bham.ac.uk/atlas`.

Notes on contributors

Brian Alspach [*alspach@math.regina.ca*] received his Ph.D. under the supervision of Paul Kelly in 1966. He spent 1966–99 at Simon Fraser University in Vancouver and is now an adjunct professor at the University of Regina. His main interests are applications of graphs, tournaments, Cayley graphs and graph decompositions. Now officially retired, he has time for writing, studying piano and poker.

Lowell Beineke [*beineke@ipfw.edu*] is Schrey Professor of Mathematics at Indiana University-Purdue University Fort Wayne, where he has been since receiving his Ph.D. from the University of Michigan under the guidance of Frank Harary. His graph theory interests are broad, and include topological graph theory, line graphs, tournaments, decompositions and vulnerability. With Robin Wilson he edited *Selected Topics in Graph Theory* (3 volumes), *Applications of Graph Theory*, and *Graph Connections*. He is currently editor of the *College Mathematics Journal*.

Richard Brualdi [*brualdi@math.wisc.edu*] is a professor of mathematics at the University of Wisconsin in Madison. He has written a number of papers on combinatorics, graphs and matrices, and is author of *Introductory Combinatorics*, *Combinatorial Matrix Theory* (with H. J. Ryser), and *Matrices of Sign-Solvable Linear Systems* (with B. L. Shader). A past president of the International Linear Algebra Society, he serves as editor-in-chief of the journals *Linear Algebra and its Applications* and the *Electronic Journal of Combinatorics*.

Peter Cameron [*p.j.cameron@qmul.ac.uk*] is a professor of mathematics at Queen Mary, University of London. His mathematical interests are in permutation groups and their operands (which may be logical, algebraic or combinatorial, and finite or infinite). He has written lecture notes on permutation groups, parallelisms, and projective and polar geometry (with J. H. van Lint), a book on graphs, codes and designs, and undergraduate textbooks on combinatorics and algebra. He is President of the British Combinatorial Committee.

Arjeh Cohen [*amc@win.tue.nl*] studied mathematics and theoretical computer science at Utrecht University. His research fields are discrete algebra and geometry, and mathematics on computers. He worked at Rijnmond Authority (Rotterdam), the University of Twente (Enschede) and CWI (Amsterdam), and at Utrecht University. Since 1992, he has been a professor of discrete mathematics at Eindhoven University of Technology. He is scientific director of RIACA, chair of the board of the Research School EIDMA, and president of the OpenMath Society. He has written about ninety papers, eleven books, and three software packages.

Dragoš Cvetković [*ecvetkod@etf.bg.ac.yu*] graduated in electrical engineering at the University of Belgrade in 1964, where he was awarded a doctoral degree in mathematics in 1971 and where, since 1986, he has been a professor of mathematics. His major field of interest is discrete mathematics, and he has published about 160 scientific papers and about 50 books in graph theory, combinatorics, linear algebra and artificial intelligence, including his best-known monograph, *Spectra of Graphs* (with M. Doob and H. Sachs). In 1985 he was admitted to the Serbian Academy of Sciences and Arts, and in 1991 he was appointed an Honorary Professor at the University of Stirling, Scotland.

Michael Doob [*mdoob@ccu.umanitoba.ca*] received his AB degree from Columbia University and his Ph.D. from the City University of New York. His doctoral supervisor was Alan Hoffman, who introduced him to graph spectra. After completing graduate school, he joined the faculty at the University of Manitoba, eventually rising to the rank of professor. He is the author of dozens of research papers and three books, two of which are directly related to spectral graph theory.

Bojan Mohar [*bojan.mohar@fmf.uni-lj.si*] obtained his Ph.D. from the University of Ljubljana, Slovenia, in 1986. He did his postdoctoral studies at Simon Fraser University (Vancouver) and Ohio State University, where he was on a Fulbright Scholarship. He is now a professor in the mathematics department of the University of Ljubljana. His areas of research include topological graph theory, graph minors, graph colouring, graph algorithms, and algebraic graph theory, and he has published over a hundred research papers (with over fifty co-authors). He has also co-authored a recent monograph, *Graphs on Surfaces*. In 1990 he was awarded the Boris Kidric prize, the Slovenian national award for exceptional achievements in science.

Cheryl Praeger [*praeger@maths.uwa.edu.au*] is a professor of mathematics at the University of Western Australia, and a Fellow of the Australian Academy of Science. She has published over 200 research articles and two monographs on permutation groups, algorithms for group computation, and applications to combinatorics, especially to vertex- and edge-transitive graphs. She initiated the modern theory of finite quasi-primitive permutation groups and the study of finite edge-transitive graphs via their normal quotients.

Peter Rowlinson [*p.rowlinson@stir.ac.uk*] is a professor of mathematics at the University of Stirling, Scotland, where he has taught since 1969. His Oxford D.Phil. thesis was concerned with finite groups, and in 1975–76 he was a visiting professor of mathematics at the California Institute of Technology. He has worked in algebraic graph theory since 1980, and is co-author (with D. Cvetković and S. Simić) of *Eigenspaces of Graphs* and *Spectral Generalizations of Line Graphs.*

Bryan Shader [*bshader@uwyo.edu*] is a professor of mathematics at the University of Wyoming in Laramie. His research interests focus on the interplay between combinatorics, graphs, and matrices. He is co-author with R. A. Brualdi of *Matrices of Sign-Solvable Linear Systems* (1995). He is currently an associate editor of the journals *Linear Algebra and its Applications*, and the *Electronic Journal of Linear Algebra*, and is an editor of *Image*, the International Linear Algebra Society's newsletter.

Leonard Soicher [*l.h.soicher@qmul.ac.uk*] is a reader in mathematics at Queen Mary, University of London. His research interests are groups, graphs, designs, algorithms and computation. He is the author of the GAP package GRAPE for computing with graphs and groups, and the co-author with Cheryl Praeger of *Low Rank Representations and Graphs for Sporadic Groups.*

Robin Wilson [*r.j.wilson@open.ac.uk*] is head of the pure mathematics department at the Open University, UK, and Gresham Professor of Geometry, London. He graduated in mathematics from Oxford University, and received his Ph.D. in number theory from the University of Pennsylvania. He has written and edited many books on graph theory and combinatorics and on the history of mathematics, including *Introduction to Graph Theory* and *Four Colours Suffice*, and his interests include graph colourings, spectral graph theory, and the history of graph theory and combinatorics.

Index of definitions